Élisée Reclus, Henry Woodward

The Earth: a Descriptive History of the Phenomena of the Life of the

Globe

Vol. 2

Élisée Reclus, Henry Woodward

The Earth: a Descriptive History of the Phenomena of the Life of the Globe
Vol. 2

ISBN/EAN: 9783337415006

Printed in Europe, USA, Canada, Australia, Japan

Cover: Foto ©berggeist007 / pixelio.de

More available books at **www.hansebooks.com**

THE EARTH

A DESCRIPTIVE HISTORY OF THE

PHENOMENA OF THE LIFE OF THE GLOBE

By ÉLISÉE RECLUS

Translated by the late B. B. WOODWARD, M.A.

AND

Edited by HENRY WOODWARD, British Museum

CONTINENTS

Illustrated by

TWO HUNDRED AND THIRTY MAPS INSERTED IN THE TEXT
AND TWENTY-FOUR PAGE MAPS PRINTED IN COLOURS

SECTION II.

LONDON
CHAPMAN AND HALL, 193, PICCADILLY
1871

CHAPTER XLIII.

SYSTEM OF SUBTERRANEAN STREAMS.—JOINTS AND FISSURES OF ROCKS.—
STALACTITES.—THE INHABITANTS OF CAVES.—THE MAMMOTH'S CAVE.—
CAVERNS OF CARNIOLA AND ISTRIA.

ABOVE the springs, the course of subterranean rivulets is generally indicated by a series of chasms, or natural wells, which disclose the stream beneath. The arches of caves not being always strong enough to support the weight of the superincumbent masses, they necessarily fall in some places, leaving above them other spaces into which the upper beds successively sink. The *débris* of the ruin is afterwards cleared away by the water, or dissolved, atom by atom, by the carbonic acid contained in the stream, and, gradually, all the loose rubbish is carried away. In this manner, above the subterranean rivulets, a kind of well is formed, which is designated in various countries by very different names. They are called *sinks* in the United States; *dolinas* in Carinthia; *catavothras* in Greece; *pots, entonnoirs,* and *creux* in the Jura; *embues, embucs, goules, gouilles, gourgs, gourgues, bétoirs, boit tout, anselmoirs, emposieu, avens, scialets, ragagés, garagaï* in Southern France;* swallow-holes, sand-pipes, sand-galls, &c., in England.

By means of these natural gulfs, it is possible to reach the subterranean streams and to give some account of their system, which is exactly like that of rivulets and rivers flowing in the open air. These streams also have their cascades, their windings, and their islands; they also erode or cover with alluvium the rocks which compose their bed, and they are subject to all the fluctuations of high and low water. The only important difference which superficial waters and subterranean currents present in their phenomena, is that these streams in some places fill the whole section of the cave, and are thus kept back by the upper sides, which compress the liquid mass. In fact, the spaces hollowed out by the waters in the interior of the earth are only in a few places formed into regular avenues which might be compared to our railway tunnels. Throughout its thick-

* Fournet, *Hydrologie Souterraine.*

ness, the rock opposes an unequal resistance to the action of the water, on account of the diversity of its fissures, its strata, and its particles. When the faults are numerous and the strata not very compact, the current gradually hollows out vast cavities, the ceilings of which fall in, and are carried away by the water almost in single grains. Where beds of hard stone oppose the flow of the rivulet, all it has done during the course of centuries has been to hew out one narrow aperture. This succession of widenings and contractions, similar to those of the valleys on the surface, forms a series of chambers, separated one from the other by partitions of rock. The water spreads widely in large cavities, then, contracting its stream, rushes through each defile as if through a sluice.

On account of these partitions it is very difficult, or even impossible, to navigate the course of subterranean rivers to any considerable distance, even at the time the water is low. When it is high, the liquid mass, detained by the partitions, rises to a very high level in the large interior cavities, and often reaches the roof above. Sometimes when, through the clefts of the rocks, a communication exists between the cave and some hollow above, the surplus water from the subterranean streams makes its appearance there. Thus the Recca, which flows beneath the adjacent plateau of Trieste, does not always find space enough to flow freely in its lower channels, and Schmidl has seen it ascend in the chasms of Trebich to a height of 341 feet. It may be understood that the pressure of such a column of water often shatters enormous pieces of rock, and thus modifies the course of underground streams.

When the water, impelled by force of gravitation, seeks a new bed in the cavernous depths of the earth, and disappears from its former channels, these are at first much easier of access than they formerly were; but, ere long, in most caves, a new agent intervenes, which seeks to contract or even completely obstruct them. This agent is the snow-water, or rain, which percolates, drop by drop, through the enormous filter of the upper strata. In passing through the calcareous mass, each one of these drops dissolves a certain quantity of carbonate of lime, which is afterwards set free on the arch or the sides of the cave. When the drop of water falls, it leaves attached to the stone a small ring of a whitish substance; this is the commencement of a stalactite. Another drop trickles down, and, trembling on this ring, lengthens it slightly by adding to its edges a thin circular deposit of lime, and then falls. Thus, drop succeeds drop in an infinite series, each depositing the particles of lime which it contains, and forming ultimately a number of frail tubes, round which the calcareous deposit slowly accumulates. But the water which drops from the stalactites has not yet lost all the lime which it held in solution; it still retains sufficient to enable it to elevate the stalagmites and all the mammillated concretions which roughen or cover the floor of the grotto. It is well known what fairy-like decoration some caverns owe to this continuous oozing through the vaults of their roofs. There are few sights in the world more astonishing than that of these subterranean galleries, with their dead-white columns, their innumerable pendants and multiform groups, like veiled statues, all yet unstained by the smoke of the visitor's torch. These stalactite caverns can only retain this primitive beauty on the condition of not being given over to idle curiosity. But yet how large is the number of those vulgar admirers who, under the pretence of loving nature, seek only to profane her!

When the action of the water is not disturbed, the needles and other deposits of the calcareous sediment continue to increase with considerable regularity. In some cases, each new layer which is added to the concretions may be studied as a kind of time-measurer, indicating the date when the running water abandoned the cave. At length, however, the soft concentric layers disappear, and are replaced by forms of a more or less crystalline character; for in every case where solid particles exist, subject to constant conditions of imbibition by water, crystals are readily produced.* Sooner or later, the stalactites, increasing gradually in a downward direction, meet and unite with the needles rising from the surface of the ground, and, forming by their number a kind of barrier, obstruct the narrower passages and close up the defiles, separating the cavern into distinct chambers. Any objects which lie on the surface of the ground in these dripping caves gradually become hidden by the calcareous concretion which thickens round them. Generally speaking, when geologists find in these grottoes the remains of men or animals—the former inhabitants of the mountain caves—they are covered with a crust of stone, slowly deposited by the dripping water. In 1816, in one of the caves of Adelsberg, a skeleton, probably that of some bewildered visitor, was discovered, which the stone had already enveloped in a white shroud; but these bones have now for some years been firmly fixed in the thickness of the rock, added to, as it constantly is, by fresh layers; indeed, the lateral cave itself will soon be filled up by stalactites, and will cease to exist. In like manner,

* Kuhlmann, *Presse Scientifique*, 1865.

the skeletons of three hundred Cretans, who were smoked to death by the Turks in 1822, in the cave of Melidhoni, are gradually disappearing under the incrustation of stone which has enveloped them with its calcareous layers.*

In the gloom of these dark recesses there is still some little manifestation of life. Since, however, plants of a higher order are unable to dispense with light, fungi form the only vegetation which we meet with, and even these growths of darkness do not always arrive at their full development; they often present monstrous and anomalous forms, which puzzle the botanist and hinder his attempts at classifying them. Some fungi never reach any further development than a mass of confusedly organised cells; others grow so as to cover a considerable surface. The Fauna, being more independent of light than the Flora, reckons a much larger number of representatives in these caves. Not only do these subterranean cavities serve as places of refuge for various birds, and as dens for several kinds of beasts of prey, such as foxes, badgers, hyænas, which carry thither the prey which they have caught (as our ancestors the troglodytes once did), but they are also inhabited by several families of animals which only exceptionally, or through accident, ever emerge from the depths of the caverns. Among the latter there is at least one mammal, a species of bat,† which is found in the caves of Istria, the Apennines, and the Algerian mountains. The subterranean pools and streams of Central Europe also contain several varieties of a strange reptile—the *Hypochthon*, or *Proteus*—the eyes of which, being useless in the darkness, are almost aborted. Insects are the class which is best represented in these subterranean regions; but none present those vivid colours which the light of the sun conveys to most of their congeners. All are clad in a dull garb which blends with the dark shades of the rock. The most curious of these insects is a species of fly (*Phora maculata*) which never uses its wings, and various Coleoptera (*Anophthalmus*), in which the eyes are entirely wanting. Then follow spiders, centipedes, crustaceans, and molluscs. M. Schiner, who has made a special study of the Fauna of caves, enumerates twenty-three species of animals which inhabit the caverns in the vicinity of Trieste alone; but these species form, doubtless, but a very small proportion of the subterranean tribes which live in the caves scattered far and wide over the whole earth. It is said that the caves in Kentucky contain a species of blind cray-fish; also whitish rats, of a very large size; and lizards, wandering gloomily in this world of darkness; and, lastly, a

* Perrot, *L'Ile de Crète*. † *Miniopterus Schreibersii*.

species of yellow cricket, which crawls like a frog, guiding its course by means of enormous antennæ.

One of these Kentucky caves, called the "Mammoth's Cave," is the largest which is at present known. The whole of its extent has not been as yet fully explored, for it may be almost called a subterranean world, having a system of lakes and rivers, and a network of galleries and passages without number, which cross and recross one another, going down to an immense depth. From the chief entrance to the further recesses of the cave, the distance is reckoned to be not less than 9¼ miles, and the whole length of the two hundred alleys that have been traced out in this enormous labyrinth is 217 miles in

Fig. 83.—Chasms of Carniola.

extent. This "Mammoth Cave" once served as a retreat for savage tribes; for skeletons of men of an unknown race have been found buried in it under layers of stalactite.

The district which is the most remarkable amongst all the calcareous countries of Europe for its caves, its subterranean streams, and its abysses, is unquestionably the region of the Carniolan and Istrian Alps, which extends to the east of the Adriatic, between Laibach and Fiume. The whole surface of the country, as in certain plateaux of the Jura in France, is everywhere pierced with deep boat-shaped cavities, at the bottom of which the water forms a kind of whirlpool, like the water flowing out of the hold of a stranded ship. Many mountains are penetrated in every direction with caverns and passages, just as if the whole rocky mass was nothing more than an accumulation of cells. On one steep cliff-side may be noticed all kinds of perforations at different heights—arched portals, and orifices of fantastic shape; on another, there are numbers of springs of blue water gushing from the caves, or from the rocks heaped up at the foot of the cliff, and forming rivulets which disappear a little further on in the fissures of the ground, as if through the holes of a sieve. The whole surface of the plateaux, whether bare or covered with forests, is scattered over with wells, or funnel-shaped holes communicating with subterranean reservoirs. The geography of the underground labyrinth of the Illyrian caves is as yet only sketched out; and yet a considerable number of *savants*, at the head of whom stands M.

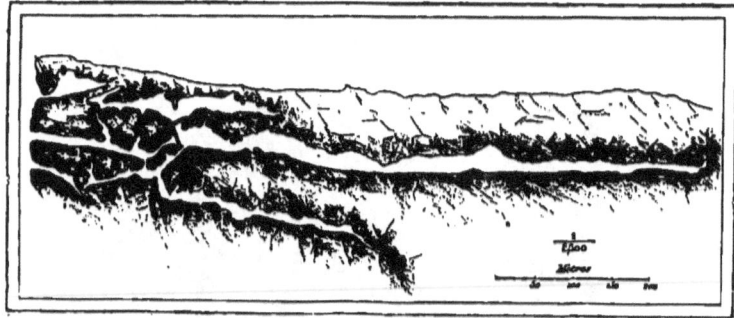

Fig. 84.—Grotto of Lueg, Illyria.

Schmidl, have devoted many years of their lives to this study. Thanks to their investigations, some of the passages in these caverns, especially those of Lueg, are almost as well known as the corridors and chambers of a palace.

One of the Istrian rivers, the subterranean course of which, although still unknown as regards a great number of points, has given rise to a most continuous course of investigations, is the celebrated Timavus (Timavo), which falls into the sea near Duino, about twelve miles to the north of Trieste. Virgil's description[*] no longer applies to the mouths of the Timavo; at present they do not reach the number of nine, because either the extermination of the woods

[*] " Fontem superare Timavi
Unde per ora novem vasto cum murmure montis
It mare proruptum, et pelago premit arva sonanti."

of the Carso has diminished the mass of the water, or the action of the stream and the alluvium of the delta have modified the form of the shore. But still it is a magnificent spectacle to see the outlet of the three principal torrents of water, which rush foaming out of the heart of the rocks, and are navigable from their mouths to their very source. A river of this importance must certainly receive the drainage of a vast basin, and yet all the neighbouring valleys seem perfectly devoid of rivulets, and their surface presents little else but the bare rock; in fact the whole of the rain and snow-water runs away through underground caverns. We do not meet with any tributary until we reach a spot 21 miles south-east of the mouth of the Timavo. This tributary, known under the name of the Recca, is lost in the rock under a high arch on which stands the village Sant-Canzian; it appears again at the foot of two precipices, and then engulfs itself in the depths of the rocks by a series of beautiful cascades, beyond which explorers have not traced it. Further on, the course of the subterranean torrent is only indicated by abysses opening here and there in the midst of the plain. In 1841, M. Lindner, who was seeking in every direction for springs of water to supply the city of Trieste, the inhabitants of which were threatened with drought, formed the idea of sending some miners down into the chasm of Trebich, situated about four miles to the north-east of the city. After eleven months of labour the miners at last reached the floor of the lower cave, 1,062 feet below the surface of the plateau, and there in fact they found the Recca of Sant-Canzian flowing at their feet. The descent into this cave is by means of ladders, and it is thus rendered accessible by the work of man.

The most remarkable network of caverns in this region of the Alps, is that which spreads out from the south-west to the north-east across the Adelsberg group of mountains between Fiume and Laibach. The principal cave is especially curious on account of its size, the variety of its calcareous concretions, and the torrent which runs roaring through it; certainly its vast compartments, its innumerable white and rose-coloured pendants, its abysses wrapt in shade, and the eternal echo of its rushing water, would produce upon visitors a much more striking effect if its proprietors had not conceived the untoward idea of decorating their property with rustic or Chinese bridges, elegant staircases, and pyramids adorned with sentimental inscriptions.

North of the town of Adelsberg, the traveller passes along the base of a hill with steep and bare sides, bringing into view the sharp edges of its highly-pitched calcareous beds. On the right, the stream of the Poik winds peaceably in the valley; and then, its course being arrested by a headland, turning suddenly, it flows into the interior of the mountain through a kind of high portal, opening between two parallel beds of rocks. Unless the water in the stream is very

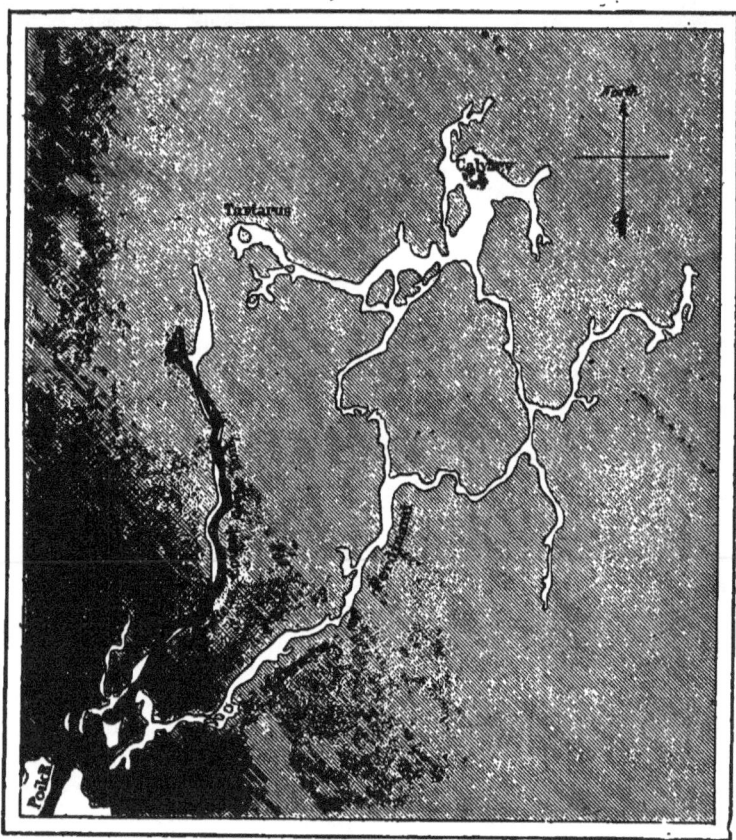

Fig. 85.—Grotto of Adelsberg.

low, it is impossible to follow it over the accumulation of rocks upon its bed; but on the right, at a height of a few yards, there is another entry, through which the traveller may descend dry-shod into a vast cavity or chamber, where the Poik again appears issuing from its narrow passage of rocks. At this point, the cave divides; on the north the stream, the depth of which varies, according to the season, from a few inches to 30 or 33 feet, buries itself in a winding avenue, which has been traversed in a boat as far as a point 1,027 yards from the entrance; on the north-east, a higher avenue, discovered only in 1818, pushes its way far into the heart of the mountain, branching out in various directions into narrow passages and wide compartments. This portion of the grotto, which appears to have been the former bed of the Poik, is the most curious part of the Adelsberg labyrinth; it affords wonderful groups of stalactites, especially in the Salle du Calvaire, the vaulted roof of which, having the enormous span of 210 yards, has dropped upon a hillock of *débris* a perfect forest of stalagmitic columns and white needles. The full length of the principal cave is not less than 2,575 yards; but very probably some other and still longer avenues may yet be discovered.

Although it is impossible to go in a boat along the subterranean portion of the Poik for a greater distance than 1,027 yards, by traversing the surface of the calcareous plateaux we can at all events trace out the subterranean stream by means of the funnel-shaped holes which open above its course. One of these gulfs, the Piuka-Jama, is situated about a mile and a half to the north of the entrance of the Adelsberg caves; the only way to descend into this is by clinging to the branches of the shrubs and sliding down by the assistance of a cord fastened to the top of the rocks. By these means the entrance to a kind of air-hole may be reached, from which the Poik is visible foaming over its bed of rocks, and only a slope of *débris* is to be descended to reach the edge of the stream. It can only be followed in the down-stream direction for about 275 yards; but it can easily be ascended for a distance of 495 yards by passing under a high portal with lofty pillars, and in this way a point can be reached which is less than a mile from the place where the stream disappeared in the cave of Adelsberg.

Further down the stream, the Poik is not visible again until it emerges from the mountain, where it is known under the name of the Planina; it rushes out through a circular arch at the base of a perpendicular bluff crowned with fir trees. It really is the Poik, as is proved by the equal temperature of water and the sudden increase of its liquid mass after a storm has burst at Adelsberg; but the stream always issues from the cave much more considerable in bulk than it is when it enters, owing to the tributaries which pour into it on both sides during its subterranean course of five to six miles. One of these rivulets, which comes down from the plateaux of

Kaltenfeld, joins the Poik at a little distance from its outlet. Above the confluence, the principal stream can be ascended in a boat to a

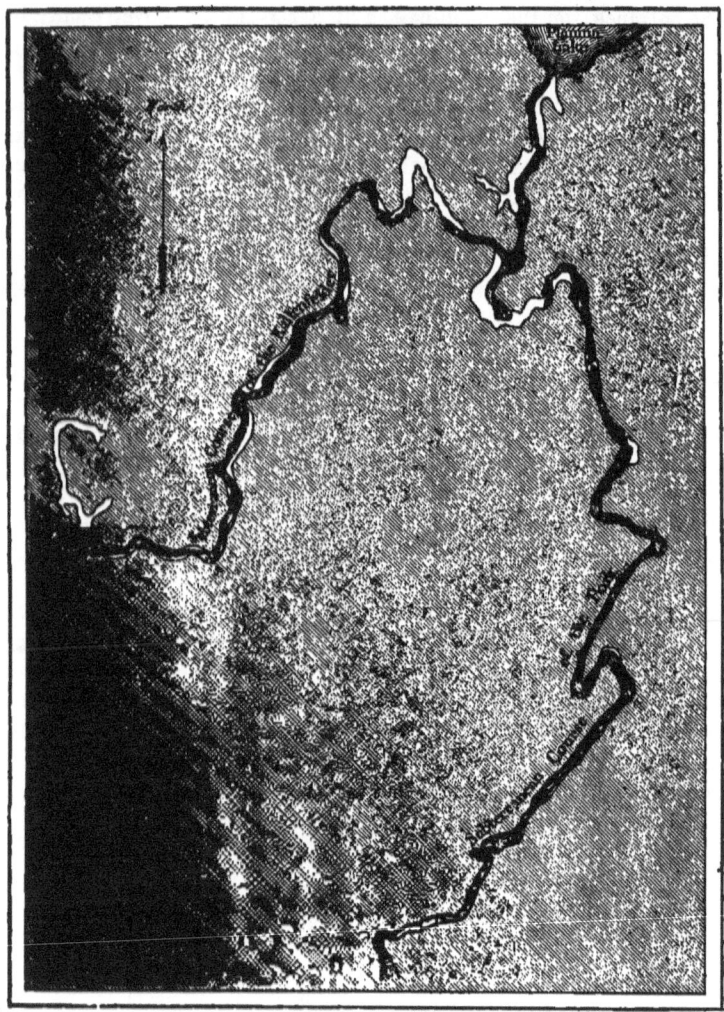

Fig. 86.—Grotto of Planina.

distance of more than 3,500 yards, which, with the other explored parts of the subterranean river, makes about three miles. Below the point of outlet, the stream is partially lost in the fissures of its bed, and then joining the Unz, goes on and empties itself into the Danubian Save.

About a dozen miles to the south-east of the Adelsberg and Planina caves, extends a large plain surrounded on all sides by high calcareous cliffs, at the base of which nestle seven villages. In this hollow, the most elevated portion of which is under cultivation, the remainder being covered with rushes and other marsh plants, there are to be found more than 400 funnel-shaped holes resembling those in other parts of Carniola. These *dolinas*, the average depth of which is from 40 to 60 feet, have each their special name, such as the "*Grand Crible*" (great sieve), the "*Crible-à-froment*" (corn sieve), the "*Tambour*" (drum), the "*Cuve*" (tub), the "*Tonneau*" (cask), pointing out the form or some remarkable peculiarity of each abyss. During extremely dry seasons, there is only one of these cavities which contains any water; but after continuous and heavy rain, the water of a stream which is swallowed up in the rocks a little above the plain, rises with a roaring noise in each of these wells. Torrents escaping from all these open "*cribles*" form in the wide space hemmed in by the cliffs a sea of blue and transparent water. This is the lake of Jessero or Zirknitz, the *lacus Lugens* of the Romans. The surface of the sheet of water extends over an area of 14,826 acres; at the time of great inundations this extraordinary temporary lake, thus vomited out by the under-ground river, is not less than 24,711 acres. The water runs away through a subterranean channel, and, further on, empties itself into the Unz, below the Planina.

Lacustrine basins of this sort, first emitted, and then again absorbed by a subterranean water-course, are rather rare; there are, however, some other remarkable instances of them in Europe. Thus, in the oriental Hartz, in the midst of a beautiful spot surrounded by fir trees, the charming lake called Bauerngraben (Peasants' Ditch) or sometimes Hungersee (Lake of Famine) sometimes makes its appearance; but when this mass of blue water has filled but for a few days its basin of gypsum rock, it is suddenly swallowed up, and flows away by subterranean channels into the stream of the Helme. The celebrated lake of Copaïs in, Beotia, may likewise be compared to the Zirknitz lake, at least as regards certain portions of its basin.

CHAPTER XLIV.

RIVERS.—VARIOUS DENOMINATIONS OF WATER-COURSES.—DETERMINATION OF THE PRINCIPAL BRANCH AMONG THE AFFLUENTS OF A RIVER.—RIVER BASINS AND WATERSHEDS.—FORKS OF CERTAIN RIVERS.

GEOGRAPHERS have long discussed, and are still discussing, the precise import of the names which are used to designate running waters. How are we to lay down any distinction between a river and a stream, or between a stream and a rivulet? Obviously, no absolute difference can exist, as all water-courses are alike composed of liquid masses impelled by their own weight over an inclined bed. The only relative difference which at first sight it seems easy to establish consists in the greater or less quantity of water which each bed contains; but even this mode of estimation must vary in every continent and in every country, according to the importance of its hydrographical system. Many an European river would seem nothing but a slender rivulet, and would scarcely be thought worthy of a name, if it were situated in the immense basin of the Amazon. Added to this, the mass of water alters its bulk according to the various seasons. Many rivers in tropical regions, which flow very abundantly during the rains, are during the dry season often entirely dried up, or changed into a series of pools.

All the phenomena of nature being full of diversity, she has omitted to furnish any fixed rule for the classification of water-courses; but some geographers, desiring, at any rate, to assume an appearance of authority in matters relating to the earth, have given the name of river (*fleuve*) to the liquid masses which empty directly into the ocean, and apply the name of stream (*rivière*) to mere affluents which are themselves fed by tributaries of the second order, or rivulets. In virtue of this purely scholastic distinction, the Argens, the Seudre, and the Leyre, would be rivers, while the Tapajoz and the gigantic Madeira would only have the right to the title of streams. Our ancestors, the Celts of Western Europe, who understood much more about nature than many of our modern *savants*, employed the same name (although variously modified by use) for distinguishing water-courses of all sizes, viz., the Rhine, the Rhône, the Arno, the Orne, and the Arnon. All running water was in their eyes a river.

The principal difficulty which systematical geographers meet with is that of determining, as regards each basin, which is the chief branch; that is, which is to be considered the river *par excellence*, all the other water-courses being mere tributaries. In some cases, certainly, it may readily be perceived to which artery of the river-basin the pre-eminence unquestionably belongs; but more generally it is difficult, or even impossible, to pronounce with any certainty on this question. Is it the Seine or the Yonne, the Adour or the Gave-de-Pau, the Rhine or the Aar, the Inn or the Danube, the Mississippi or the Missouri, the Marañon or the Apurimac (Ucayali), which has the best right to impose its name on the principal artery which bears onward to the sea the mingled water of the two rivers? Does the point in question chiefly depend upon length of course? If it does, the Saône and the Rhône are only tributaries of the Doubs, which has a total development, from Mont-Rizoux to the Gulf of Lyons, exceeding that of the Rhône by 93 miles. In like manner, the Mississippi would thus become a tributary to the Missouri, which has a course more than 1615 miles longer—an excess which is equal to three times the length of the Seine. In deciding which of the upper tributaries is the principal water-course, would it be more to the point to compare the quantities of the liquid supply which each brings to the common fund? In this case, the Yonne, the Aar, and the Inn, are rivers which are fed by the Seine, Rhine, and Danube respectively. Ought we not rather to consider the more or less rectilinear direction, and the comparative geological unity of the valley of each affluent, as the principal signs which should determine the real *river*? Then, the Rhône and the Seine are nothing but secondary water-courses in comparison with the Saône and the Yonne, and the Yonne itself must yield its pre-eminence to the Cure.

The *savant* who devotes himself to the unthankful task of seeking out the principal branch in a river system has, therefore, to take account of the most diversified points of detail: the average mass of water, the length of the course, the general direction of the valley, and the geological nature of the soil; but, whatever may be the result of his investigations, he must ultimately yield to the allpowerful authority of tradition. For it is tradition, and not science, which has invested rivers with their titles and dignity; it is the voice of our ancestors, founded on a thousand circumstances in connection with mythology, the history of conquests or colonisation, agriculture, navigation, or even on various natural phenomena, which has arbitrarily decided to give the pre-eminence to some particular watercourse over the other rivers of the same system. It is now too late to try to change the hydrographical nomenclature.

But even were it possible, this alteration would be almost entirely inefficient, for the vitality of nature will not accommodate itself to the strict classifications to which pedants would seek to restrict it. It is only by pure abstraction that we come to consider a river as an isolated existence. In reality it is the aggregate of the streams and rivulets which flow into it from all the points of its basin; it unites the millions of rills which are set free from the ice, or trickle from the crevices of the rocks; it is made up of the innumerable springs which ooze out from the ground saturated with rain or covered with snow. A river is in a constant process of change, and every tributary takes its share in this work of transformation. The entire drainage area, and not any particular affluent, ought, therefore, to be considered as the real river. We must take into account the Missouri, the Ohio, and the Red River, no less than the Mississippi, extending its long and constantly-increasing peninsula of mud into the Gulf of Mexico; also the Tapajoz, the Rio-Negro, and the Madeira flowing with the Solimoës into the vast estuary of the Amazon. In like manner, to use the language of the sailors of the Bay of Biscay, the "two seas" of Garonne and Dordogne unite their waters to compose the "Sea" of Gironde.

Those names of rivers which are formed by the contraction of the designations of their chief tributaries are, indeed, the only terms which are geographically correct. We may mention, as examples, the names of the Somme-Soude, the Thames (Thame, Isis), the stream of Gyronde (Gyr, Onde) in the Upper Alps, and, better still, that of the Virginian river Mattapony (Mat, Ta, Po, Ny). The aggregate of the arteries of a river system may be compared to the branches and twigs of an immense tree. The Rhine and the Mississippi remind one of the oak by the majesty of their shape and the magnitude of their branches, thrown out at right angles to the parent stem. The Nile, with its long trunk devoid of lower boughs, and crowned with its plume-like terminal branches, recalls to our mind the palm tree of the oasis. These comparisons, it is true, have no scientific element in them, but still they do not fail to present themselves to the eye, and geographers, as well as artists, must be to some extent struck by them.

Almost all those portions of continents on which the humidity of the atmosphere falls in the shape of snow and rain have their system of rivers, into which all the water is emptied which is not, immediately after its fall, absorbed by the earth or sucked up by the roots of plants. But when the surface of the ground is almost or quite horizontal, the rain-water cannot find a sufficient amount of slope to enable it to flow down towards the sea, and, consequently, spreads out in stagnant pools. Thus, in the *pampas* of the Argentine Republic—where however the annual rainfall is greater than that of France—the prairies are dotted over with lagunes having no outlet, and the Vermejo, the Salado, and the Pilcomayo—the great rivers descending from the mountains of the north-west—are not replenished by a single tributary from the plains through which they pass.*

Lofty mountain chains, the peaks of which tower up into the sky, crossing the very tracks of the clouds, collect in proportion a much larger share of moisture than the plains, and consequently give rise to the most abundant streams of water. Nevertheless, as low-lying countries, or those possessing a moderately elevated vertical outline, embrace an area much more extended than that of mountainous districts, these flat regions are the localities where rivulets spring from the earth in the greatest number. In a general way, the ravines or dells in the plains in which the water of river-sources is collected are representations in miniature of the deep gorges and the hollows of erosion existing in high mountains. But amongst the incipient river affluents there are some which take their rise on level plateaux, or in some trifling depression of the ground; there are others, especially in the great plains of Russia, which issue from lakes or marshes, which spread out in vast sheets in the centre of the country. Thus, the watershed, that is, a ridge separating two slopes, or perhaps a mere ideal winding line, on each side of which the water flows in an opposite direction, is developed under the most diversified conditions. A river basin, that is, the area which is traversed by all its affluents, may be bounded on one side by the jagged ridge of some mountain chain, on the other by the gentle undulations of a range of hills, further on by an almost imperceptible rising in some low-lying plain. In certain localities, indeed, it is necessary to level the soil in order to ascertain the exact spot where the "divorce of the waters," as the ancients used to call it, actually takes place. Added to this, even in the mountains, the ridge line, or the highest elevation, is very far from uniformly coinciding with the watershed which separates two drainage areas. Mountains exhibit such infinite variety in their

* Martin de Moussy, *Confédération Argentine*.

original form, and the agents which denude them have hollowed out their sides in ways so diversified, that some rivers actually take their rise on the contrary side of the mountain to that which they are about to water.

The most remarkable instance which is to be found on the earth of sudden interruption in a mountain system is probably the astonishing cut of Riñihue, situated in the Chilian Andes, near the fortieth degree of latitude. According to the unanimous evidence of the aborigines and the Chilian peasants of the country, the stream of Huahuum, which takes its rise in the high *pampas* of Buenos Ayres, empties its waters into the Pacific, after having crossed the chain of

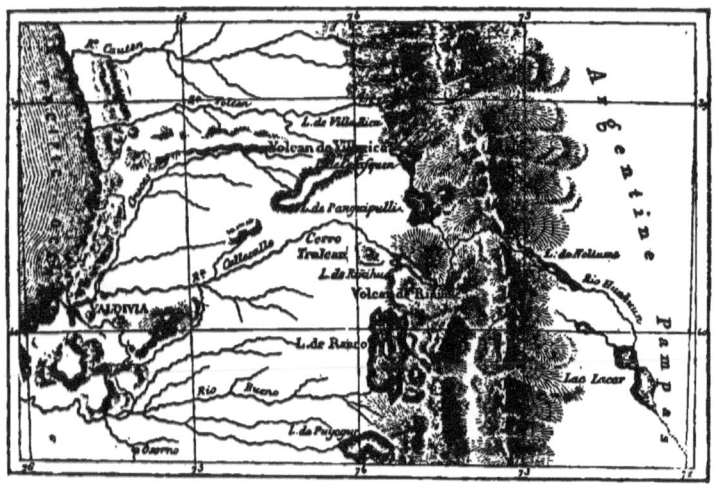

Fig. 87.—Cut of Riñihue in the Chilian Andes.

the Cordilleras. Issuing from the lake Neltume, the stream then penetrates into a defile, where it is known under the name of the Caillitue. It is certain that further down it is replenished by a stream flowing from the Andine lakes of Panguipulli and Cafalquen, and that the combined liquid mass flows on and empties itself into the lake of Riñihue, which is a tributary of the Pacific Ocean. The Indians assert that the whole extent of the Huahuum and Caillitue are navigable, and are only interrupted by one single rapid. Unfortunately, the scientific explorations of this region have not yet gone beyond the banks of the lake Riñihue. It appears that in this part of its development the chain of the Andes owes its form to the action of the subterranean agents, which have raised cones of eruption at intervals along the volcanic line of fault.*

Several river-basins exhibit rather a curious phenomenon. The watershed line, traversing high mountain chains, plateaux, and marshes, and separating two hydrographical systems, is interrupted by breaches or gaps, through which the water can flow out of one basin into another. On reaching this breach the flow of water, being attracted by two inclines, forks out into two streams running in contrary directions, and sometimes towards two different seas. Thus, in

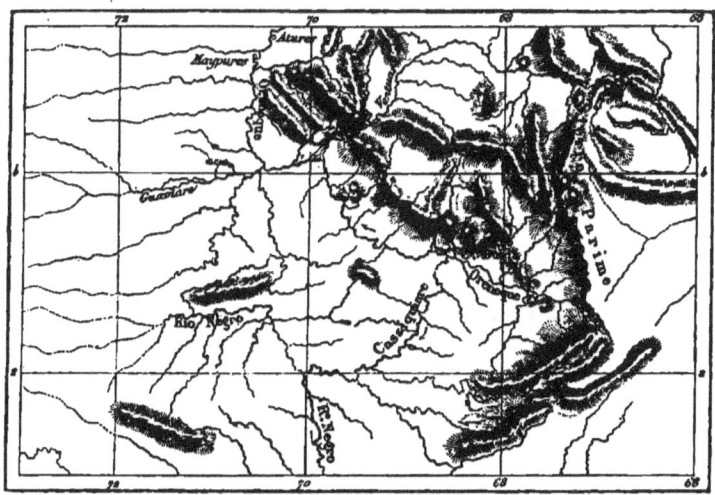

Fig. 68.—Bifurcation of the Orinoco.

Columbia the Upper Orinoco divides into two rivers, one of which empties into the Atlantic immediately to the south of the Antilles, whilst the other, known by the name of the Cassiquiare, runs to the south-west towards the Rio Negro, a tributary of the Amazon. The river, therefore, which collects the waters of the upper basin of the Orinoco is a tributary of two seas at the same time; it assists in turning the whole of the Guianas into a great island, surrounded on one side by the ocean, on the other by a channel navigable along a double incline, having its summit level at the foot of the high mountain of Duida.

This phenomenon of bifurcation, which has been rendered famous

* Frick, *Mittheilungen von Petermann*, ii., 1864.

x

by the journey of Humboldt and Bonpland, is also found, although certainly on a less magnificent scale, in several other countries of the earth, some mountainous, and others only slightly undulating. In some places, owing to the kind of indecision which is produced in the liquid mass by the double attraction of the two inclines, man has been enabled to regulate at his will the course of the two diverging streams, or even entirely to do away with the bifurcation by means of a dam, or some other hydraulic works. But to make up for it, in a multiplicity of other cases human ingenuity has been able to utilise the depressions of the surface so as to draw off laterally an arm of a river, and thus create an artificial fork.*

In Europe only, numerous instances may be mentioned of natural bifurcations. In Sweden a small lake, which is situated at a height of more than 3,300 feet at the foot of the lofty mountain of Sneehättan, simultaneously feeds the stream of Lougen, which descends towards Christiania, and that of Romsdal, which empties itself into the Molde-Fjord, between Bergen and Trondjhem. Added to this, the marsh of Kol, on the plateau of Hardanger, gives rise to eight rivulets, each diverging in its own particular direction.† In like manner, on a rocky plateau situated at a height of about 2,640 feet to the east of Puy de Carlitte, in the Eastern Pyrenees, we find the little pool of Las Dous (the Two) emptying its waters simultaneously into an affluent of the Têt du Roussillon and into the rivulet of Angoustrine, a tributary of the Sègre and the Ebro. Central Italy affords a still more curious instance of bifurcation. It appears unquestionable that, at the time of the Romans and during the first centuries of the middle ages, the Arno was divided into two branches, one of which emptied itself directly into the sea, whilst the other, crossing on the south the valley of Chiana, fell into the Paglia, a tributary of the Tiber.‡ When the River Arno, gradually sinking its northern bed, ceased to flow into the valley of Chiana, the water which descended from the lateral ravines in this almost horizontal depression flowed to a very slight extent on one side into the Tiber, and on the other into the Arno; but more often it stagnated in wretched marshes, which were a constant source of fever. These marshes have now disappeared, thanks to the splendid hydraulic works undertaken since Torricelli's time by the Tuscan engineers for the amelioration of the valley. By

---

* *Vide* vol. ii., the chapter on "The Labours of Man."
† Fritsch, *Mittheilungen von Petermann*, vol. xi., 1866.
‡ Salvagnoli Marchetti.

means of the alluvium brought by the torrents on both sides into the settling basins, an artificial watershed has been created in the middle of the valley, giving the water two very perceptible slopes, inclined in contrary directions.* One of the tributaries of the basin of the Seine also once offered an instance of constant bifurcation; at Mœurs the Grand Morin divides into two streams, one of which flows down to the Marne, and the other feeds the Superbe—an affluent of the Seine. But lately, owing to the destruction of the woods, the sources have become diminished; the double communication of the water only takes place in an artificial way by means of a dam.†

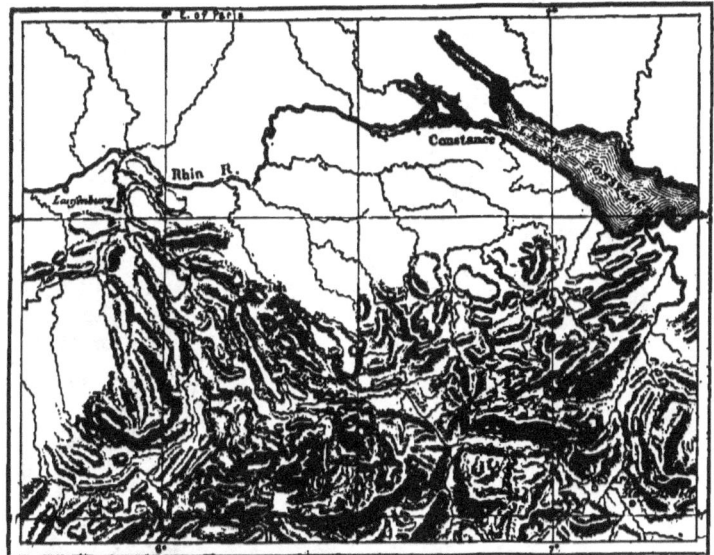

Fig. 89.—Bifurcation of the Valleys of the Rhine.

Amongst phenomena of a like nature, we must also class the division of the contents of a river into two branches, which, flowing separately each in its own valley, ultimately reunite at a considerable distance below the point of bifurcation. It is not improbable that, at a recent geological period, the Rhine was thus divided into two branches, embracing in its course an immense island of rocks and mountains comprehended between the lakes of Wallenstadt, Zurich,

* Simonin, *L'Etrurie et les Etrusques.*
† Plessier, *Formation des Plateaux et des Vallées de la Brie.*

and Constance, and the present confluence of the Aar and the Rhine. In the earth's history the two valleys may be looked upon as having an equal title to be considered the axis of the river-basin, as they have both served as the river-bed, either simultaneously or in turn. Between Meyenfeld and Sargans, at a height of 1,580 feet, the Rhine doubles round suddenly to the north-east, and, penetrating a narrow defile, runs down to the Lake of Constance, which it crosses, and flows on to join the waters of the Aar and the Limmat at about 93 miles

Fig. 90.—Threshold of Sargans.

below Sargans. The latter town is situated on an isthmus of pebbles and peat, which divides the present bed of the Rhine from its former bed, which tends towards the north-west. If this isthmus, which is only about 16 feet high, were to disappear, the river would again divide into a fork, and one of its arms would flow on to empty itself into the Lake of Wallenstadt, and thence into the Lake of Zurich and the valley of the Aar. Various hypotheses have been propounded to explain the formation of this isthmus which has severed the river basin into two parts, and forced the whole body of the Rhine to flow into the Lake of Constance. It is probable that this mass of pebbles is a portion of a slope of *débris* brought down by the torrent of Seez from the recesses of the gorge of Weiss-tannen (White Firs), and deposited at the outlet of the lateral valley.* So long as the river was able to clear a way through these heaps of stones a portion of it followed its old course towards the Lake of Wallenstadt; but, being constantly impeded by the ever-growing barrier, it was ultimately compelled to open out a new outlet towards the north.

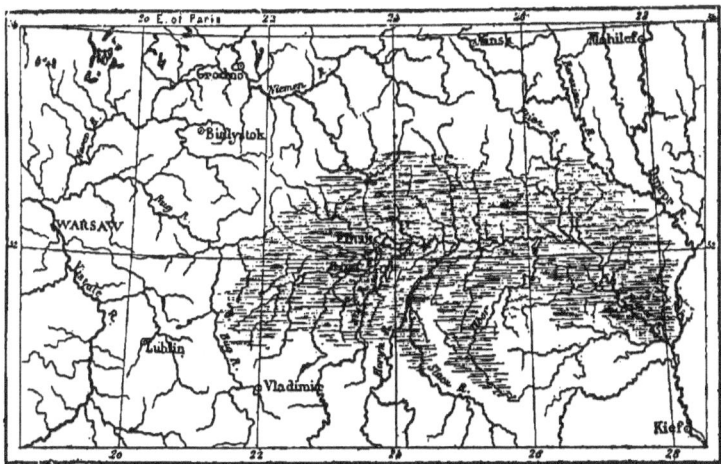

Fig. 91.—Marshes of Pinsk.

A great number of examples of this double flow of portions of one mass of water towards two different basins are afforded by low and marshy plains. The marshes of Pinsk, in Volhynia, serve as a common source to various affluents both of the Vistula and the Dnieper, thus forming a link between the Baltic and Black Seas. In spring, when the snow melts, and towards the end of autumn, after the heavy rains, a series of lakes, wet marshes, and temporary rivulets connect the inland Caspian with the Sea of Azof and the Euxine. The water of the Kalaous, coming down from one of the rugged valleys of the Caucasus, divides, and forms a temporary channel between the two basins, which were, indeed, once united in one and the same ocean.

* W. Huber, *Report of the Geographical Society*, February and March, 1866.

The two principal river systems of North America—those of the Mississippi and the St. Lawrence—are likewise blended together for a few days after a prolonged rainfall. Even before the construction of the canal which at present unites the two rivers, small boats could sometimes pass from the Chicago River into the Illinois, and thus cross the scarcely-indicated watershed which divides the basin of Newfoundland from that of Mexico. In a recent period—that is, about 4,500 or 5,000 years ago—the union of the two river-basins, which has now become but temporary, appears to have been of a permanent character. The calculations and observations of Sir Charles Lyell, Schoolcraft, and many American geologists, render it very probable that, at this epoch, all the upper affluents of the

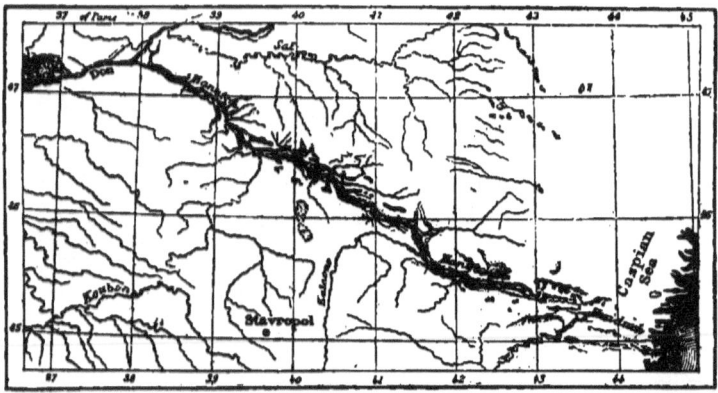

Fig. 92.—The Ponto-Caspian Isthmus.

Mississippi and the St. Lawrence fed a lacustrine reservoir, the vast sheet of which, situated about 600 feet above the level of the ocean, stretched towards the north as far as the mouth of the Wisconsin, and on the east joined the Lake Michigan, covering all the intervening peninsulas. The centre of the continent was occupied by a sea as large as our Mediterranean, which emptied itself into the ocean by an immense delta, each arm of which was one of the greatest rivers of the earth.*

The bifurcations of water-courses do not, however, all take place on the surface of the ground; and if the deeper layers could be disclosed to our view, it is probable that we should find the majority of river-basins would afford instances of subterranean derivations. In a

* Humphreys and Abbot, *Report on the Mississippi River.*

country like France, in which geological exploration has seriously commenced, a considerable number of these curious phenomena have been discovered, although they are in general but little noticed. Thus, in the Basses Pyrénées, the Gave d'Ossau forms a fork at the foot of the high hill of the Sévignac. One arm, running to the north-west, flows on to join the Gave d'Aspe, and forms the Gave d'Oloron; but the other buries itself under the rocks, and reappears about five miles to the north, in two very strong springs, the stream resulting from which, called the Neez, empties itself into the Gave of the same name, not far from Pau. In like manner, in the centre of France, the Haute-Vézère sends one of its arms under the ground, for a distance of about three miles, to feed the stream of the Isle, which meanders through its deep valley in long parallel windings.

CHAPTER XLV.

THE HYDROGRAPHICAL SYSTEMS OF VARIOUS PARTS OF THE WORLD.

THE great difference which exists between continents as regards both their vertical outline and their extent of area, gives to the watercourses in each part of the world the most diversified directions and characteristics. In every place the general plan of the hydrographical system varies in proportion to the height and bearing of the mountain chains, the length and inclination of their slopes, the geological nature of the regions which it waters, and the annual quantity and distribution of the rainfall. But, since the continental masses, both in their general outline and in their different parts, present an evident equipoise in their forms; since the clouds and winds are in full obedience to constant laws; the result is, that the rivers themselves are arranged on the surface of the earth with a remarkable degree of order, which is all the more beautiful in that it so considerably deviates from any symmetrical regularity. The graceful windings of a river, its long and almost quivering curves, and the intricate bends of its innumerable tributaries, prevent our noticing the rhythm of its system, and how this system prevails from one end of the world to the other. On our earth physical laws are but rarely manifested in all their inflexible simplicity. Owing to the vitality which pervades everything, they often assume a character of beauty, and through this very beauty they not unfrequently evade the notice of man.

A study of the map with regard to the distribution of rivers over the surface of the earth, brings before our view, at a glance, this fact—that the water-courses which are tributaries of the Atlantic exceed considerably, both in number and importance, those which belong to the great Pacific Ocean. This sea, the greatest of all seas, receives directly only five considerable rivers—the Cambodin, the Yantse-kiang, the Hoang-ho, the Amoor, and the Columbia; but the comparatively narrow channel of the Atlantic is the reservoir into which the most enormous rivers of the earth pour their contents—the Uruguay and the Parana, the River of the Amazons, the Orinoco, the Mississippi, and the St. Lawrence, without reckoning the Congo, Niger, and Gambia, all the water-courses of Western Europe, and, by the intervention of the Mediterranean, the Nile and the Danube —the two great rivers of the ancients. This unequal distribution of rivers is a result of the semicircular arrangement of the Andes, the Californian Mountains, and those of Kamtschatka and Siberia, all round the basin of the Pacific. The western side of South America is excessively poor in rivers. All over this narrow belt, which is on the average not one-tenth as wide as the opposite Atlantic side, and is, besides, rarely visited with rain, there are at most but two or three rivers which are navigable. The streams of Chili and Western Columbia would scarcely merit the name of rivulets in the basin of the gigantic Marañon.

In a hydrographical point of view, the continent of Asia may be divided into three entirely distinct systems—those of the north, the centre, and the south. The first is the great plain of Siberia, which is gently inclined towards the Frozen Ocean, and the whole extent of which is crossed by three parallel rivers, certainly among the largest, but also, perhaps, the least used by man, of all the watercourses in the world. In the centre of the continent there are several closed basins, consisting of plateaux more or less desert, the streams of which are lost in some lake, or evaporate during their course. The southern and eastern countries of Asia are the portions of the continent which show a genuine vitality—thanks to the sea which bathes them, the deeply indented shape of their peninsulas, the varied productions of their soil, and, above all, to the numerous watercourses which traverse them.

The most remarkable of these rivers are arranged in pairs, so as to constitute three groups of twin currents. These are the Tigris and Euphrates, the Ganges and Brahmapootra, the Yantse-kiang and Hoang-ho. In each of these pairs, the two rivers take their rise side by side in the bosom of the same system of mountains, and, bending their course in opposite directions, each describes a vast semicircular line all across the continent, and ultimately again unite before they empty themselves into the sea through the same delta. There is another point which still further augments the analogy between these double fluviatile arteries, viz., that each empties its waters into one of the three seas situated to the east of the three southern peninsulas of Asia. The Shat-el-Arab flows into the Persian Gulf, to the east of Arabia; the Ganges into the Bay of Bengal, to the east of India; the Chinese rivers into the Pacific

Ocean, which stretches to the north and east of the Indo-Chinese peninsula.

In order to understand the general features of the river system of the Asiatic continent, there is a fourth group of allied rivers which we must also notice—the Indus-Sutlej. Certainly these two rivers of the western regions of Hindostan unite their waters at a rather considerable distance from their mouth; but their lower course has entirely the character of a delta. The Indus and the Sutlej were probably once separated, and became united in consequence of the alteration of their course, and the considerable elongation of the delta common to both which received their alluvium. In like manner, at the time of Alexander the Great, the mouths of the Tigris and Euphrates were situated at a good day's march from each other; but at the present day the two river-arms coalesce at some considerable distance from the sea, and form together the Shat-el-Arab. The Indus and the Sutlej may, therefore, be classed among the double rivers, as their sources lie very close to one another; their courses take an entirely different direction, and they also have a common outlet. As the waters of this fourth group of rivers descend from the same mountain range which gives rise to the Ganges and the Brahmapootra, we might even say that in the north of Hindostan there is a double system of allied rivers which at their sources are almost joined. The four most considerable currents of water in India, taking their departure from nearly the same point, flow away in opposite directions, and, after describing enormous circuits, unite in pairs, as if to obey some double law of harmony and contrast—the Indus and the Sutlej to the east, the Ganges and the Brahmapootra to the west. They are the four animals of the Indian legend—the elephant, the stag, the cow, and the tiger—which spring down from the same mountain peak into the green plains of Hindostan.

The contrast offered by Europe proper—so rich in mountains, peninsulas, and deep indentations of the coast—to the vast plain of an almost Asiatic character which distinguishes Eastern Europe, shows itself equally in the river systems of the two halves of the continent. In Western Europe the Alps and the other chains of mountains radiating from them determine the characteristics of the water system. In the Sclavonic countries, inhabited as they are by peoples hardly emerged from barbarism, the great rivers, such as the Volga, the southern Dwina, the Niemen, the Bug, and the Dnieper, all take their rise in the marshy or slightly undulating regions which occupy the interior of Russia. Certainly they roll down a very considerable mass of water; but in their historical importance they are very inferior to the rivers which spring from the Alps, and, flowing in every direction, water the various countries of Western Europe—the principal theatre of modern civilisation. The Alpine group of streams is that to which it is chiefly material to devote a separate study. From the sides of the St. Gothard, the centre of the Alps, three rivers, not counting the Reuss, take their rise—the Rhine, the Rhône, and the Tessin—falling respectively into the North Sea, the Mediterranean, and the Gulf of Venice. Two other water-courses, which do not precisely descend from the St. Gothard itself, take their rise in its vicinity. These are the Aar, the principal tributary of the Rhine and the Inn, a stream more important than the Danube—the name which it assumes below the point of their confluence. Here, then, are five rivers which radiate towards four seas from one single group of the Alps; but as isolated rivers, and not in the form of double systems, like those of India and China. However, these distinct water-courses, especially the Rhône and the Rhine, present some remarkable peculiarities. These two great rivers, nearly equal in volume, flow each in a diametrically opposite direction; then, turning suddenly towards the north by an abrupt bend, and crossing a lake of considerable dimensions—one the Lake of Geneva, the other the Lake of Constance—cross the parallel chains of the Jura either in rapids or cataracts, and, finally emerging from the mountainous regions, flow, the one directly to the north, towards the German Ocean, the other directly to the south, towards the Mediterranean.*

Other groups of the same mountain chain, such as those of the Viso and the Levanna, near Mont Cenis, form secondary centres for the radiation of streams; but as regards their hydrographical importance, none of them can be compared to the central group of the St. Gothard.

The great rivers of peninsular Europe which are not fed by the Alpine snows flow to the north of the almost continuous line of mountains which is formed across the continent by the chains of the Pyrenees, the Cevennes, the Jura, the Alps, and the Carpathians. The rivers which descend to the south are smaller, on account of the more contracted area which is afforded them in Europe by the Mediterranean slope. But it must be remarked that the line of summits does not exactly mark out the watershed where the waters divide, some flowing to the north, the others to the south; there is, in fact, a complete mutual invasion of the opposite basins, and their

* W. Huber, *Bulletin de la Société de Géographie*, 1866.

respective interpenetrations fit, as it were, one into the other. A river flowing to the north receives affluents from the southern side of the mountains, and another flowing to the south receives those from the north. Thus, on the Tatra (Carpathians) the watershed is far from coinciding with the line of summits, and cuts across the chain of mountains. The Arva, coming from the north, penetrates the mountain chain, and flows on into the Theiss; whilst the Poprat, taking its rise in the south, hollows out a bed for itself through the gorges, and runs on to join the Vistula.* In like manner, the Garonne rises in the glaciers of the Maladetta, to the south of the principal chain of the Pyrenees, and makes its way into the district of Aran and the plains of France; but to effect this it is compelled to cross the base of the mountain of Poumero through a subterranean gulf 4,376 yards long. The water, which disappears on the Spanish side in the high

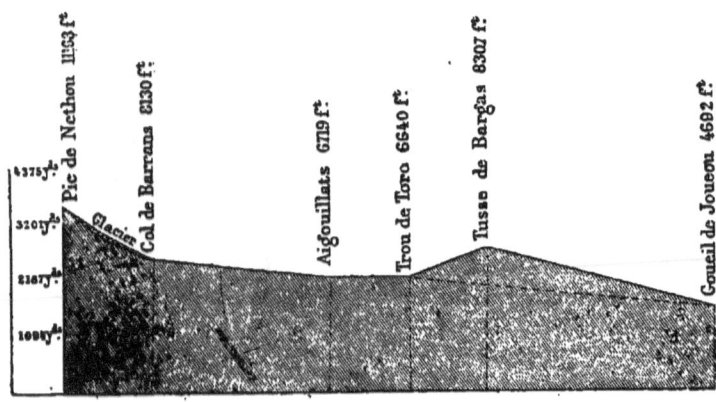

Fig. 93.—Sources of the Garonne.

valley of Essera, reappears on the other slope of the mountain at a point 1,980 feet lower down. The rising spring, the water of which thus pierces right through the rocks of Poumero, was once held sacred; it is called the "Goueil de Joueou" (Jupiter's eye).

In North America the same radiation of rivers exists as in Europe, but it spreads round three centres, two of which are mountain groups, and the other a merely gradual and imperceptible rising of the plain. In the territory of Idaho, between the 43rd and 44th degree of north latitude, a great peak towers up to a height of 13,779 feet, to which Lieutenant Reynolds has given the name of

* Carl Ritter.

"Union Peak," because the water from its melted snows, being soon increased and converted into important rivers, flows towards the Colorado on the south, the Missouri on the north, and the Columbia on the west.* More to the south, but still in the angle formed by the valley of the Colorado and those of the tributaries of the Missouri, the Rio Grande del Norte takes its rise, thus completing the system of radiation of large rivers round an elevated group of the Rocky Mountains. Nine degrees further north, in the vicinity of Murchison Peak, several of the more important springs rise, which feed the Fraser River, the Columbia, the Saskatchevan, the Athapasca, and the Mackenzie. According to Antisell, three of these rivers are fed by the snow of the same mountain. The sources of the Mackenzie and the Columbia take their rise at a distance of about 200 yards from each other; and in fourteen paces or so a man may walk from the origin of the Columbia to that of the Saskatchevan. These, then, are the spots whence the radiation takes place of the great rivers on the north-west of the continent. The radiating centre of the rivers of the plain is situated a little to the west of Lake Superior, in the vicinity of the Red Lake, Lake Itasca, Lake of the Woods, and several sheets of fresh water which are scattered over the highest part of the lower plateaux of North America. Thence spring forth the sources of the Mississippi proper, those of the St. Lawrence, and the Northern Red River, a tributary of the great Lake Winnipeg, which communicates with the Mackenzie River and the Frozen Ocean by a series of sheets of water. The radiating centre of the river system of the plains serves to link together the two centres of the Rocky Mountain chain. It forms the complement of them.

The three regions of the American river sources are mutually linked together by the two principal affluents of one great river. Thus, the gigantic development of the upper branches of the Mississippi connects the lofty groups of the Idaho mountains with the marshy plains of the Minnesota; as the Missouri it is classed as a mountain current, and as the Upper Mississippi it is a stream of the plains. The river, therefore, which unites all these waters is essentially double in its character. The Mackenzie River also presents this appearance of duality, although in a less degree, as it receives affluents both from the lake region and also from the chain of the Rocky Mountains. In like manner, the two principal branches of the Columbia, the Serpent River, and the Columbia proper, take their

* Humphreys and Abbot; Antisell, &c.

rise respectively in the two groups of summits, whence the streams radiate towards various points of the continent.

South America is *par excellence* the country of rivers. There roll down the immense Amazon, navigable for more than 3,000 miles; the mighty Parana, signifying by its name "The River" pre-eminently; and the Orinoco, surnamed "the Father of Waters," the drainage area of which is not one-third so extensive as that of the Mississippi, although the latter river pours down a much less considerable body of water. On account of the narrowness of the Pacific slope, all the great water-courses of South America flow over the plains situated to the east of the continent; but they do not all take their rise in the chain of the Cordilleras. The Orinoco takes its rise in the mountains of Guiana, the Marañon in the Andes, the Parana and the greater part of its tributaries spring from the high plateaux in the interior of Brazil. These rivers, therefore, do not radiate round the same centre; on the contrary, they belong to two basins which are perfectly distinct, and, indeed, cross one another at right angles. The basin of the Amazon tends, in fact, from west to east, whilst the plateaux and plains in the middle of the continent, forming a basin in the direction of the meridian transversal to that of the Amazon, are watered on the north by the Orinoco and the Rio Negro, on the south by the Tapajoz, the Madeira, the Paraguay, and the Parana. The distinguishing feature of the river system of South America is in the fact that the three principal rivers are interwoven by means of an almost continuous line of running water, which extends from north to south—from the mouth of the Dragon to the estuary of the Plata. More than half a century ago Humboldt placed the matter beyond all doubt that the Cassiquiare empties its water both into the Orinoco and into the Rio Negro. The communications between the Tapajoz and the Paraguay are not so perfect, but they nevertheless exist in several places. According to M. de Castelnau, the proprietor of the Estivado farm irrigates his garden by turning the water from an affluent of the Paraguay into the bed of the Tapajoz, and makes the little channels flow at his will towards either the northern or southern side of the continent. In like manner, there is a stream near Macu which at the time of inundations is divided into two currents, one forming a part of the Plata system, and the other belonging to that of the Amazon. Further to the east, the Rio Guaporé, an affluent of the Madeira, and the Jauru, a tributary of the Paraguay, take their rise in a plain which is periodically inundated during the rainy season. At the foot of the Bolivian

Andes a similar intermingling of basins takes place, as regards the Marmoré and the Pilcomayo. Thus, the Caribbean Sea and the mouth of the Orinoco are connected with the estuary of the Plata by a series of rivers, streams, and marshes.

The numerous water-courses which proceed from the central plateau of the continent are all set in an aspect parallel to the Tapajoz and the Madeira. The chief affluents of the Orinoco, on the contrary, follow the same direction as the River of the Amazons. We are, therefore, correct in saying that the river system of South America comprehends two basins crossing one another. The Rio Magdalena, the Atrato, and the other streams of Guiana, are all rivers

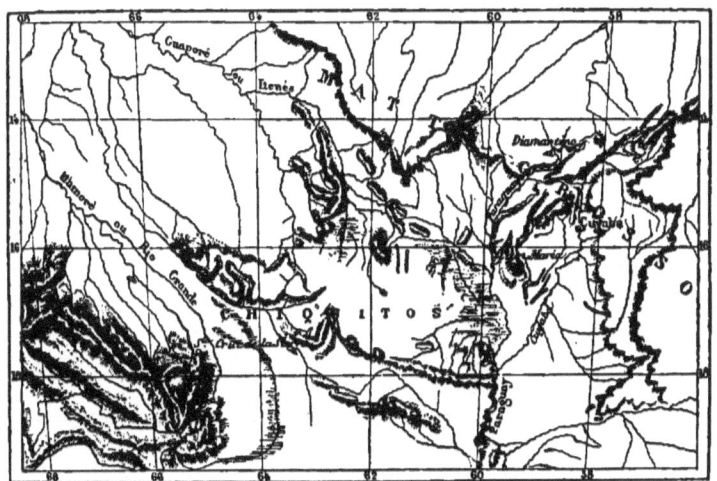

Fig. 94.—Interlacing Basins of the Amazon and the La Plata.

with distinctly limited basins; but it must be remarked that they all flow from the south to the north, in the same direction as the southern tributaries of the Amazon.

In that portion of the earth which is the most massive and the least articulated in its shape an harmonious correspondence is found between the water-courses and the continent itself. As long as the greatest part of Africa was an unknown region, geographers were able to attribute to its rivers all kinds of imaginary courses; they could, as their fancy dictated, make the Nile, the Niger, and the Congo take their rise from one common source, or interweave in a complete net-work all the tributaries of these great rivers. But the discoveries of modern travellers will now warrant us in forming some general idea of the African river systems. This land, so devoid as it is of peninsulas and of deep indentations in its coasts, does not, probably, present more than one centre of radiation for its waters, which centre is situated about the middle of the continent. From this point descend the Chary, the Binué—a tributary of the Niger—various streams falling into the Congo, and some important affluents of the Nile. Still, the principal branches of the large rivers take their rise at enormous distances from one another, and in the general features of their courses exhibit only some transient and slight similarities. The basin of the Nile is partly separated from that of the Niger by a great depression, the centre of which is occupied by the Lake Tchad. In like manner, several lakes and their affluents are interposed between the three basins of the Nile, the Zambesi, and the Congo; lastly, a small independent inland sea—the Lake N'gami—having its own special system of tributaries, fills up the space between the basins of the Zambesi, the Orange River, and the Limpopo. There is another point which distinguishes African rivers from those of other countries; this is an absence of any extent of ramifications. In this characteristic they resemble their mother-continent — a gigantic trunk without peninsular branches. From Assouan to Rosetta, a length of seven degrees, the Nile does not receive a single visible affluent; nevertheless, it must necessarily be replenished by several underground tributaries, for its liquid mass is much more considerable in Egypt than in Nubia.*

Australia is even poorer in rivers than the east of the African continent itself. With the exception of the Murray, its affluent, the Darling, and a few other rivers that are navigable at all times, the greater part of the water-courses in Australia can scarcely be said to exist except during the rainy season, and in summer their beds are only indicated by pools of stagnant water at intervals. Their special characteristic appears to be periodicity.

The general features of the river systems of each part of the world may thus be shortly summed up:—

Northern Asia is distinguished by rivers of simple character. In the south and east they are allied.

Europe is distinguished by two centres from which the streams radiate—one situated in the midst of vast plains, the other in the heart of the highest mountains of the continent.

North America is characterised by a radiation of the rivers from

* Elia Lombardini, *Essai sur l'Hydrologie du Nil.*

RIVER SYSTEMS. 321

three centres, two of which, being elevated groups in a mountain chain, are linked together by the third, occupying a marshy rising in the plains.

South America is characterised by the crossing of two mutually transverse basins and the continuous union of the river systems.

Africa is distinguished by the comparative independence of its water-courses and their poverty in tributaries.

Australia, by the small number of its rivers and the periodicity of their existence.

The form of each continent, and the phenomena of climate peculiar to them, have thus determined the rise of rivers which are modelled on a particular type in each division of the world. As all continental masses differ one from another, the circulating system of each naturally harmonises with the general features of the regions which the running waters traverse and vivify.

CHAPTER XLVI.

THE RIVER OF THE AMAZONS.—DIVERSITY IN THE CHARACTER OF WATER-COURSES.
—UNITY OF THE LAW WHICH GOVERNS THEM.—EQUALISATION OF THEIR
SLOPES.—UPPER, MIDDLE, AND LOWER COURSES OF RIVERS.

In like manner as the hydrographical systems of each continent present in their special features the most marked contrasts, so the rivers of each country and the various tributaries of each river. They are distinguished by the length of their system, the winding of their course, the abundance of their water, the nature of the soil which they pass through, the colour and character of their alluvium, the general inclination of their bed, and the shape and number of their meanderings. Thus, only mentioning the basin of one single river, we may reckon among the tributaries of the Mississippi the Clear-water River, the Mud River, and the Blue, Green, Yellow, Red, Black, and White rivers. Names designating other physical properties besides that of the colour or purity of the water are also very numerous in the tributary valleys of the American rivers. The same kind of names occur in most river systems; and, indeed, nothing would be more easy than to give to every water-course some name relating to its general aspect, its characteristics, or some of the local circumstances which distinguish it, such as gulfs, cascades, or defiles. Like the trees of a forest, so also an infinite diversity is shown in all the running waters which moisten the surface of the earth. The chief cause, however, for this infinite variety in rivers must be sought for in the geological constitution of the soil through which they flow. Thus, among the old schistose and gneissose rocks, rivers are more often characterised by the abundance of their liquid mass and the winding of their bed. In calcareous districts the water-courses are less richly supplied, more rectilinear, and generally bounded on each side by steep escarpments. Any sudden bend in the course of a river usually indicates some important modification in the geological nature of the strata. We may mention as examples the elbows of the Rhine at Basle and Bingen, those of the Rhône at Lyons, of the Danube at Ratisbon, of the Elbe at the outlet of Saxon Switzerland. In South America all the great rivers flowing into the Atlantic describe a broad curve towards the east, when they leave the valleys of the Andes and make their way into the Tertiary formations of the continent.*

The river *par excellence*, the glory of our planet, is the great stream of the Amazons, which, next to the great upheaval of the chain of the Andes, forms the principal feature of the Columbian continent. This moving fresh-water sea, which takes its rise at a short distance from the Pacific, and empties itself into the Atlantic through an estuary measuring 186 miles from promontory to promontory, serves as a line of division between the two halves of South America, and, like a visible equator, separates the northern hemisphere from the southern along a length of about 3,000 miles. Everything belonging to this great central artery is on a colossal scale. In its immense basin, embracing an area of 2,700,000 square miles, it collects two or three thousand times as much water as the Seine. In different parts of its course this immense river is known under various names, as if it were composed of distinct streams set end to end, and, together with its tributaries, its *furos*, or false rivers, its *igarapés*, or lateral arms, offers scope for steam navigation of more than 30,000 miles. It is so deep that sounding lines of 150, 200, or even 300 feet, have failed to measure its depths, and frigates can ascend it for more than 1,000 leagues. Its width is so great that in some places it is impossible to see the opposite bank, and at the mouths of the Madeira, the Tapajoz, the Rio Negro, and some other of its great affluents, the distant horizon closes in upon the water just as in the open sea. It is replenished by dozens of rivers which scarcely find their equals in Europe, and many of them, being yet unexplored, still belong to the realms of fable. In several places its banks serve as limits to two distinct Faunas, and many species of birds will not venture to cross the broad sheet. Like the sea, it is inhabited by cetaceans; like the sea, too, it has its storms, and during a tempest the waves will rise to several feet in height. When we sail over the grey water of the estuary at the mouth of the river, we feel tempted to ask, says M. Avé-Lallemant,† whether the sea itself does not owe its existence to the enormous tribute which the rolling current is incessantly bringing down to it. The difference in the motion produced by the movement of the waves or by the force of the current is the only thing which points out on which domain a voyager is sailing—that of the fresh or salt water. Even in late years, the greater part of

* Ami Boué.
† *Reise durch Nord-Brasilien.*

the inhabitants of the shores of the Amazons—white, black, or red men alike—are in the habit of fancying that the great river surrounds the whole universe, and that all the nations of the earth are denizens of its banks.*

Certainly, the difference is considerable between the mighty South American river and some slender stream; as, for instance, the Argens, which is crossed by a bridge with a single arch, and can readily be waded through by travellers. But whatever may be the comparative importance and the discrepancy of aspect in these rivers, they are none the less governed by the same laws. The geographer can describe them all together by forming an outline of an ideal river, the course of which would afford the combined phenomena of all the streams which traverse the globe.

The function of rivers in the plan of nature is incessantly to renovate the surface of continents, to convey the life and the alluvium of lofty mountains down to the plains and the coasts of the ocean. It has often been said that a landscape cannot be really beautiful when it is destitute of the rippling motion of a lake, or the presence of running water. The fact is, that man, whose life is so short, and, in consequence, so restless, has an instinctive horror of immobility. To make him fully appreciate the vitality of nature, it is requisite that motion and sound should bring it home to his senses. Only by a course of long reflection can he duly estimate the long-protracted movements of the terrestrial crust; he therefore needs to view the rapid bounds of the water leaping down in cascade after cascade, or the harmonious undulations of the waves. More than this, he also demands the contrast between the stable and the unstable, between restlessness and rest. This is the cause why a field of snow as far as the eye can reach, a desert without water, a sky without clouds, or a shoreless ocean fail to excite in him anything better than a gloomy or melancholy admiration. In the presence of these spectacles man feels himself crushed, whilst in a narrow valley, with its streams of running water, he is fully conscious of his own vitality.

On our earth, water is, *par excellence*, the symbol of motion. It flows and flows on for ever, without rest, and without fatigue. The lapse of centuries cannot dry up the slender rill of water trickling from the fissure of a rock, and fails to silence its soft and clear murmur. It leaps down joyously, in cascade after cascade, to mingle with the impetuous torrent; then, blended with the calm and mighty river, it flows on, and loses itself at last in the immense and

* Bates, *The Naturalist on the River Amazon.*

mysterious ocean—that tomb in which every water-borne fragment finds a temporary grave till the resolved elements enter again into the vast bosom of nature and reassume fresh forms of vitality. Motion is only another word for action. Water does not merely flow through a bed hollowed out ready for it; it is incessantly eating away, undermining, corroding, washing away, and moving the earth and the rocks which hem it in or oppose its course. Pebble by pebble, and grain by grain, it is carrying the mountains into the sea. Water, as Pascal says, is "not merely a road in motion, it is also a travelling continental mass which, in the centuries of yesterday, was covered with the eternal mountain snow, and will in the ages of to-morrow be fixed on the sea-shore, to augment the domain of man." Rivers carry out the circulation of solid as well as of liquid matter; they are, like the blood, ever-flowing life-renewers. It is, then, requisite that we should study carefully the mode of operation which rivers adopt in their renovating action on the continents they traverse.

Every current of water is constantly tending to equalise its slope, to increase it where it is almost imperceptible, and to diminish it where it is too rapid. The whole course of the river, from its mountain source down to its junction with the sea, may be compared to an avalanche falling from the heights of some snow-clad peak. The masses which sink down into the valleys modify gradually in their fall the outline of the cliffs. The projections are broken down, the fissures are filled up, a gracefully curved slope of *débris* abuts against the vertical walls, and extends in a gentle incline down into the plain. Owing to all these excavations and fillings up, the passage through which the avalanche makes its way ultimately assumes an outline of considerable regularity. Although less abrupt in its progress, less violent in its effects, and gliding over a gentler slope than the avalanche, still the river adopts a very similar course of action; it clears away the obstacles before it, and fills up any depressions, appearing as if it endeavoured to provide for itself a uniform incline down to the sea.

The portions of a river's course where this equalisation of its incline chiefly takes place are naturally those where the declivity of the bed is most rapid, and where the waters consequently attain their highest rate of speed. It may be generally asserted that those portions of the river-beds which are distinguished by the most abrupt incline are also the most elevated; for in almost all the countries of the earth the plains lie round the circumference of the land, and the mountains rise far in the interior. Most rivulets and streams take their rise thousands of feet above the level of the sea, and descend first through a very steep bed, sometimes intersected by precipices, or even interrupted by lacustrine basins. On reaching the lower plains, the running

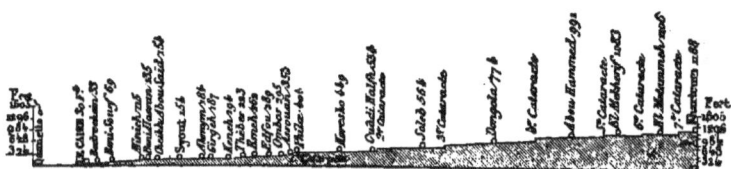

Fig. 95.—Inclination of the Nile from Khartoum to Damietta.

water, now converted into a considerable river by the tributaries which have joined it on both sides from the valleys of the mountain system, extends in long and peaceful windings across the more or less sloping ground which serves as a pedestal for the mountain chain. This is its middle course, during which the river receives its principal affluents descending from other mountain chains, or the high ground which commands it laterally. Then, below the last hills, its lower course begins; the fresh water descends slowly down to the

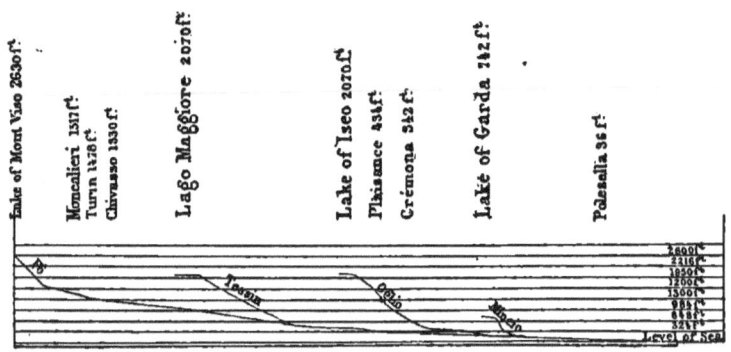

Fig. 96.—Slope fo the Pó, the Tessin, the Oglio, and the Mincio.

sea, and, not far from the mouth of the river, is arrested in its course twice every day by the salt tide which meets it.

The Rhine is a magnificent example of a river in which the three divisions of its course are regularly developed.* The upper course,

* Carl Ritter, *Europa*.

the whole of which is included in the Alpine regions, bends round in a vast semicircle to Laufenburg and Basle, where the rapids cease. The middle course, remarkable for its regularity, rolls on uniformly to a point below Mayence, where the Rhine is compelled to open a passage across the Odenwald and other hills; then, below the Siebengebirge, between two low banks of alluvial origin, commences the lower course, which ultimately terminates in the muddy estuaries of Holland. But for one river where the three divisions of its course are marked with so much distinctness, how many there are which exhibit no marked difference between the various portions of their bed! How many there are, indeed, which are even calmer and less inclined on the plateaux of the interior than in the vicinity of the sea! How many there are, especially, which—as represented by some of their affluents—are entirely rivers of the plain, whilst in other tributaries which descend from the mountains they exhibit all the characteristics of torrents! These are differences essential to the fluviatile system and to its geological operations.

CHAPTER XLVII.

MOUNTAIN TORRENTS.—INEQUALITIES OF THEIR BEDS AND OF THEIR DISCHARGE OF WATER.—TEMPORARY STREAMS.—FILLING UP OF LAKES.—EROSIONS, GORGES, AND SLOPES.—TORRENTS OF THE FRENCH ALPS.

The principal features which distinguish the mountain torrent from the water-course in the plain is the irregularity of its bed, its mode of action, its discharge of water, and its sedimentary matter. Among the gentler features of the plain, the stream runs but slowly, and all the changes of slope, curve, and level take place in gradual transitions; but, on the contrary, in narrow winding gorges it is violent, impulsive, and uncertain. Rocky angles project abruptly across the water; the declivity is intersected with precipices; the liquid mass poured down by the torrent may sometimes be compared to that of a river, but at other times it forms only a slender rivulet, or even dries up altogether. Lastly, most mountain streams are sometimes as pure as crystal, and at others are loaded with so large a quantity of alluvium that they are more like avalanches of *débris*. The turns and twists of the gorges are so much the more sudden as the rocks through which they are cut are higher, harder, and more irregular in their stratification and fissures. The water dashing against some projection springs back at right angles on the opposite rock, to be again driven back, and thus descends towards the valley in a series of zigzag falls. In these rugged gorges, where the pathway seems suspended from the ledges of the opposing cliffs, on either side overhead may be seen the abrupt fissures where the torrent has cut a passage; and not only is this mass of water and foam incessantly cast from one side to the other by the obstacles which hem it in, but it is very often temporarily kept back by the barriers of *débris* which crumble down across its course. When the dam, composed of stones and blocks of rock, affords no interstices through which the water can glide, the latter gradually rises in the form of a lake, and then makes its way as a cascade over the wall of rubbish, which by degrees it hollows out down to the level of its old bed. But usually the avalanche which pens back the torrent consists of a mass of snow, dust, and broken stones; the water kept back by this more plastic dam slowly converts it into a kind of pasty mass, and forces its way through a subterranean outlet. In the spring, when a good many avalanches are falling from the sides of the Alps, it is curious to trace the course of the torrent, visible here and there, in the gorges. The water may be seen diving down under some greyish or dark mass, joining with its graceful curve the two opposite sides of the ravine. The entrance of the gulf forms a kind of porch ornamented with icicles, down which the melted snow trickles or falls drop by drop. Above the torrent which is roaring in the depths below, the mass of *débris* is intersected in some places with *crevasses*, and the closely-packed snow presents a bluish edge, like ice; wells open in it at intervals, at the bottom of which the foaming waves may be indistinctly seen careering along.

In ravines and defiles where the slope is uniform the torrent-water affords some degree of regularity in its volume; but when the declivity is unequal and broken, and especially when, as is the case in most of the calcareous districts, it is composed of horizontal layers intersected by precipices, the liquid mass is incessantly changing in width and depth. In the level or gently inclined portions of its bed the water, flowing slowly, spreads out into a wide stream, until, reaching the edge of the cliff, it suddenly tumbles over, and, losing in volume what it gains in speed, seems nothing but a slender thread of foam gliding over the face of the rock. Below the fall a new basin opens out, often hollowed in the shape of a tub, in which the water, now to all appearance unstirred by the slightest current, reposes quietly as in a lake. A great number of the valleys in the Alps, the Jura, and all mountainous countries, owe their picturesque beauty to this succession of pools of quiet water and graceful cascades. This series of gradations constitute the successive planes of elevated valleys.*

The variations which are found in the discharge of a torrent stream are really enormous, even in those mountainous countries where, owing to the accumulation of the winter snow upon the heights, the water never entirely dries up. During severe cold, when the snow above is frozen on the ground, and numbers of rivulets are converted into solid ice, the main stream of the valley sends down only an inconsiderable liquid volume, and a traveller may easily cross it by jumping from stone to stone; but on the arrival of the earliest warm weather, when the rain and the sun,

* See above, Chapter on Valleys, p. 144.

assisted by the south wind, melt the snow and cause it to slide down in avalanches, the masses of water which are discharged into the torrent from all sides change it into a formidable river, running sometimes, Surell tells us, at a speed of 46 feet a second—more than 30 miles an hour. It spreads out widely over its basins, flows over the meadows, and often washes away farm-houses, trees, and even the vegetable mould. In the defiles, on the contrary, it is compelled to gain the requisite space in height, as it cannot find it in width, and its level suddenly rises 60, 80, or even 120 feet. All this may easily be noticed in the narrow Italian valley-streams fed by the snow from

Fig. 97.—Circle of the Valley of Lys.

the Mont Blanc and Monte Rosa groups. The Sesia, the Dora, and many of their affluents, before they empty into the plain, pass through dark gorges, where the liquid mass of the flood-water, ten times deeper than its width, descends with the rapidity of an avalanche. Looking forward to these rushes of water, the mountaineers, in many places, have dug out their paths more than 150 feet above the bed of the torrent.

The Var may be mentioned as an instance of this astonishing fluctuation in the discharge of its torrent-waters. At its outlet, the liquid mass of this river varies from 37 to 5,240 cubic yards of water in a second; this difference is as 1 to 143, and the proportion would be still larger if the fluctuations were measured above the confluence of the Vaire, the Tinée and the Vésubie.* In the level countries of Western Europe, the difference presented between the high and low-water levels is, on the average, scarcely one-tenth of that afforded by the Var. In great rivers, such as the Mississippi, the difference between high and low water is as 1 to 4 only. As a standard of comparison between the floods of a torrent and those of a lowland river in the same climate, we may mention the Upper Loire and the Somme. Above Roanne, the basin of the Upper Loire, at its first outlet from the mountains, comprises an area of 2,470 square miles, and the stream discharges during exceptional floods 9,549 cubic yards of water a second—rather less than four yards for each mile of surface. In its highest floods, the Somme sends down 117 cubic yards of water—a quantity which, if it was spread over its drainage area, would render the floods of the Upper Loire 84 times more considerable than those of the Somme; † and doubtless a comparative study of the inundations of all the water-courses in France would disclose still greater variations between the system of torrents and that of the lowland rivers.

In the tropical regions, where the rainy season is succeeded by the season of drought, the greater part of the mountain rivers only run during half the year; they are alternately considerable rivers and dry ravines.

Thus, some valleys, those, for instance, of the Sierra-Nevada de Santa Marta, exhibit a daily fluctuation in the discharge of their streams, owing to the storms which the gusts of the trade-winds rarely fail every afternoon to dash against the heights. In the evening, all the gorges are filled with masses of raging water, and the traveller finds himself compelled to put a stop to his journey; he bivouacs on the edge of the river, and is lulled to sleep by the noise of the cataracts roaring over the rocks; when he wakes up at dawn next day, all he sees is a slender rivulet of water, only visible here and there among the masses of gravel.

But the torrents which must be instanced as the most striking types of the merely temporary water-course are the *ouadys* in the

* Villeneuve Flayosc.
† Belgrand, *Annales des Ponts et Chaussées*, 1854.

Sahara and the plateaux of Arabia, and the liquid masses which sometimes roll down the *quebradas* of Bolivia and the Argentine *pampas*. All round the Red Sea, embracing an extent of more than 1,550 miles of coast-line, there does not exist one permanent stream. All the *ouadys* which, during heavy rains, flow into the sea, convey to it only the surplus of the surface-water which the sand of the desert was not able to absorb. In a general way, before the complete disappearance of these streams, most of which run over a bed of subterranean rock, they ooze up imperceptibly through the sand, and show themselves in pools stagnating in the passes of the defiles. Instances of streams thus converted into a chain of ponds are very numerous in deserts all over the world—in Arabia and Algeria, in the Caspian Steppes, and in the North American solitudes.

In these regions the ground on the plateaux and in the plains is furrowed with valleys exactly like those found in a country that is well watered with rivulets, streams, and rivers. The river system exists in full force, and for hundreds of miles the traveller may trace wide hollows, perfectly developed, which would contain rivers like the Danube or the Rhine, and on either side debouch the stony stream-beds of the lateral valleys. Nevertheless, these deep and winding depressions, hollowed out by the temporary water-courses, generally contain nothing but pebbles and sand; water is altogether wanting except during the season of the periodical rains. One of these waterless rivers, the Roumah, which connects its bed with the Euphrates, not far from the mouths of the Chat-el-Arab, is not less than 750 miles in length. The only permanent elements, so to speak, of its vast drainage area, are a few springs and rivulets flowing from the mountain-sides round the circumference of its basin.

In the upper part of their course, these torrent-waters assist, as we have said, in modifying the relief of the terrestrial surface; but these magnificent operations of erosion, which crumble away mountains or, at least, by enlarging the clefts, ultimately convert mere fissures into openings of such important dimensions both in width and depth, are not the work of the torrents alone. The latter, in fact, are scarcely the chief agents in the work; they do little else than clear away the stones and *débris* fallen from the heights above. All the meteoric phenomena of the atmosphere—among which, however, snow and rain may certainly be considered as the real origin of torrents—contribute to the work of destruction, and detach from the mountain-sides masses of *débris*, which accumulate at the foot of the rocks in more or less inclined slopes. The torrent into which this *débris* crumbles down washes away all the sand and lighter matter, until the time when, swelled by rain and melted snow, it rolls down towards the valley the great blocks of rock that have fallen into its bed. It

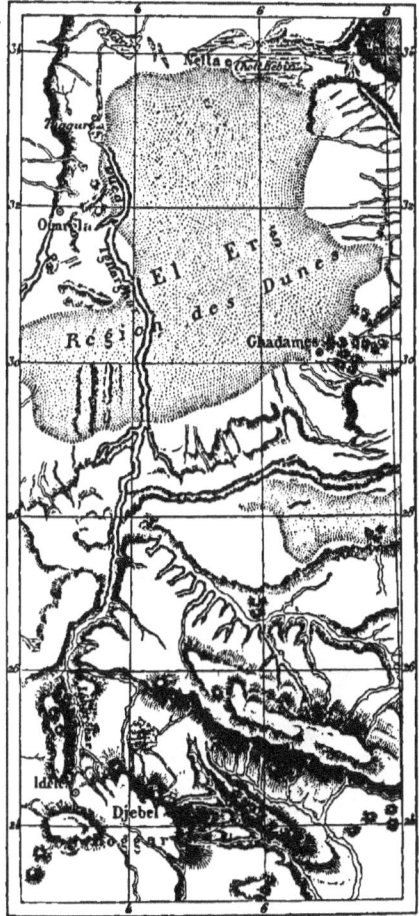

Fig. 98.—The Igharghar.

is difficult to restrain a feeling of dread when we pass along the bank of a flooded torrent and hear, above all the uproar of the water, the dull thunder of the masses of stone dashing one against the other as they are hurried along under the rushing water, yellow with the earth which it washes away.

Thus, year after year and century after century, the torrent clears away whole mountain-sides which have crumbled down into it rock

Fig. 99.—Valley of Cogne.

by rock, and this great work of erosion is incessantly going on. In some mountain groups, where the rocks are easily shifted by the action of the weather, nothing is left but a mere skeleton of

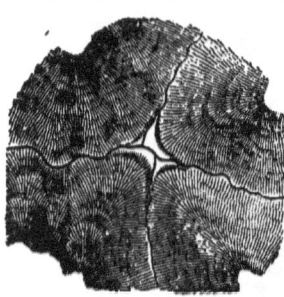

Fig. 100.—Quadrangular basin of erosion; after Sonklar.

those former proud heights which once towered up towards heaven. But in the regions where the mountain strata are of a compact formation, and the water consequently takes some considerable time to penetrate them; all that we notice in the way of dilapidation are large holes which the torrents have gradually hollowed out in the body of the rock. Where two mountain rivulets form a junction, it is very seldom that the three headlands which overlook the confluence do not leave at their base a

small triangular valley, whence the water leaps down into the lower gorge. In like manner, when two streams proceeding from directly opposite ravines fall into the main stream of the valley at the same spot, the little plain of erosion which is found at their confluence generally assumes a quadrangular form. It must, however, be understood that the dimensions and the outlines of these basins must vary infinitely according to the force of the torrents, the hardness of the rocks, and the energy of the agents that attack them. Ultimately, the surface of the country having been carved out by the water for an unknown number of centuries, completely changes its aspect; the mountains and the plateaux are swept down by the rivers, and little else remains but the isolated land-marks of the ancient piles.

There is probably no country in the world where this devastation goes on more rapidly than in the French Alps. The mountains of this region, and especially those which enclose the basins of the Durance and its tributaries, are in general composed of very hard rocks alternating with other beds, which easily give way under the action of the water; in every place we may notice immense cliffs resting upon bases without any solid consistence. The marls, the disintegrated schists, and the other friable matter are gradually washed away, and their fall precipitates that of the compact layers at the summit, which suddenly fall down or glide slowly into the valleys.* It is, however, the improvidence of the inhabitants, and not so much the geological constitution of the soil, which is the principal cause of the devastating action of the streams. In the mountains of Dauphiny and Provence, the slopes, most of which are now so bare, were once covered with trees and various plants which kept back the surface-water resulting from the rain or melting of the snow, by absorbing a great part of the falling moisture and thus retaining the coating of vegetable earth over the beds of crumbling rock. During the course of centuries, the trees have been cut down by greedy speculators, and by senseless farmers who wished to add some little strips of land to the fields in the valleys and to the pastures on the summits; but when they destroyed the forest they also destroyed the very ground it stood on.

The rain or snow, being now no longer kept back upon the slopes by the roots of the trees, descends rapidly into the valley, driving before it all the *débris* torn away from the sides of the mountain. The tooth of the goat and the sheep helps to lay bare the rootlets of the herbaceous plants and the brushwood; bit by bit, the whole of the thin coating of vegetable earth is removed, the bare rock shows itself, and deep ravines are hollowed out in the cliffs and are traversed in the rainy seasons by furious torrents which once did not

* Scipion Gras; Rozet; De Ladoucette; De Ribbe.

336   THE EARTH.

exist. The water which used slowly to penetrate the earth, con-

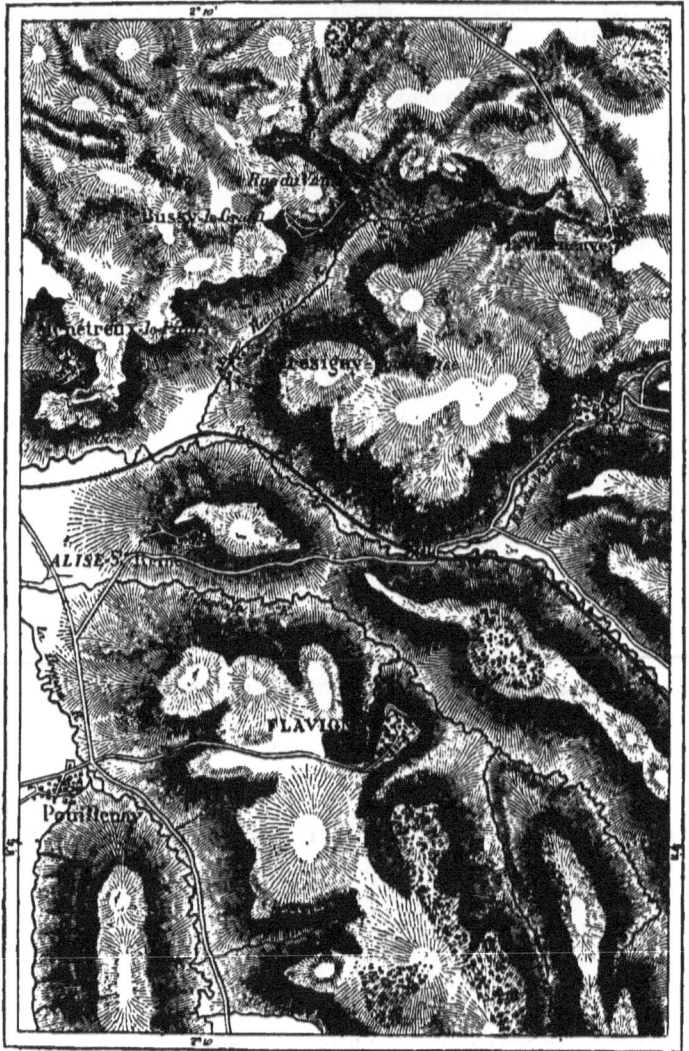

Fig. 101.—Valleys of Erosion of the Bourgogne.

veying fertilising salts to the roots of the trees, now serves no other purpose than that of devastation. When the forests are gone, great furrows of erosion may be noticed opening out at intervals on the slopes; these furrows often correspond to ravines situated on the other side of the mountain, and in a comparatively short space of time, they ultimately sever the ridge of the mountain into distinct peaks, uniformly surrounded by a slope of rocks or fallen earth: summits of this kind are being formed every year. In some localities, there is not a single green bush over a space of several leagues in extent; the scanty grey-coloured pasturage is scarcely visible here and there on the slopes, and ruined houses blend with the crumbling rocks that surround them. The stream in the valley is generally nothing but a scanty rill of water winding among the heaps of stones; but these very heaps of shingle and rock have been carried down by the torrent itself in the days of its fury. In many parts of its course, the Haute Durance, which is generally not more than 30 feet wide, seems lost in the midst of an immense bed of stones, a mile and a quarter wide from bank to bank. The Mississippi itself does not equal it in dimensions.

The devastating action of the streams in the French Alps is a very curious phenomenon in an historical point of view; for it explains why so many of the districts of Syria, Greece, Asia Minor, Africa, and Spain have been forsaken by their inhabitants. The men have disappeared along with the trees; the axe of the woodman no less than the sword of the conqueror have put an end to or transplanted entire populations. At the present time, the valleys of the Southern Alps are becoming more and more deserted, and the precise date might be approximately estimated at which the Departments of the Upper and Lower Alps will no longer have any home-born inhabitants. During the three centuries that have elapsed between 1471 and 1776, the *vigneries* of these mountainous regions have lost a third, a half, or even as much as three-quarters of their cultivated ground; and the men have disappeared from the impoverished soil in the same proportion. From 1836 to 1866, the Upper and Lower Alps have lost 25,090 inhabitants, or nearly a tenth of their population. At the present time, in an area of 3,860 square miles, embraced between Mont Thabor and the Alps of Nice, there is not a single group of inhabitants which exceed the number of 2,000 individuals. Barcelonnette, the most considerable place, has more than once been in danger of being carried away by the stream, the bed of which is higher than the streets of the town; the latter certainly would be still less populous were it not that the numerous functionaries necessary in every sub-prefecture tend to give it an artificial life. Without the *employés* and the custom-house officers, who almost consider themselves as exiles, the whole extent of a great portion of these mountainous regions would be nothing more than a gloomy solitude. It is the mountaineers themselves who have made and are seeking to extend this desert, which separates the tributary valleys of the Rhone from the populous plains of Piedmont. If some modern Attila, traversing the Alps, made it his business to desolate these valleys for ever, the first thing he should do would be to encourage the inhabitants in their senseless work of destruction. Is it necessary that man must ultimately rid the mountains of his odious presence, so that the latter, left to the kind offices of beneficent nature, may again some day recover their forests of fir trees and their thick carpet of flower-studded turf?

Although the torrents lower the mountains, on the other hand they elevate the plains; but their deposits, not being pulverised into clays and sand, are often the means of bringing another disaster on

Fig. 102.—Talus of *Débris* in the Valley of the Adige.

the inhabitants, who find their fertile land covered beneath enormous masses of rocks and pebbles. In fact, when a stream empties itself into a valley which has a moderately inclined slope, and the former consequently experiences a sudden check in its progress, it deposits over a long extent of descent all the *débris* which it conveys in its water or rolls down before it. The masses of rough alluvium accumulate on both sides of its course, so as to form a rising, with regular slopes abutting against the escarpments of the mountain. Even in places where the stream once rushed down into the valley in rapids or cascades, its tendency always is to conceal gradually every irregularity in its old bed under the ever-increasing slope of rocks, pebbles, and sand. The deep ravine of the upper valley is succeeded by a long embankment, which, continuing the incline, pushes out far into the principal valley, and forces the stream to describe a considerable bend round the base of the cone of *débris*. Some of these

TALUS FORMED BY TORRENTS. 339

banks attain very important dimensions; they accumulate to an enormous extent at the outlet of each lateral ravine opening into the elevated valley of the Adige, to the south of the Œtzthal group. One, that of the Litznerthal, is 1,036 feet in height at the outlet of

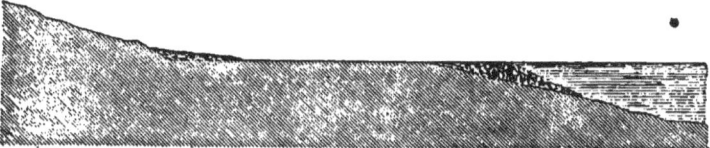

Fig. 103.—Talus formed by Torrents.

the ravine, and extends 4,148 yards in length as far as the Adige, with a mean slope of 4° 46′; the curve of the river which winds round its base is not less than five miles in length.

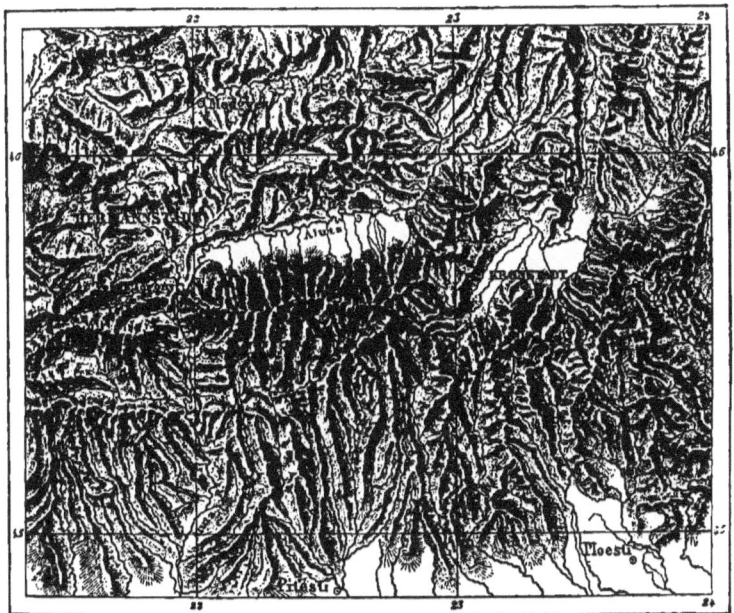

Fig. 104.—Ancient Lakes and Defiles of Aluta.

When the streams empty their waters into a mountain lake, and not into a valley, the *débris* which they carry down accumulates at the upper end of the lacustral basin, forming a slope much more abrupt

z 2

than the mass of stones deposited at the entry of a ravine. In fact, at the outlet of the latter the water of the torrent continues to flow over the masses which it has heaped up; fresh materials are continually being brought down, some of a small, others of a large size, which serve both to prolong the slope and to render it more and more uniform with that of the plain below. In lakes, on the contrary, a separation immediately takes place in the various *débris* brought down by the current. The blocks of stone and pebbles fall by their own weight into the depths of the water, and form a kind of *moraine*, which incessantly pushes on into the quiet water. The lighter alluvium, which is held in suspension by the liquid mass, is partially carried on by the current towards the middle of the lake;

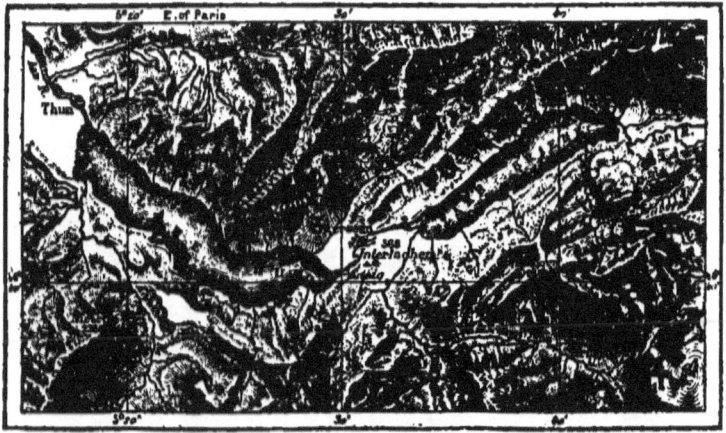

Fig. 105.—Lakes of Thun and Brienz.

but the greater part of this matter is soon dropped on each side of the *embouchure*, and ultimately extends in horizontal promontories above the accumulated mass of heavier rubbish. Thus, the bed of the stream, with its steep slope of stones in front, bordered by its layers of lighter alluvium, incessantly encroaches on the lake.

A large number of lakes have thus been gradually filled up altogether; in several high mountain valleys, where lakes exist at intervals one above another, all the basins have in turn been filled up. In other places the upper pools only are choked, and the work is going on in one of the lower lakes, which, sooner or later, will ultimately be converted into a horizontal plain. By very carefully measuring the annual deposits of a torrent, and ascertaining, by poring, the depth of the former lakes which they have filled up, the number of centuries might be approximately estimated which this immense work has taken. Also, sounding the depths of the basins which are still full of water would show the duration of ages which will be required to fill up their abysses. At the foot of the great group of the Bernese Alps, on the isthmus of Interlachen, so well known to travellers, it would be comparatively easy to make the experiments necessary for the solution of this problem, which would also inform us approximately as to the duration of the geological period during which the streams have flowed down from the mighty group over which towers the Jungfrau. For this calculation it would be necessary to measure the present deposits of the furious Lutschine, and to estimate the enormous solid mass of the isthmus of Interlachen which has been thrown down by the stream as a kind of dam between the two lakes of Brienz and Thun, which once formed only one lacustrine basin.

CHAPTER XLVIII.

EROSION OF LACUSTRINE DIKES.—CATARACTS AND RAPIDS.

WHILST crossing the lakes situated at the bases of the mountains, the waters of the torrent become tranquillised, and their course regulated;* they emerge from the basin in streams of a less turbulent shape, and flowing on to join other water-courses, descend with them quietly to the sea.

But even the outlet-stream of the lake, although usually more peaceable than the water-course above, accomplishes its special geological labour, and is also employed in the task of doing away with the lacustrine basin. The water, impelled by its own weight, constantly wears away the layers which form the lower margin of the lake. The edge of this margin being gradually destroyed by the liquid mass, sinks by slow degrees, and the average level of the water in the lake sinks also in the same proportion. Thus, at the

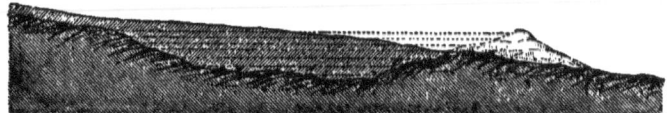

Fig. 106.—Filling up of a Lake-basin.

two extremities of the basin the river is carrying on two kinds of work, contrary in appearance, but which have both an equivalent result in reducing the area of the lake which the river crosses. Up above, it gradually elevates its bed, and gains on the lake by filling it up with alluvium; down below, it lowers the brink, and, by this constantly increasing waste-gate, gradually drains out the water. The two stream-beds, the upper and the lower, will ultimately meet in the middle of the lake, and the latter will cease to exist. This is the double phenomenon which has been going on for ages in the Lake of Geneva. This crescent-shaped sheet of water certainly once

* *Vide* the chapter on "Lakes."

FILLING UP OF LAKES.

extended as high up the stream as the place where the town of Bex now stands, 11¼ miles from the end of the lake; it also extended down the stream in narrow basins as far as Ecluse, 9⅔ miles from the outlet of the Rhone.

It must, however, be understood that the outlets of lacustrine reservoirs are not the only places where the rapids and cataracts of a river crumble away the rocks so as to lower the up-stream and elevate the down-stream beds. However hard may be the strata which form the bed of a rapid, the eddying waters ultimately penetrate the stone, and deposit the *débris* below the gulf that the furious shock of the torrent has hollowed out at the foot of the rocks. In like manner,

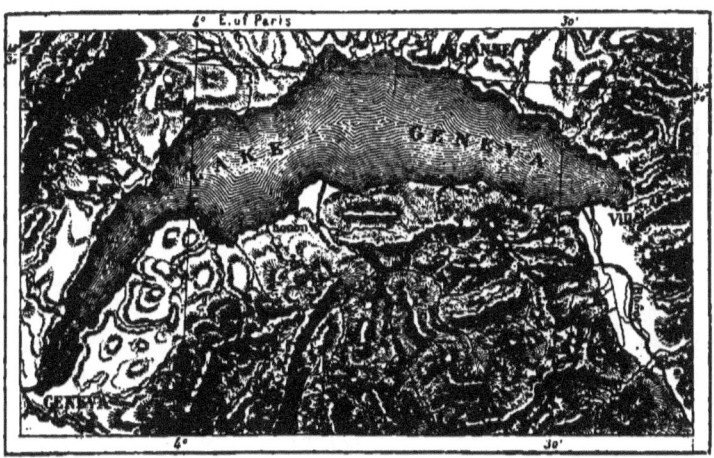

Fig. 107.—Alluvial Deposits of the Rhone and the Dranse.

cascades and cataracts incessantly wear away the ledges from which the mass of their water pours down to the bottom of the abyss, carrying the great stones with them in their fall, and, destroying layer after layer, they continually retrograde towards the source of the river, and tend to convert themselves into mere rapids, which, in some thousands of years or perhaps centuries, are destined to assume a perfectly uniform inclination. This is the ideal, so to speak, of every river—to do away with the irregularities of its course, and to flow down towards the sea, describing a regular parabolic curve. This ideal, however, is never perfectly attained, on account of the diversity of rocks in its course, the changes in its bed, the disturbances or elevations of the ground, and other circumstances of various kinds which may cause a deviation in its current. But whatever may be the obstacles which oppose the levelling of the declivity, still every river intersected by falls and rapids is constantly at work in effecting the general uniformity of its slope.

In their magnificent beauty, cataracts and rapids only yield the pre-eminence to hurricanes and volcanic eruptions. Of the former, there are some in Europe which are very remarkable, such as the falls of the Rhine at Schaffhausen; the four cataracts of the Gotha-Elf at Trollhäta (dwelling of sorcerers); the Hjommel-saska (the hare's leap), where the river Lulea plunges over in a body from a height of 264 feet; and the Riukan-fos (roaring cascade), which falls at the outlet of the Norwegian lake Mjösvand in a single jet of 885 feet. The most celebrated waterfall in the whole world is that of Niagara—"the falling sea"—the constant thunder of which may sometimes be heard 12 miles off. Above the cataract the river, which discharges on the average 1,300 to 1,400 cubic yards of water a second, breaks against the shore of Goat Island, and divides into two rapidly inclined currents. Even at this point the mass

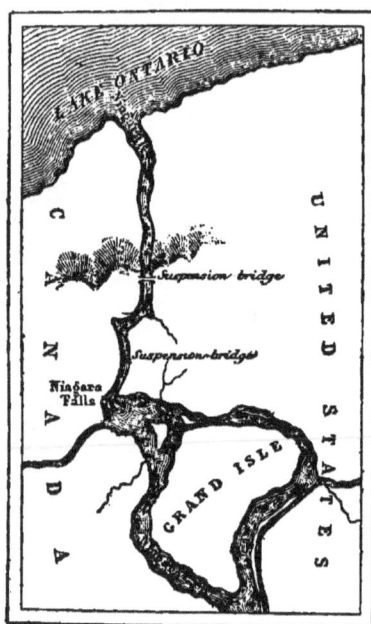

Fig. 108.—Course of Niagara.

of water is impelled by such velocity of movement that engineers have not yet been able to sound its depth, and they have similarly failed to do so below the cataract. On reaching the edge of the cliff, the two halves of the river—one 655, and the other 295 yards wide—take their final leap, and describe their vast parabola, 147 feet and 160 feet in height. A gloomy passage, penetrated by furious gusts of wind, opens between the wall of rock and a sheet of water, 18 to 33 feet in thickness, which curves widely overhead like an immense arch of crystal. Columns of iridescent vapour spring from the whirlpool of the roaring waters, and half hide the two white masses of the cataracts. At every instant of the day, following the path of the sun, the great rainbow painted on the wavering and misty spray shifts its position, and thus modifies the aspect of the fall. The various seasons, each in their turn, add some feature of beauty to the magnificence of the spectacle. The trees still left on Goat Island and the cliffs contrast with the whiteness of the water—in summer by their verdure, in autumn by the more varied colours of their foliage. In winter, stalactites, glittering in the sunlight like immense strings of diamonds, hang down from all parts of the rock, and serve as a framework to the two great plunging sheets of water. In spring, when the ice breaks up, a formidable spectacle is presented by the blocks of ice, like mountain fragments, crowding together at the edge of the cataract, and crashing against one another as they glide over the enormous curve of water which is sweeping them along.

Other great waterfalls in different parts of the earth afford similar phenomena, and several of their number may even rival Niagara in their beauty. Among these we may mention, in North America, the magnificent falls of the Missouri, the Columbia, and the Montmorency. Of like beauty, also, there is in Brazil, not far from Bahia, the wonderful cataract of San Francisco, known by the name of Paulo Affonso. At the foot of a long slope over which it glides in rapids, the river, one of the most considerable of the South American continent, whirls round and round as it enters a kind of funnel-shaped cavity roughened with rocks, and, suddenly contracting its width, dashes against three rocky masses reared up like towers at the edge of the abyss; then, dividing into four vast columns of water, plunges down into a gulf 246 feet in depth. The principal column, being confined in a perpendicular passage, is scarcely 66 feet in width, but it must be of an enormous thickness, as it forms almost the whole body of the river. Half way up, the channel which contains it bends to the left, and the falling mass, changing its direction, passes under a vertical column of water, which penetrates through it from one side to the other, and breaking it up into a chaos of surges, converts it into a sea of foam. Sometimes the white misty vapour may be seen, and the thunder of the water may be heard, at a distance of more than 15 miles.*

This turbulent cataract is very different in character from the majestic falls of the Zambesi, the existence of which Livingstone has

* Avé-Lallemánt, *Reise durch Nord-Brasilien.*

made known to the world. Above the precipice the river is calm, and flows over a gently inclined bed; some islets, covered with cocoa-nut trees, are reflected in the clear water. A large island, called "the Garden," on account of its rich vegetation, divides the Zambesi into two branches, and the general features of the landscape are full of grace. All on a sudden, without the least transition, the ground comes to an end beneath the water, and the two liquid masses, one of which is 1,858 and the other 546 yards wide, plunge down to a depth of 348 feet into the gaping fissure of a vast mass of basalt. They then escape by a narrow and winding channel, which the river itself has hewn out of the rock during the lapse of many centuries. Ten columns of vapour, answering to ten great projections on which the body of water dashes itself to pieces, rise in

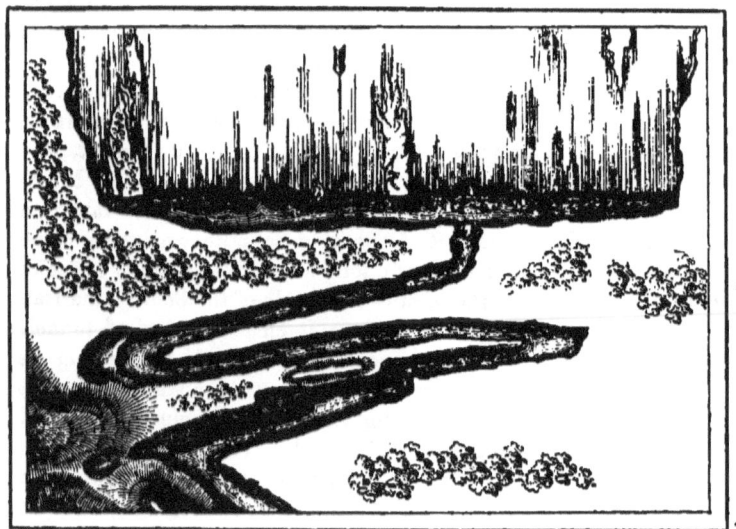

Fig. 109.—The Falls of the Zambesi.

eddies from the foot of the precipice, and float, like the smoke of a conflagration, far away above the surface of the river. They vary in height according to the state of the water and the atmosphere; but, generally speaking, they do not rise less than 1,000 or 1,150 feet above the brink of the gulf.* On account of these clouds of spray and vapour, the natives have given to the cataract of the Zambesi the name of *Mosi-oa-Tounya*, or "Thundering Smoke."

* Baines, *Exploration in South-West Africa*.

FALLS AND RAPIDS. 347

With regard to rapids, we find them on most rivers at different points of their course; either at spots where cataracts once existed, or at the mouths of streams which carry down with them large quantities of *débris*, and pile them up like dikes across the current. The

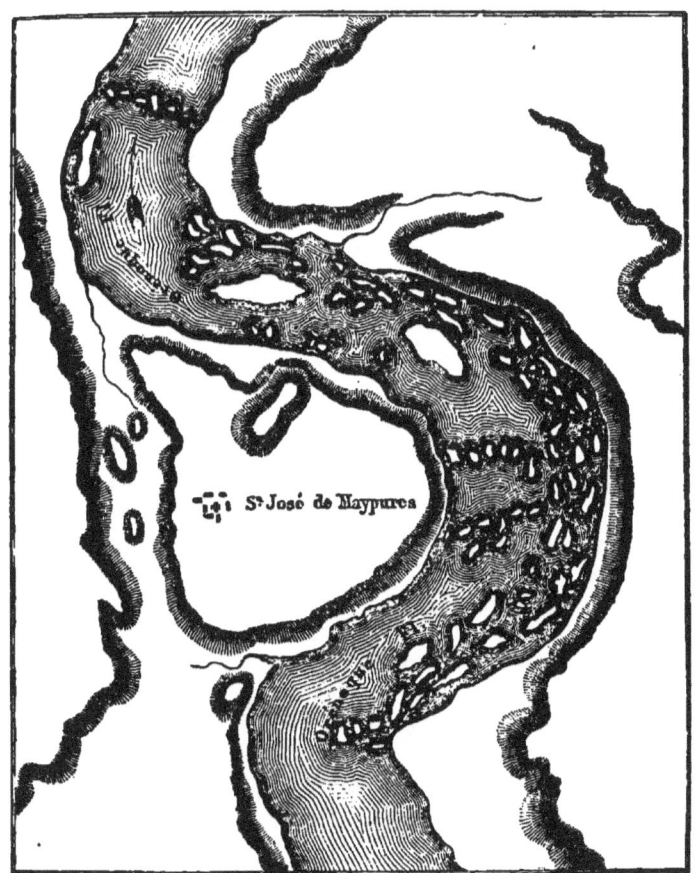

Fig. 110.—Rapids of Maypures on the Orinoco.

American rivers are the principal localities where these rapids may be contemplated in their full beauty. Humboldt was the first to describe the *raudales* of Atures and Maypures, where the Orinoco, changed into a mass of foam, pours down innumerable cascades over a chaos of rocks and banks with dark sides crowned with foliage and verdure. Each mass of granite, resembling in its shape some ruined tower or castle, is surmounted by a group of palms or densely-foliaged trees. Every stone below the level which the river reaches during flood-time is covered with alluvium, on which the mimosa, with its delicate leaves, grows abundantly; also ferns and orchids, with their charming flowers. They are perfect little gardens surrounded with foam, reminding one of the rocks covered with flower-studded turf which spring up in the midst of some of the glaciers in Switzerland. A cloud of vapour hovers over the river, and the rainbow shines through the verdant hues of innumerable bowers of foliage. This is the lovely spectacle which the Orinoco affords for a distance of several miles along each of its two rapids. The fall is not considerable, that of the *raudal* of Maypures being scarcely 30 feet; but still the slope is very difficult to overcome, and in a width of 2,841 yards the navigable channel is sometimes not more than 18 feet.*

About the same time as that when Humboldt and his friend Bonpland visited the rapids of Atures and Maypures, Azara examined the great *salto* of Maracayu, where the river Parana, which, just above, is 4,590 yards wide, is suddenly contracted into a deep channel only 66 yards across, and sliding over an inclined plane of 60°, forms a fall of 56 feet of vertical height. The narratives of travellers have also made us acquainted with the rapids of the Madeira, the Huallaga, the Ucayali, and several other rivers, down which the canoes of the savages used to glance like arrows in the midst of the foam. In North America the most celebrated rapids are those which the St. Lawrence forms at its issue from Lake Ontario; but all-powerful steam has succeeded in overcoming them. The European rapids are not so imposing, on account of the inferior quantity of the river discharge, and also because the general relief of the continent is much more gentle than that of the New World. We may, however, mention the rapids of the Shannon, above Limerick; the *porogs* of the Dnieper; and the whirlpools (*strudeln*) of Bingen, which were so dangerous before the rocks were blown up, which impeded the course of the Rhine. Among the most imposing rapids in France, both on account of their bulk and the fury of their foaming water, and also of the calm solemnity of the surrounding landscape, are those of the Gratusse, formed by the Dordogne, some miles above Bergerac.

In surveying both falls and rapids, there is one point that especially impresses the mind; it is that, in a general way, immediately the water has emerged from its state of turbulent effervescence,

* Humboldt, *Voyage aux Régions Equinoxiales*.

CATARACTS AND RAPIDS. 349

it assumes an unbroken surface, and spreads out into wide calm sheets, known in Spanish America under the name of *remansos*. On one side we look down on the giddy chaos of the liquid masses dashing against one another as they rush along; on the other we see a pool of water almost still, or at most slowly rotating. Here, the long gentle eddies seem unable even to move the straws and twigs which incessantly float round and round in the same circle; higher up the stream, the river in its impetuous career sweeps away trunks of trees, tears up the stones of its bed, and notches out the edge of the cliff over which it falls. This contrast becomes still more striking when we reflect that the cataract once descended at the very spot where

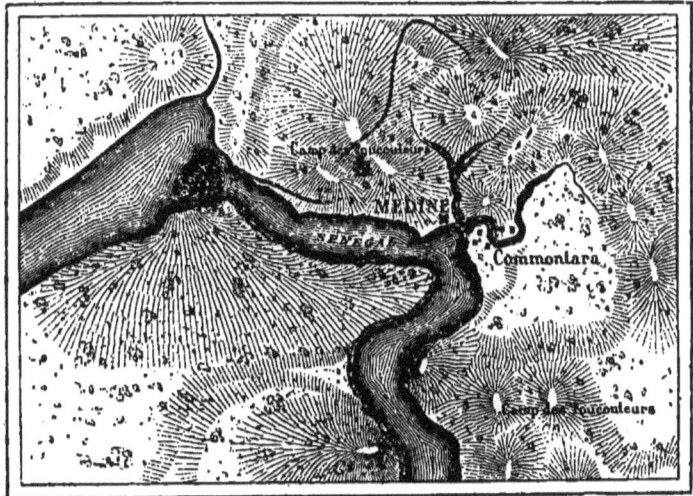

Fig. 111.—Cataract of Felou, Senegal.

this tranquil sheet of water now lies, and that during a long course of ages the fall has continually retrograded. The high vertical rocks which hem in the two banks of the river belong to the same geological formation, and the parallel lines of their strata exactly correspond on both sides. The traces of the current which has eaten away the stone are still visible, and the marks of the work slowly accomplished by the water can be distinctly traced out by the eye. The immense cavity which extends like a dark passage below the fall has been hollowed out by the cataract, scooped out, so to speak, grain by grain.

The rate of speed at which the fall shifts its position might serve to estimate approximately the age of the river itself. If geologists had studied this retrograde movement for a sufficient number of years, they would know the exact degree of resistance afforded by the rocks throughout the whole length of the cavity; they would be able to say with certainty how many centuries the present system has lasted with regard to every river which is interrupted in its course by a cataract. But this comparative study of waterfalls has scarcely commenced, except, perhaps, in the case of Niagara and some other of the great watercourses of North America. According to Hall, Lyell, and other geologists, the Falls of Niagara have receded three miles and a half in the space of about 35,000 years. The erosion of the edge of the precipice is now taking place at the average rate of 12·183 inches a year.* This is a tolerably rapid movement of retrogression, which, however, is explained by the nature of the rocks;

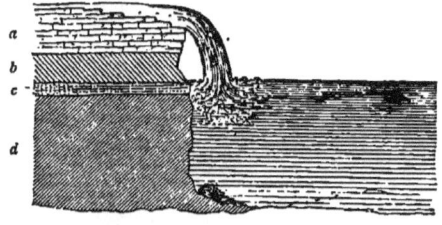

a. Niagara Limestone.   c. Niagara Marl.
b. Clinton Group.   d. Medina Sandstone.
Fig. 112.—Profile of Cataract of Niagara; after Marcou.

these latter are composed of calcareous strata resting on beds of soft and friable marl. The water penetrates into these layers, and slowly undermining them, washes them away, thus throwing down the upper strata in massive blocks, which are carried away by the cataract. The observations of M. Marcou have established the fact that the volume of water is constantly diminishing in the fall on the American side, and that, in consequence, the rocks there have scarcely been encroached upon for some twenty years. To make up for it, the great cataract is rapidly receding up stream, and even now it no longer assumes the graceful semicircular form which obtained for it the name of the "Horse-Shoe Fall." In an interval of time, which may be estimated at eight or ten centuries, the cliff of the cataract will probably be lowered as far back as the little islets of the Three Sisters; the whole liquid mass will then rush down the current which

* According to Lyell. Mr. Bakewell says that the annual retrogression of the falls has been nearly a yard since 1790.

runs along the Canadian shore, and the branch on the American side, no longer receiving any water, will gradually dry up; Goat Island will become united to the mainland, and the River Niagara, constantly receding towards Lake Erie, will pour down the whole of its water in one formidable fall.

It may likewise be presumed that the height of the cataract will tend to increase; for the calcareous strata which gave way under the weight of the water gradually augment in thickness in the up-stream direction.* But we are scarcely warranted in estimating, even approximately, the time which the Niagara will take in receding to Lake Erie; for, as M. Marcou remarks, the prodigious manufacturing activity of the Americans may much modify matters in this respect. A canal, which is, in fact, a perfect river, already turns a large number of mills on the American side; and if the river is tapped by thirty or forty conduits of this importance, the mighty Niagara will become nothing but a humble rivulet. "Arts and manufactures will have disarmed the thundering Jupiter." But Lake Erie itself, which, according to Ellet, contains at present more water than the fall could run off in six or eight years, will be perhaps filled up with alluvium before Niagara has been able to wear away the lower ledge of rocks which prevents the lake from rushing down bodily into the basin of Ontario.

That which the future will perhaps accomplish as regards Niagara, has already taken place in the Mississippi. Nearly halfway between St. Louis and Cairo, the river penetrates a defile which cuts through the chain of the Ozark Mountains; rocks, 300 feet in height, rise between the two banks, and on their perpendicular sides may be clearly distinguished the lines of erosion which were once traced out by the current of the Mississippi. In former days these rocks formed a barrier over which fell a cataract like that of Niagara, which, too, like the former falls, constantly wore away the strata which served as its bed. Above this barrier of hills, the water of all the upper tributaries united in a vast lake, which extended north as far as the mouth of the Wisconsin, and joining Lake Michigan on the east, covered all the immense prairies of the intervening peninsulas.† In like manner, the Rhine, the Danube, and a great many other rivers, the course of which at the present day is tolerably uniform, presented a succession of lacustrine ponds, placed in gradation one above another, and united by cascades. The

---

* Marcou, *Bull. Soc. Géol. de France*, 2nd Series, vol. xxii.
† Humphreys and Abbot, *Report on the Mississippi River*.

rocky barriers situated between the ponds have been gradually demolished and washed away by the water; some of them have even been pierced through at their base, and this is the origin of the natural bridges which throw their arches above a great number of streams and rivulets. The Pont-de-l'Arc, which the water of the Ardèche has slowly bored out during the course of geological ages, has a span of not less than 177 feet. The famous natural bridge of Virginia is only 111 feet in width.

By operations of this kind, rivers gradually regulate their slope and effect a communication between plateaux of different height, which sink in successive gradations from the base of the mountains to the sea-coast. Whatever may be the irregularities of the continental surface, running waters cut out their beds in the form of inclined planes, and give them a more or less regular slope, which is followed by merchandise, travellers, and even civilisation itself, in order to penetrate into the separate basins of the river-system. Every cut made by a river across a chain of hills or the side of a plateau, may be considered as a gate opened through a wall dividing two distinct regions. Thus, in studying the monography of each river, it is necessary to study specially the apertures which the water has made through the barriers which once opposed its free course. By a succession of victories obtained over enormous masses of rock, the river has succeeded in emerging from the lacustrine reservoirs where its waters once lay dead, and has gradually constituted itself as a living individuality ever at work shaping anew with its waves, its alluvium, and the bars at its mouth. The Danube obtained its hydrological importance from the time when its waters ceased to be lost in the former lakes which have now become the plains of Hungary, Austria, and Wallachia.

Oftentimes, when the river thus cuts away a passage through a rocky barrier, it leaves standing erect, as an evidence of the former state of things, an islet of hard stone which it has failed in washing away. In the most picturesque parts of their course, almost all large rivers exhibit some of these solid masses which continue to resist the pressure of the water some centuries after the destruction of the surrounding strata. Thus, on the Danube, we find those proud rocks, with their perpendicular sides towering up, like enormous pillars, as high as the level of the rising ground by the river-side, and crowned on their summits, some with a feudal fortress, some with a hermitage, and some with nothing but a clump of bushes or brushwood. Thus, too, in the Mississippi, not far from the spot where the whole body of its water once poured over a precipice in a mighty cataract, we notice the fine rock which, from its form and majestic aspect, has obtained the name of the "Great Tower." This rock still bears, at a height of 132 feet, the circular line of erosion which was once traced out by the current. But although we still find a pretty considerable number of these natural water-girt "towers," the greater part of those which once existed have gradually disappeared under the action of the elements, and their place is now marked only by hidden reefs or rocks on a level with the stream.

CHAPTER XLIX.

FORMATION OF ISLANDS.—RECIPROCITY OF CURVES.—WINDINGS AND CUTTINGS.—
SHIFTING OF THE COURSES OF AFFLUENTS.

It is, therefore, a fact that rivers, like all other natural agents, never cease in their work of destruction; but they destroy only to reconstruct in another place. They are continually eating away rocky islets, and employing the *débris* in the formations of islands of sand. Wherever some obstacle exists in mid-current, such as a bank of rock, the trunk of a fallen tree, or some construction of human industry, the water, arrested suddenly in its course, divides into two flows as if before the cut-water of a ship, which gliding round the opposite sides of the obstacle, rush, one on the right and the other on the left, either against the remainder of the water flowing regularly down the current, or, in small streams, against the banks themselves. From this results a double encounter; the two-curved flows, being more or less inflected and retarded by a thousand local circumstances, are thrown back towards the middle of the river, where they meet, each having described its parabola. There, one portion of each of these flows continues to descend, describing a more elongated parabola; whilst another portion flows into the comparatively tranquil space which lies below the obstacle, and gradually deposits on the bottom the sediment with which it is charged. Thus is formed the first islet, which is destined to increase by degrees, and to serve as a starting-point for a series of other islands and sand-banks which make their appearance in turn in the extent of tranquil water embraced between the parabolas of the two curved flows. The Germans give these islands the name of "Werder" (from *werden*, "to grow" [?]), to point out their slow and gradual mode of formation.

In conformity to this same law of the *seriation of islands*, banks of alluvium ought likewise to emerge regularly at the confluence of two rivers; for there, too, the masses of water come in collision, and being mutually repelled, again approach one another in elongated curves. In fact, the tongue of land which separates the two water-courses is often continued by a series of islets below the confluence;

* *Vide* vol. ii., the chapter on "Clouds and Rain."

FORMATION OF ISLANDS. 355

but the force of the current being considerably increased by the doubling of the liquid volume, and the joint bed being always much less in width than the sum of the two beds together, the stream must naturally gain in depth all that it loses in surface. The water, being confined in a narrower channel, hollows out the bed with increased energy, and thus tends to prevent the formation of sand-banks.* The alluvium is deposited in the interval between the two currents, and helps to lengthen, in a down-stream direction, the "bec" or tongue of land which separates the two streams.

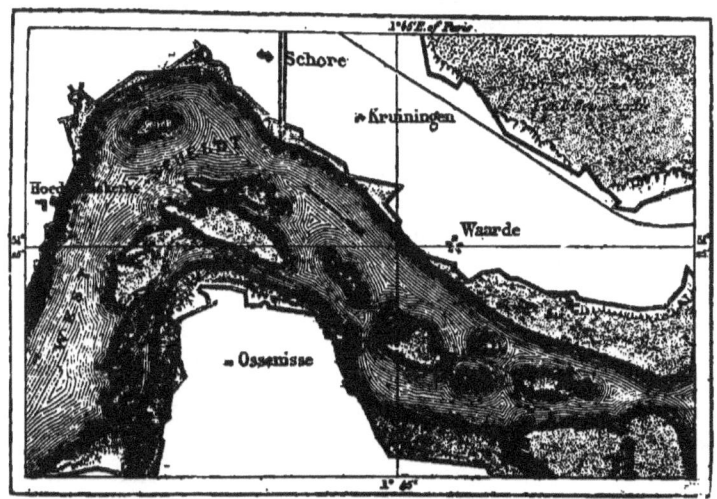

Fig. 113.—Series of Islands in the Western Scheldt.

These chains of sandy islets would always be deposited with the greatest regularity if the river descended to the sea in a straight line. It is true that every watercourse, in obedience to the law of gravity, seeks to scoop out for itself a rectilinear channel, so as to gain the ocean by the most rapid incline. But the irregularities of its bottom and banks considerably modify the direction of the river, and cause it to describe a series of curves or windings, thus lengthening the total extent of its course. Thus, another law, that of the *reciprocity of curves*, is combined with the *succession of islands* in beautifying the surface and contour of a river, and in causing it to incessantly remodel the ground of the valley through

* Von Hoff, *Veränderungen der Erdoberfläche*, vol. iii.

A A 2

which it flows by hollowing away the ground, sometimes on one side, and sometimes on the other.

Some lateral impulse communicated to the liquid mass is all that is required to throw the current of the river either to the right or to the left. If the water strikes against a wall of rock, or any other obstacle placed across the regular direction of the stream, the latter rebounds so as to form an angle of reflection equal to the angle of incidence, and, induced both by its impelling force and the general slope of the bed, it becomes more and more inflected, and describes a parabolic curve towards the opposite bank. There its current is again turned back, and again takes an oblique course across the bed of the river. When the first seriation is once brought about, the current must necessarily form a succession of windings, in conformity with the law of the *reciprocity of curves*, which is, in fact, nothing more than the law of the pendulum. Each oscillation calls forth an equal and isochronous oscillation in a contrary direction; each curve calls forth another curve of an equal radius and equal velocity. If the fluviatile economy was not constantly modified by the varied composition of the soil and the immense diversity of the obstacles of every kind which it meets with, the river would flow down towards the sea, always forming a series of zigzags as regular as the oscillations of a pendulum.

But the mass of the current does not confine itself to merely striking against the two banks in turn; it also continually wears them away, and modifies their outline. When the water dashes against the bank with all the impetus which is communicated to it by the current and the action of centrifugal force, it tears away the earth, dissolves some of the solid particles, washes away the sand, and has a constant tendency to penetrate farther. Being then driven back towards the opposite side, there, too, it destroys and washes away the soil before it is repelled afresh to continue on each shore alternately its work of destruction. Thus by a law of equilibrium the current undermines each bank in turn, whilst its alluvium is deposited at the points of the two bends. In consequence of the succession of bends and points, the windings are sometimes almost perfectly annular. A boat leaving the upper bend describes a long curve in following the river, and when it arrives at last at the lower bend it is sometimes actually in sight of the starting-point that it quitted long before. In the greater part of its middle course the Mississippi forms a series of windings so exactly like one another that the Red Skin Indians and the earliest European colonists were

WINDINGS OF RIVERS. 357

in the habit of estimating distances by the number of curves which

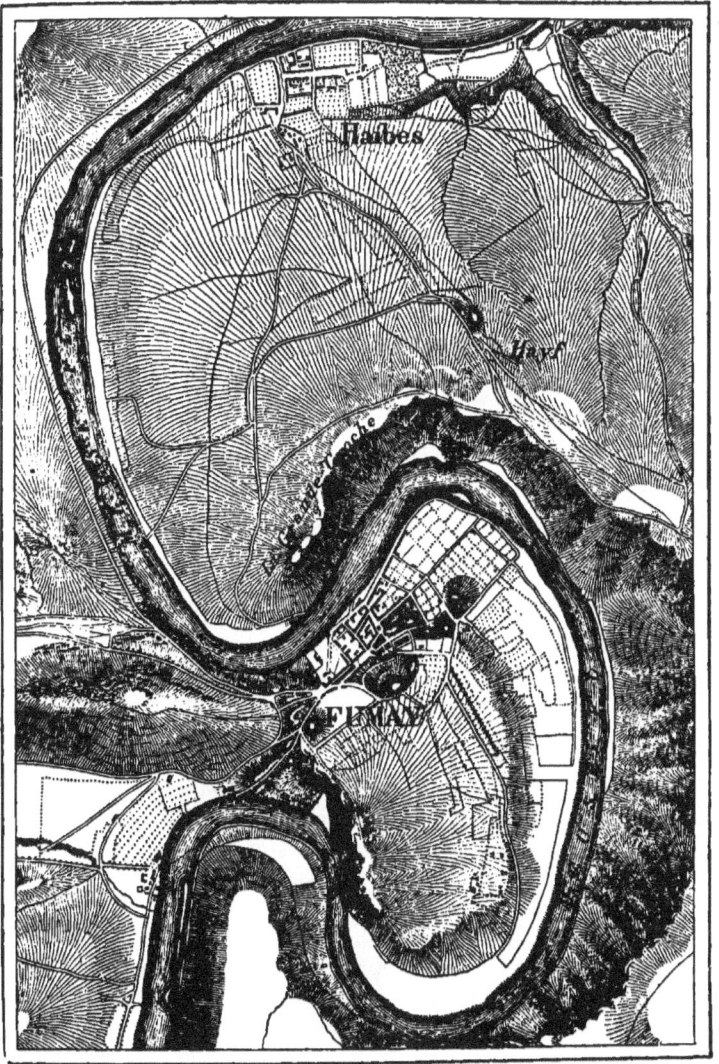

Fig. 114.—Meandering of the Meuse, at Fumay.

the river described. These windings, however, in a certain point of view, are of the very greatest utility for navigation. Every bend has the effect of moderating the slope, and thus retarding the velocity of the current, proportionately augments the mass and the depth of the water.

By dint of gradually washing away both the upper and the lower bend in a contrary direction to one another, the river constantly tends to diminish the isthmus of necks which still connects the little peninsular with the surrounding plains; thus the time will ultimately come when, the isthmus having disappeared, the two bends will be united, and the winding of the stream will be converted

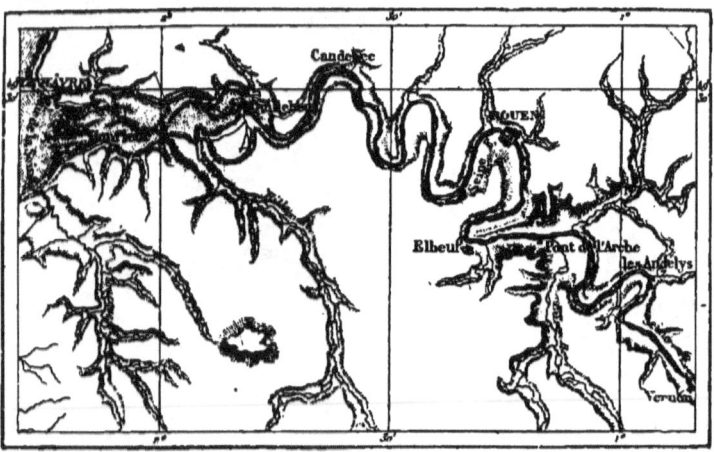

Fig. 115.—Meanderings of the Seine.

into a perfect ellipse. Then, unless the labour of man offers any opposition, the whole liquid mass will flow on in a straight line along the rapid slope formed by the junction of the two bends, whilst the water still remaining in the old beds will become sluggish and dead on account of the slight slope which is afforded to it by the enormous curve of the circuit in comparison to the newly-formed passage. The rapid waters of the upper bed striking against the still water in the former winding are suddenly arrested in their course, or even driven back; they then deposit the earthy *débris* that they hold in suspension, and thus gradually form natural embankments of sand and mud between the two old beds of the river. It is not long before a similar embankment likewise separates the two beds of the lower bend, so that the forsaken winding is ultimately

WINDINGS OF RIVERS. 359

left without any communication with the new current of the river; its water becomes stagnant, and it is, in fact, converted into a lake. In the basins of the Mississippi, the Amazons, the Ganges, the Rhone, and the Po, there are a considerable number of these circular lakes. We may trace out with the eye, as it were, three rivers, one of which, active and living, flows without interruption

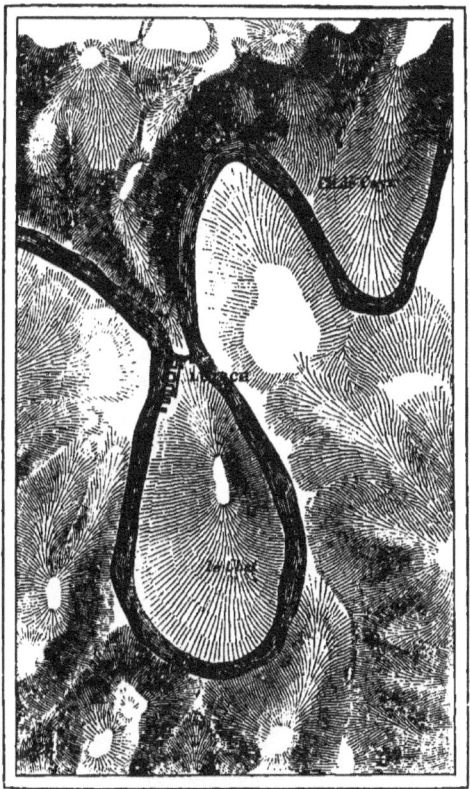

Fig. 116.—Meandering at Luzech.

from its source to the sea; whilst the two others on either side are become "dead waters." The remains of them, scattered all along the existing river, still point out the spots where once extended its ring-like windings. In consequence of these alternate shiftings of position, the valley is always much wider than its river; and along its circuitous path winds the continually changing bed of the existing stream. In some parts of its course the Po only

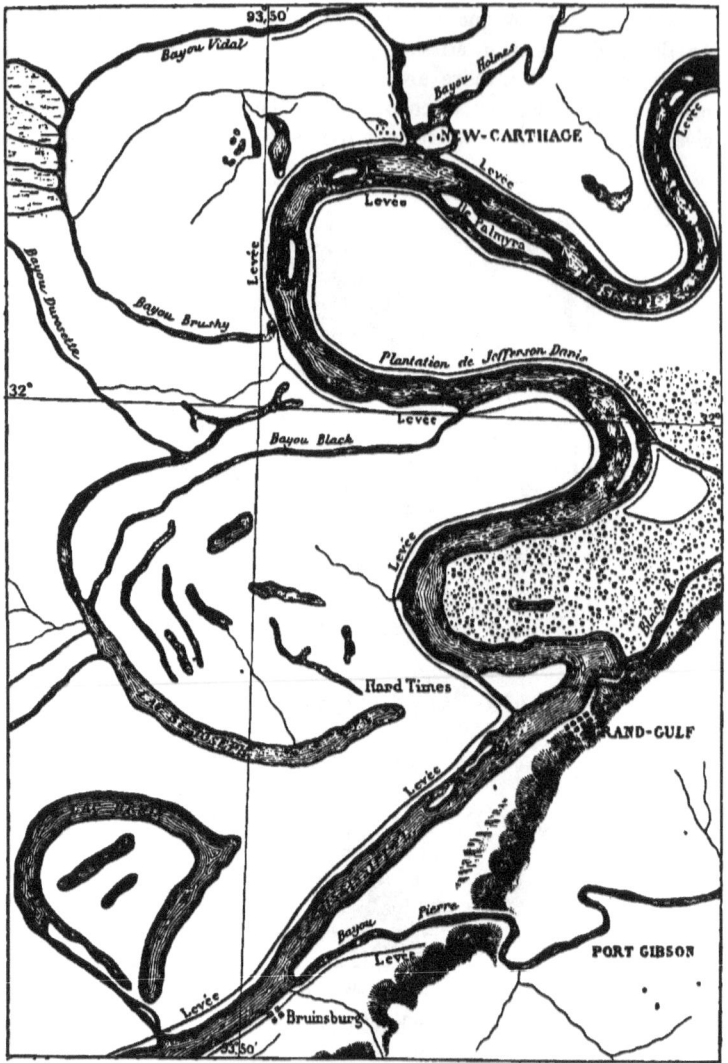

Fig. 117.—Old Channels of the Mississippi.

takes about thirty years in forming and destroying each of its meanders.*

* Elia Lombardini, *Dei Cangiamenti del Po*.

MIDDLE COURSE OF THE MISSISSIPPI

PL. XVI

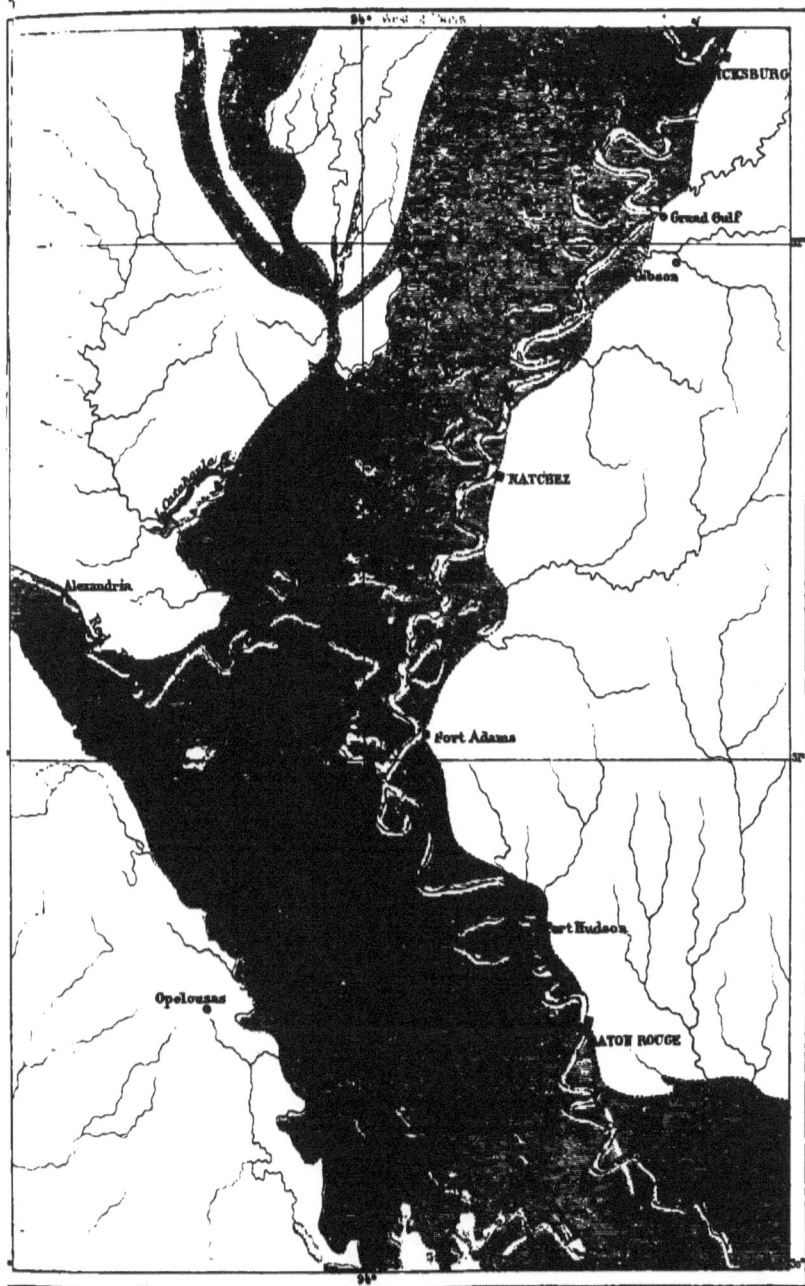

Eng^d by Erhard.   Vincent Brooks, Day & Son, Lith. London W.C.   Drawn by A.Vuillemin after A.Humphreys & Abbot.

CHAPMAN & HALL LONDON.

The perforation of these river isthmuses is not always brought about by the sole action of nature; many channels uniting two river-

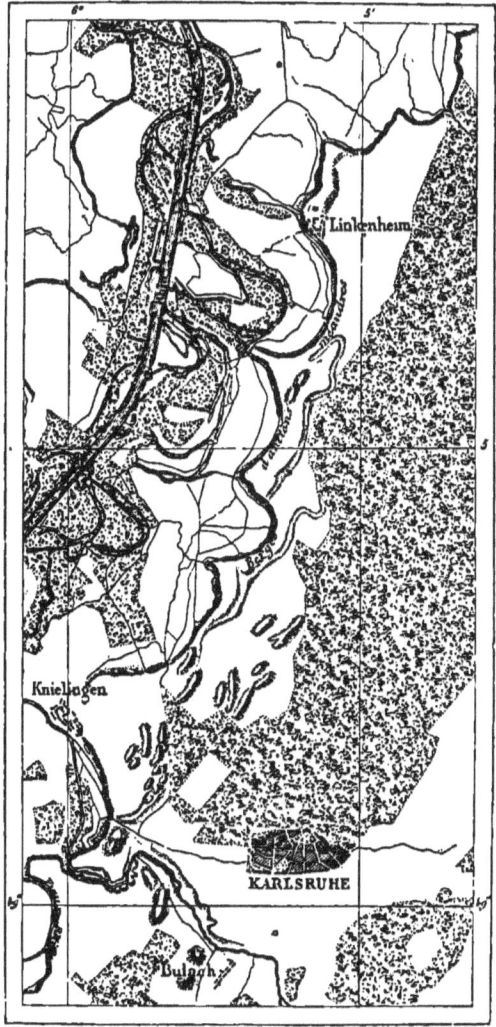

Fig. 118.—Old Meanderings of the Rhine.

beds have been dug out by the hand of man, and, thanks to the currents which have deepened them, they have ultimately replaced the former beds. Some engineers have gone so far as to propose to carry out a systematic series of operations as regards the whole line of the Mississippi, and thus to rectify the bed of the river from Cairo to New Orleans. Since the colonization of Louisiana the labour of man, assisting the action of the currents, has already rectified several beds; in this way have been formed the "cuts-off" of Bunch, Needham, Shrieve, Point Coupée, and Fer-à-Cheval. Above these different points the isthmuses are much more difficult to cut through, on account of the strata of compact and hard clay which extend immediately below the superficial bed of the modern alluvium, and are not easily washed away by the water. Thus it was that in front of Vicksburg a portion of General Grant's army worked in vain for several months endeavouring to cause the current of the Mississippi to pass through a channel cut across the narrow isthmus of the right bank.

Nevertheless, all these "cuts-off" dug out by the hand of man cannot fail ultimately to become obliterated; for, in conformity with the law of the reciprocity of bends, a river, when deprived of its windings, is not long before it forms new ones. This was the case above Compiègne, where it was vainly attempted to straighten the course of the Oise. In a very short time the river made fresh windings, the development of which was found exactly to equal those which had been done away with. It was managed better in fixing the course of the Midouze, in the Landes, for there the ingenious idea was adopted—and followed by success—of giving to the river a series of meanders of perfect regularity. When man attempts to meddle with nature, he can only succeed in permanently modifying its aspect by studying the constant laws of its phenomena, and by making his work conform to these.

The idea of digging out and maintaining a straight channel between two parabolic bends of the Mississippi, is in no way more absurd than the idea of constructing piles, perpendicular to the current of the stream, in order to limit the bed of the river, and to throw the waters into a regular channel. By operations of this kind, contrary to every principle of hydraulics, our French engineers have entirely ruined the system of the Loire, the Garonne, and several other watercourses. The Loire—a river which is the despair of engineers, and still more of boatmen—is distinguished above all the streams in France by the inconstancy of its current, and the continual shifting of its navigable channels. There is a very great difference between the mass of water which a river rolls down in

CUTTING OF CHANNELS. 363

flood-time, and the slender rivulets which slowly make their way through the sand in the dry season. Now, as a lateral shifting of the current takes place at the time of every fluctuation in the level, the result is that several temporary channels are formed, and obli-

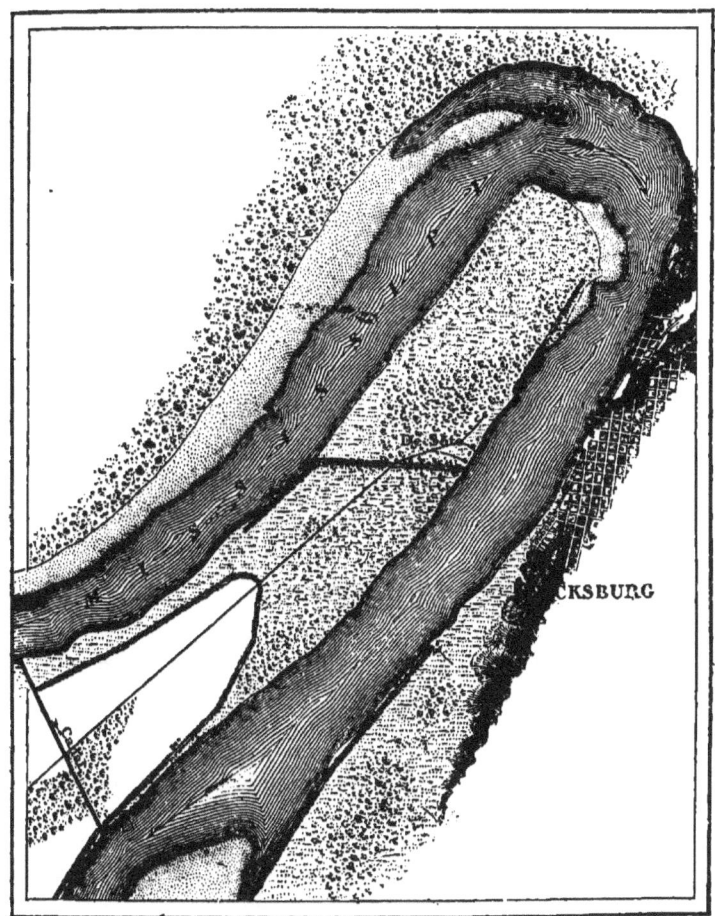

Fig. 119.—Channel of Vicksburg.

terated in turn. Some *mouilles*, or comparatively deep holes, are certainly to be met with almost constantly at the concave extremity of the bends, where the partial currents unite; but everywhere else the bed rises more or less over its whole extent, so as to form ridges (râcles), and navigation becomes impossible during a great part of the year. This fatal interruption to commerce, representing an annual loss of many thousands of pounds, would not take place if it was decided to adopt the system of "guiding banks," proposed by M. Edmond Laporte,* and subsequently by M. de Vézian. Instead of rectilinear rows of piles constructed across the bed of the river, embankments should be raised formed with a parabolic curve, against which the principal current might strike at all seasons, so as to describe unhindered its regular series of serpentine curves; this is the only plan for insuring the greatest possible depth to the channel for navigation. The annexed plate, borrowed from M. de Vézian's work,† points out the position which the embankments ought to occupy, and the direction that they would communicate to the current of the river.

Even if the mass of water were to remain the same from year to year

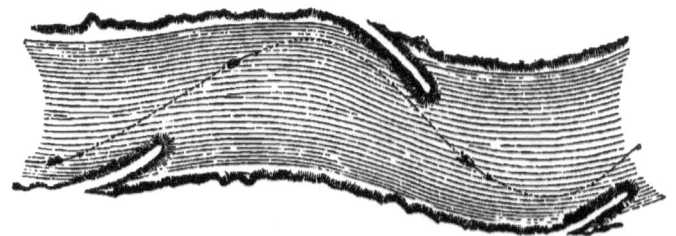

Fig. 120.—Diagram to show "Guiding Banks."

and from century to century, it is certain that the mere action of the current striking each bank in turn, would in the long run be sufficient to alter the curves of the river, and gradually to remodel the ground in the valley. But the liquid mass of every stream is incessantly varying from the commencement of spring until the end of winter. It increases during the rainy season, and when the snow melts; it diminishes, on the contrary, when the supply from the clouds, the snow, and the glaciers, is not equivalent to the water which is absorbed by the innumerable rootlets of the vegetation by the river-side, and by the continual evaporation caused by winds and heat. Under the influence of these various phenomena which either increase or abate, the level of every river constantly fluctuates between *flood* and *low-water*. The current of the stream is consequently shifted, first to one

* *Gironde*, September, 1864.
† *Annales du Génie Civil*, May, 1863.

NAVIGATION CHANNELS. 365

side and then to the other, and thus every day contributes in a different

Fig. 121.—Middle Course of the Rhine.

way to the erosion or consolidation of each of its banks. The quantity discharged by the river during flood-times being five, ten, fifty, or even a hundred times as much as at low-water seasons, it is hardly to be wondered at that the erosion accomplished by the current should also vary in very considerable proportions.

By dint of manipulating the small particles which it has itself been the means of conveying to the alluvial plains, the river ultimately succeeds in completely altering the direction of its own tributaries. The short promontories which are situated at the confluence of the principal river and the streams which run into it, are constantly lengthened in a down-stream direction by all the sandy or muddy *débris* which is deposited by the two currents. The two masses of water, which are ultimately to encounter one another, tend to take a direction more and more parallel on each side of this increasing promontory, and developing their windings on both sides of their axis of descent, thus make their way side by side through the plains. A magnificent example of this inflection of river-beds may be noticed in the valley of the Rhine between Basle and Mayence. All the affluents that the Vosges and Black Forest send down to the great river, bend to the north as soon as they have emerged from their natal valley, and wind through the plain, tending in the same direction as the current of the Rhine. Above and below this wide plain of alluvium, in which nature has afforded no obstacle to the free passage of the water, the lateral rivers do not double round in this way before they join the Rhine. Being kept back by the mountains or hills which command them, they fall directly into the river, nearly at a right angle to it.

CHAPTER L.

PERIODICAL RISING OF STREAMS.—"EMBARRAS" OF FLOATING TREES.—ICE-FLOODS IN THE NORTHERN RIVERS.—INUNDATIONS.

A LARGE supply of rain-water being the principal cause of the swelling of rivers, the rainy seasons must necessarily be the times when floods are generally produced. In tropical regions, where the zones of clouds and showers shift regularly from north to south and from south to north during the course of the year, the fluctuations in river-levels can be calculated and predicted beforehand, just as the seasons themselves, according to the passage of the sun over the ecliptic.* When the luminary shines above the northern hemisphere, and dry seasons prevail on the north of the equator, the water-courses in the northern tropical zone become low, and many are completely dried up. During the winter season, on the contrary, when the sun has brought back to the north the rain-clouds and tempests, then the rivulets, streams, and rivers, again swell and flow brimful of water. The same phenomena take place in a contrary order in the southern hemisphere. Thus the level of running waters on the north and on the south of the equator fluctuates in turn, so as to form a kind of annual tide, which in its regularity may be compared to the daily tides of the ocean. We must, however, add that in all tropical regions the periodicity of the annual floods is variously modified both by the relief of the ground and also by aërial eddies and other phenomena which have an influence on the falls of rain.

Amongst all the rivers of the intertropical zone, the floods of the Nile have obtained the most world-wide celebrity. Herodotus and other historians of Greek antiquity have told, with a sort of religious astonishment, of this periodical swelling of the sacred river which conveys to Lower Egypt the soil which nourishes it. To the agriculturists by the river-side, this beneficent flood seemed like a miracle, and their priests never failed to take advantage of it so as to increase their power among the people. So long as the valleys of the Upper Nile and its tributaries were unknown, it was really difficult, when surveying the annual inundations of the river, not to consider it as a

* *Vide* vol. ii., the chapter on " Clouds and Rain."

prodigy. The course of the Lower Nile is not fed by a single tributary; it traverses an arid country rarely watered by the rain of heaven; a burning sun evaporates its water, and yet, all of a sudden, about the beginning of July, the river-level rises, without any apparent cause, in its wide isle-studded bed. The water rises, and goes on rising, and from August to October it covers the sand-banks, flows over its brink, and, inundating the banks, pours itself out in strata no less regular than the annual rings in the trunks of trees. At the very highest flood the river often contains a mass of water twenty times* as great as that which it conveys to the sea when at its very lowest, and yet perhaps the Egyptian sky has not for several months yielded a single drop of rain. This prodigy, incomprehensible enough to our ancestors, may nowadays be easily explained. The enormous mass of water, which serves to irrigate the cultivated districts of the delta, proceeds from the snow and rain which the clouds so abundantly shed on the mountains of Ethiopia and on the other countries of equatorial Africa.

There are many water-courses in the intertropical zone which afford the phenomenon of periodical floods with as much regularity as those of the Nile; but there are none which are more curious in this respect than the great river of the Amazon basin. This "Father of Waters" flows nearly under the equator, and receives simultaneously the affluents of two hemispheres. Owing to this arrangement of its river system, the floods of the northern rivers take place in summer and autumn, whilst the southern tributaries overflow during the winter. The principal river and the Madeira are mostly swelled by the equinoctial rains, and their floods take place in spring and summer. An actual system of compensation is thus established in the lower bed of the Amazon between the tributaries flowing in on the right bank and those on the left. When the Pastaza, the Japura, and the Rio Negro are at low water, the Ucayali, the Madeira, and the Tapajoz are running brimful; when the latter begin to get low, the northern affluents increase their mass of water.

Beyond the tropical zone, rivers must necessarily manifest less regularity in their annual floods, the rains themselves being more irregularly distributed among the various seasons. Nevertheless, an unquestionable system of order never fails to show itself each year in the fall of atmospheric moisture, and this system is again met with in the corresponding fluctuation of the river-levels. This is a fact which may be proved by the study of various water-courses. In

* Elia Lombardini, *Essai sur l'Hydrologie du Nil.*

regions like the north of France, which are favoured with rains in winter, spring, and summer, the floods generally take place between the 15th of October and the 15th of May, and it is only due to the rapid evaporation which takes effect during the hot weather, that summer floods are so very rare.* In the Mediterranean districts, where autumn rains predominate, the water-courses begin to swell towards the end of the year. In those river-basins which, from the vastness of their area, extend into several meteorological regions, the fluctuations of level, which succeed one another with more or less regularity in each of the different affluents, are combined so as to form a fresh series of floods as regards the principal artery, and the general course of these floods may be easily foreseen. The most striking example which can be mentioned is that of the Mississippi, a river which unites in its vast bed, the water coming from the great western

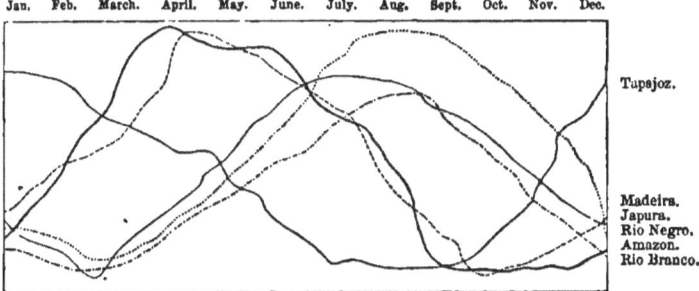

Fig. 122.—Compensation of Floods in the Basin of the Amazon.

deserts and the streams which flow down the pleasant valleys of the Alleghanies. At New Orleans the river commences to rise about the 1st of December, and its mass of water increases until about the middle of January, which is the time of the first flood. Then the level slowly sinks, and afterwards remains nearly stationary during the months of February and March. In April and May the river swells afresh, and in the course of the month of June, it forms the great flood so dreaded by the planters. Immediately after it sinks rapidly until the end of September, and its lowest level coincides very generally with the commencement of November.

Several water-courses in the temperate zone exhibit, in the fluctuations of their level, a phenomenon of compensation similar to that of the Amazon. These are rivers which are replenished simultaneously by streams fed by rain-water, and also by torrents increased by the

* Belgrand.

melting of snow and glaciers. The variations of lowland streams being, as regards the seasons, precisely contrary to the variations to which mountain tributaries are subject, the level of the main river remains at a nearly regular height. The rain-water tributaries diminish in bulk at the very time when the affluents, which have come down from the glaciers, are increasing—that is to say, in summer; in winter and spring, on the contrary, the glaciers supply very little water, whilst the plains are inundated with rain, and their streams are filled to the brink. Thus the abundance of one affluent balances the poverty of another. As an instance of this, the Rhone and the Saone have often been brought forward. During the heat of summer the latter brings down only one-fifth of its winter discharge. On the other hand, the Upper Rhone rises much higher during the same season; but below its junction with the Saone the average height of its water is nearly the same during every season of the year. A compensation of a similar kind likewise takes place between streams of surface-water and those which are fed by springs. The rivulets which traverse the subterranean passages of rocks, cannot descend into the plain so rapidly as the water-courses which flow on the surface of the ground.

The grandeur of the geological operations accomplished by floodwaters, are best to be appreciated on the banks of rivers which have been placed by the labour of man in a state of defence against the watery enemy. When the river Amazon overflows, it forms in some places, with the marshes on its banks, a perfect sea of 100 or even 200 miles in width. The animals seek a refuge in the tree-tops, and the Indians who live by the sides of the river make a kind of encampment on rafts. About the 8th of July, when the river begins to sink, the water, returning to its original bed, undermines the thoroughly soaked banks and slowly washes them away. A sudden fall then takes place, and masses of earth, amounting to hundreds or thousands of cubic yards in bulk, sink down into the water, carrying with them the trees and animals existing upon them. The very islands are exposed to sudden destruction: when the entangled masses of fallen trees, which serve as a breakwater to them, give way before the violence of the current, a few hours, or even a few minutes, are quite sufficient for their disappearance; they are literally washed away by the flood. They may be observed visibly melting away; and the Indians, who are quietly at work upon them collecting turtle-eggs or drying the produce of their fisheries, are suddenly compelled to fly for their lives. Then it is that the current of the stream is encumbered with long floating piles of entangled trees, which hitch together only to break away again, and accumulating round some headland, are heaped up one above another all along the banks. All round these immense trains of trees, which roll and plunge heavily under the impetus of the current like great marine monsters or drifting wrecks, great masses of the plant *Canna rana*, float on the surface of the water, giving to some parts of it a resemblance to broad meadows.* We may thus readily comprehend the almost religious awe which has been experienced by travellers who have made their way up the river of the Amazons, and viewing these whirlpools yellow with sand, have been eye-witnesses of their destructive operations in tearing away the river-banks, throwing down trees, washing away islands in one place to form them again in another, and drifting down the current long trains of trunks and branches. "The great river was terrible to look on," says Herndon, the American traveller, "as it rolled through the solitudes with a solemn and majestic air. Its waters seemed to wear a wrathful, malevolent, and pitiless aspect. The entire landscape had the effect of stirring up in the mind a feeling of horror and dread similar to that produced by the imposing solemnities of a funeral at sea, by the minute-gun firing at intervals, the howling of the tempest, and the wild uproar of the waves, when the crew assemble on the deck to bury their dead in the bosom of a troubled sea."

The Mississippi presents a remarkable instance of a great watercourse which man has recently annexed to his domain, and has succeeded in modifying considerably, as regards its geological action, during the course of a few years. In 1782, and even at the time of the great inundation of 1828, the whole of the region embraced between the left bank of the Mississippi and the course of the Yazoo —that is, an area of more than 30 miles in width on the average— was completely covered with water, as is proved by the bones of wild animals which have subsequently been found on the artificial mounds raised by the red-skin Indians. At the present day the river is confined on both sides by lateral embankments, and no longer floods the whole basin of the Yazoo. Now it only tears away narrow strips of the vast forests by the river-side; and even in the very highest floods, the masses of trees which drift down the current do not form, as before, long floating trains.

Even at the beginning of the present century, these floating trains, or *embarras*, rendered the navigation almost impossible in some

* Avé Lallemant, *Reise in Nord-Brasilien*.

reaches of the Mississippi and its tributaries. A great portion of the courses of the Atchafalaya and the Ouachita were completely choked up by heaps of trees. In many places a person might cross them without any idea that he was going over a river. Bushes and even large trees grew upon some of these floating masses.* One of these entanglements of drift-wood, known by the Americans under the name of the "Great Raft," always obstructs the bed of the Red River. This immense agglomeration of trees, under which the water disappears in a mass as if under a movable arch, gradually gets higher up the course of the river as the trees at the lower end break away, and the annual floods bring down fresh drift-wood to the upper extremity. The obstruction was probably first formed at the confluence of the Red River and the Mississippi, and has since gradually advanced 391 miles from the mouth, gaining a mile or two every year. In 1833 the Federal Government undertook some important operations for the removal of the obstruction, which had then attained a length of 124 miles; but whilst a flotilla of boats was occupied in pulling out the trees which formed the lower extremity of the "raft," the upper end was constantly increasing by means of the fresh drift. In 1855, after twenty-two years devoted to this "labour of Sisyphus," the question was raised whether it was not better worth while to abandon this ungrateful labour, and to apply the funds at disposal to the improvement of the *bayous*, or lateral channels. "The Great Raft" being thus abandoned in the marshes, which formed the old bed of the river, will be gradually converted into a great peat-bed, destined perhaps to become coal at some future geological period, unless human ingenuity should otherwise dispose of it.

In cold countries, such as British America, Russia, or Siberia, the water-courses carry down to the sea a far less quantity of vegetable *débris* than the rivers in tropical countries; but, to make up for it, they are loaded with enormous blocks of ice at the time of thaw, a period which often coincides with the highest floods. It is a wonderful sight, especially in rivers adorned with cataracts like the Niagara, when the rocks of ice, dashing against one another, and breaking up in the midst of the watery columns, give one the idea of a cataclysm, in which lakes and continent were all being simultaneously swallowed up in the abyss. The icy sheet which extends over the surface is shattered with a sharp, grinding noise, and the broken fragments are caught by the current, and dashed violently

* Lyell, *Second Visit to the United States.*

against each other; their sharp angles are broken off in the collisions, and they are whirled round and round in long eddies. In the curves of the headlands, at the points of the islands and sand-banks, and also in those portions of the river where the icy barrier still remains firm, the broken masses gradually accumulate, and, mounting up one upon another, owing to the force of their impetus, butt against the banks like battering-rams, and thus often clear away an outlet into the plains for the flood-water. Sometimes they rear themselves up like dams, and drive back the body of the river up-stream again. For this reason, dikes, embankments, and other hydraulic ramparts, built along the course of a river subject to these annual breakings-up of the ice, must be constructed with the utmost solidity. Amongst other constructions of this kind, we may mention the enormous buttresses with which the piles are furnished to support the bridge of Montreal on the St. Lawrence, and the defensive icebreakers built in the Vistula on the up-stream side of each pier of the bridge of Dirschau. At St. Petersburg the granite quays and the edifices they protect would be all carried away by the ice-flood, if at the same time violent tempests from the west were to drive the waves of the gulf into the mouth of the Neva.

In temperate Europe the breaking up of the river-ice is attended by little or no danger; but the mere inundations are very much dreaded on account of the towns, villages, and richly-cultivated districts with which the banks are covered. The inhabitants on the edges of the Loire still recall to mind with horror the disasters which have been caused by the great though exceptional floods which in one year only (1856) carried away roads and defensive embankments, causing damage to a most enormous amount. In the same year the calamity was but little less disastrous in the valley of the Rhone, which was covered in some places, especially the Camargue, by an inundation almost like one of the floods of the Amazon. The inhabitants of the banks of the rivers are now worse off in this respect than their ancestors. The extraordinary rains caused by atmospheric changes are not all they have to dread. They have now to look for greater irregularity in the action of the streams and still more sudden inundations, in consequence of so many of the marshes and pools being drained dry, and the mountain slopes being cleared of wood by the axe of the woodman, or laid bare by the feeding of the goats. They also have to fear the immediate effects of the drainage channels which pour down the rain-water so rapidly into the streams. Lastly, the surface-water is every year precipitated into the plains more and more suddenly on account of the increasing number of ditches, which are carefully kept up along the roads and paths, into which the boundary trenches of the various properties all empty.* On the other hand, the extension of cultivation on the edge of a river, without the application of drainage, enables the earth to absorb the water to a lower depth in the soil, and thus diminishes the height of the floods. This fact is proved by the example of the Lake of Aragua, in Venezuela. At the commencement of the century, when the greater part of the neighbouring plains were under cultivation, the level of its water was comparatively low; but during the War of Independence it gradually rose, owing to the devastation of the country by the contending armies, and the consequent return of the plains to their

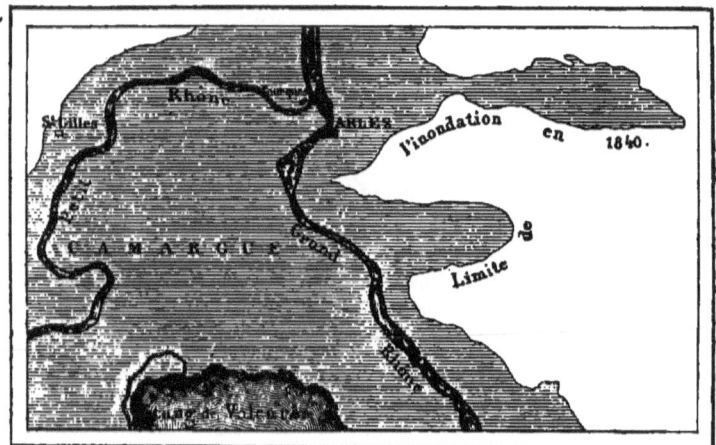

Fig. 123.—Limits of the Inundation of the Rhone in 1840.

original condition of virgin forest. Latterly, fresh clearings have for the second time sunk the water of the lake.

Under the action of all the causes which so variously influence the fluviatile economy, some rivers, such as the Oder, since 1778, and the Elbe, since 1828, have diminished in volume, although it is certain, from the meteorological registers, that the amount of rain falling into their basins has not lessened. Other rivers, as the Rhone and the Loire, do not appear to have at all decreased in the quantity of their water; but, on the other hand, their inundations are much more dangerous than formerly. The Seine, which, according to the

* Becquerel, *Comptes Rendus de l'Académie des Sciences*, November, 1866.

testimony of the Emperor Julian, poured through Paris, some fifteen hundred years ago, nearly the same quantity of water in every season of the year, shows, at the present time, a difference of about 33 feet between the high and low water levels. Some rivers indeed, such as the Garonne, appear to have been more formidable in days gone by than they now are. The highest inundation of the Garonne which is on record, is that of April, 1770. At Castets, the point at which the tide stops, the flood-level attained a height of 42½ feet above the low-water mark. This is 6½ feet more than in the largest floods of the present century.*

However this may be, some of these inundations assume such proportions that they become perfect cataclysms for all the river-side districts. The example of three little streams, the Doux, the Erieux, and the Ardèche, all three confined to the limits of a single department, may give an idea of the rapid swellings of these high floods. On the 10th of September, 1857, the three water-courses, which usually flow peaceably enough over their rocky and pebbly beds, poured down into the Rhone a combined mass of more than 18,000 cubic yards of water, instead of the 20 to 25 yards which was their ordinary discharge in the same time. This flood was equivalent to the body of water which the Euphrates and Ganges together pour into the sea. Spreading over their respective valleys to a height of 50 to 60 feet above their low-water mark, the flooded rivers overthrew the houses, washed away the cultivated ground, and uprooted the trees. So many thousand trunks of trees were carried away in one day that, below the Erieux and the Doux, the whole surface of the Rhone seemed nothing but a train of drift-wood, over which, as it appeared, a bold man might well have ventured to cross the river. Still, even these inundations have been exceeded, for, on the 9th of October, 1837, the Ardèche rose, at the bridge of Gournier, to a height of 70 feet above low-water mark, at least 10 feet higher than in 1857.† Above the Iron Gates some of the floods of the Danube have caused the river to swell to a height of more than 60 feet above low-water mark.

It is a fortunate thing that, in most river-basins, it is very seldom the case that the floods of the various affluents exactly coincide, and that all the tributaries are seen to swell at the same time. In fact, whenever a rain-cloud passes through a valley, it discharges its moisture sometimes on one side and sometimes on the other, and the various water-courses which it swells overflow in turn after the rain-

* Raulin, *Géographie Girondine*.
† Marcheguy, *Annales des Ponts et Chaussées*, vol. i. p. 861.

cloud has passed over. Thus, in the valley of the Rhone, when the damp winds encounter the Cevennes, the slopes of the Alps which are turned towards the river are sheltered from the storm, and it is only gradually that the series of showers makes its way from the Cevennes towards the mountains of Annonay. If all the tributaries of the Rhone were to swell at one time, it would roll down a most formidable mass of water, amounting to more than 130,000 cubic yards a second. It would be another Amazon. Even when the Rhone discharges into the sea only 16,000 to 20,000 yards a second, the havoc which it makes upon its banks is most frightful.

CHAPTER LI.

MEANS OF PREVENTING FLOODS.—NATURAL AND ARTIFICIAL RESERVOIRS.—IRRIGATION CHANNELS.—EMBANKMENTS, AND CRACKS IN THEM.

IT is evident that it would not do for man to remain constantly under the apprehension of these inundations, and that it was necessary to find some means of preventing them. For hundreds and thousands of years, and especially during this century of industrial activity, plenty of plans of protection against river-floods have been both projected and put into execution; but too often these works have remained useless, or have even produced entirely contrary effects to those which were expected by the engineers and the inhabitants. The fact is, that in going to work, they did not always pay sufficient attention to the laws of hydrology. If man wishes to become master of the forces of nature, and to make them work to his advantage, the first condition is, that he shall thoroughly understand them.

It must be remarked, in the first place, that the mass of surplus water forming a flood is not actuated with the same speed over all its width. The nearer the liquid particles are to the bank, the slower they move. This phenomenon, caused by the friction of the fluid against its banks and the bottom of its bed, may, it is true, be observed to some extent at low-water seasons as well; but it is when the level of the river is at the highest that the various portions of the liquid mass present the greatest differences in speed. The *thread of the current*, the mathematical line of the greatest rapidity, which varies every day and in every stream, according to the quantity of water and the section of its bed, exceeds by about a fifth the average speed of the river.* In flood-times this line gradually rises above the bottom, and by thus ascending towards the surface of the river, so as to keep—according to the direction or forces of the wind—sometimes on the surface of the river, sometimes a few feet below, it leaves the solid walls which constitute the sides of the river, and the medial part of the water, of which it is the ideal axis, and

* According to M. de Prony it is 0·1835.

moves consequently with greater facility. In great rivers, such as the Amazon, the Mississippi, or the Rhone, the current sometimes descends with the speed of seven or eight miles an hour. Whilst the central part of the current thus hastens down towards the sea, the water at the side, kept back by the irregularity of the bed, remains behind, and flows more slowly along the banks. Thanks to this difference in speed, which increases according to the height of the river, floods are sometimes lessened or even entirely prevented. In fact, when mighty masses of water, descending either from the clouds or the mountains, fall simultaneously into the basin of a stream, these liquid avalanches would certainly produce formidable inundations if they were not immediately carried away by the centre of the current, and did not distribute in succession a portion of their bulk over all the points that they traverse. Forming, so to speak, a river in the middle of a river, this rapid flow weakens the flood by dividing it

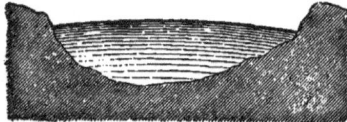
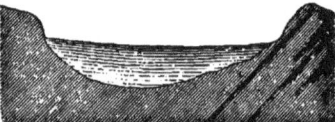

Fig. 124.—Growth by Floods.   Fig. 125.—Subsidences of the Waters.

over a vast length of bank. In the river of the Ohio the mid-flow has rolled down for a distance of five miles, when, at a spot two miles and a half from the very place where the rain fell, the high banks are scarcely touched by the rising water.

In consequence of the speed communicated to the mid-flow of the flood, the liquid mass that it carries along is perceptibly higher than the mean level of the river. It forms a kind of convexity, from the top of which the water spreads out in light sheets towards the two banks; but, on the other hand, when the flood-wave has disappeared, the middle of the river exhibits a considerable depression, and the water which has gradually accumulated near the two edges has to flow back towards the centre of the current so as to re-establish by degrees the fluviatile level. It has been ascertained that on the Mississippi the central convexity of the flood-wave is on the average about three feet. When the river sinks, almost as considerable alteration of the level takes place in a contrary direction. The woodcutters of Maine and Canada are not ignorant of this hydrological fact. They are well aware that logs of timber thrown into the river in flood-time are thrown up on the banks, whilst they float regularly down the middle of the current when the river-level sinks.*

The depression which is formed in the middle of the river during the period of subsidence is, however, obliterated as soon as the liquid mass ceases to diminish; the water then commences to bulge up again in the axis of the current, owing to the greater facility of movement possessed by it in that part. The surface of those great Russian rivers which are covered with ice for several months of the year, exhibits a remarkable instance of the bulging up of the liquid mass in the central line of the current. At the conclusion of the winter, when the water from the melting snow runs down off the banks towards the bed of the river, and the sheet of ice stretched over the river is not yet broken, it is ascertained that the surface-water collects in elongated pools on those portions of the field of ice which are nearest to the edges, whilst the medial part bulges up in an arch above the current, and remains constantly dry. On the Volga, the difference of the level between the edges and the middle of the ice amounts sometimes to more than three feet.†

The current is not the only regulating force which weakens the action of floods, and gives more certainty to the height of the water. There are other agents which assist in equalising the discharge of a river by receiving the overflow during the rainy seasons, and afterwards emptying it into the principal current. These regulating agents are the surface or underground reservoirs which exist on each side of water-courses which are still left in a state of nature. Thus, according to Humboldt, the Upper Maranon pours into the caverns of the *pongo* of Manseriche a portion of its waters, and also all the drift-wood which it brings down from the higher valleys. Many streams lose a considerable quantity of water by the mere process of filtration, through the spongy soil of their valleys. It is stated that in some places the water of the Nile penetrates laterally as far as fifty miles from the bed of the river. ‡ In like manner, during floods, the Seine feeds the land-springs which extend under Paris, and all the wells are then filled by the water of the river. §

Next to lakes, which are the chief regulators of running waters, ‖ the marshes lying close to the edges of a river take the principal

---

* Marsh, *Man and Nature*.
† De Baer, *Bulletin de l'Académie de St. Pétersbourg*, vol. vii. No. 4.
‡ Marsh, *Man and Nature*.
§ Delesse, *Carte Hydrologique*.
‖ *Vide* the chapter on "Lakes."

share in modifying its discharge. During inundations the lagoon and swamps on both sides temporarily store up a large quantity of flood-water, which is only set free after the sinking of the river. The marshy regions through which the Mississippi runs in its middle course affords a remarkable instance of this fact. Thus, in 1858, the great American river which, below the mouth of the Ohio, sent down 52,039 cubic yards of water, only discharged 45,915 yards at Bâton-Rouge, after it had received the contents of the Arkansas, the Yazoo, and other less important rivers. A mass of water, amounting to 6,124 cubic yards a second—equivalent to nineteen times the bulk of the Seine—must, therefore, have been lost on the way.* Just in the same way, the Rhone, in its great inundations, makes its way over the side embankments opposite Culoz, and, covering the whole of the vast marsh of the Chautagna, pours its surplus waters into the Lake of Bourget. It has been calculated that during the flood of 1863, this reservoir absorbed from the Rhone a mass which altogether amounted to 71,900,000 cubic yards of water, the effects of which would have been most disastrous on the plains below.†

In places where the marshes beside a river have been drained by the operations of man, the water-level of the stream rises to a much more considerable height in flood-time, and the plains around are inundated. But the inundations themselves become new regulators of the discharge of the water, and that, indeed, by means of their very irregularity. The liquid sheet which covers the fields is hindered in its flow by the inequalities of the ground and by clumps of trees. Being unable to follow the river in its impetuous course, it remains behind, like a temporary lake, until the river is low enough for it to return into its natural bed. Thus the flow of an inundation always decreases in height as it gets nearer to the sea, and ultimately it completely disappears. The inundation of the Nile diminishes as it flows on from Assouan, where it is from 53 to 56 feet in height, to Rosetta and Damietta, where it is not more than 3 feet in height. A similar decrease in the flood-wave may be observed on all other rivers. We must not, however, lose sight of the fact that this gradual waste of the water proceeds partly from several other causes, such as the porous nature of the ground bathed by the river, the activity of the vegetation growing by its side, and the amount of evaporation. This last cause of the exhaustion of the water is probably the most important in all hot countries like Egypt and Guinea.

* Humphreys and Abbot, *Report on the Mississippi River.*
† Gobin, *Commission Hydrométrique de Lyon*, 1862.

It should be man's part to complete the work of nature by imitating in his operations some of those means which she employs for storing up surplus waters, and afterwards distributing them equally over vast areas, thus insuring a regular discharge. Man should make it his task to watch the drop of rain as it falls from the sky, to follow it in its course, to arrest it in its progress when it would help to swell a dreaded flood, and to employ it for the benefit of agriculture, navigation, and manufactures. On every mountain-side and elevated plateau he may avail himself of a powerful remedy for the prevention of floods, by replanting them with trees; for, as M. Becquerel's experiments have proved, the quantity of water which drops during heavy rain on wooded ground is only six-tenths of that which falls on the bare soil.* In a great number of the upper valleys reservoirs might be constructed, where the liquid mass would accumulate in times of rain, and be subsequently emptied over the slopes in innumerable irrigating conduits. On cultivated declivities, as Provence and the Maritime Alps, man should enlarge and consolidate the flat stages which rise one above another along the mountain-sides, forming, as it were, so many staircases, each step of which should keep back its share of rain-water. In the valleys he should tap the river in order to feed irrigation ditches and mill streams. Finally, in the lowland plains, it would be easy to line each side of the river with reservoirs, where the stream might deposit the sediment with which it is charged.

The streams, on the sides of which water-mills and manufactories have been established, are, as it were, disciplined by means of the waste water channels and the reservoirs where the water is stored up, and especially through the mill-dams and other obstacles which convert the river or stream into a regular canal, with its dammed-up levels. Inundations are therefore very rare, or even quite unknown, in a great number of the manufacturing valleys of England, Scotland, and the United States. Still, the gratuitous power afforded by water is not by any means generally utilized at present; and even the inhabitants of manufacturing countries allow a very considerable quantity of available water-power to run away to waste. Thus—to select an instance among the French streams which turn the greatest number of mill-wheels—the Doubs itself, flowing through the manufacturing districts *par excellence*, scarcely does one quarter of the work which might be obtained from it. From Vougeaucourt to Besançon, a distance of 43 miles, the total fall being 248 feet, only

* *Comptes Rendus de l'Académie des Sciences*, November 5, 1866.

900,000 horse-power was utilized in 1860, out of the total amount of 3,400,000 horse-power which might have been employed.

Although manufacturing operations can only assist exceptionally in moderating and gradually doing away with floods, the agricultural processes which are going on in all the valleys inhabited by man, ought to exercise a direct and decisive influence in regulating the flow of streams and rivulets. The husbandman ought not to allow the waste of a single drop of the beneficent water, which, by a widely-extended system of irrigation, might double, or even increase tenfold, his crops, and convert a wilderness into a garden. In the intelligent employment of running water in the fertilization of a district, our agriculturists have much to learn from the example of the uncients. As far back as the time of the early Egyptians, works of irrigation of really colossal dimensions had been accomplished; and, perhaps, among all the undertakings of this kind due to modern industry, there is not one which, in boldness of plan or practical utility, can be said to surpass the *meri* (basin), or Lake Mœris, which was opened to the waters of the Nile in the reign of Pharaoh Amenemha III., more than 4,500 years ago, according to the chronology of M. Brugsch.

From the topographical details which have been left by ancient authors as to this wonder of the world, we know that the site of Lake Mœris must be looked for in the present province of Fayoum, the name of which is derived from the Coptic, and signifies *sea*. Now, a considerable lake exists at the present time—the Birket el Keyroun—in the lowest part of the province; and so long as the geography of this part of Egypt was but partially known, it was very natural to look upon this lake as the ancient excavation of the Pharaohs. A study of the localities has proved that this is not the case. In fact, the Birket el Keyroun is situated in a deep depression, nearly on the level with the sea, and 53 feet below the average waters of the Nile. This basin, therefore, cannot be the one which alternately received the surplus flood-water of the river and emptied it out again through two wide gates, as Strabo tells us, into the plains by the side of the Nile. Besides, the position of this lake differs much from that which the ancient geographers assigned to the Mœris. According to the discoveries of M. Linant de Bellefonds, the engineer, the site of the great reservoir was just in the very highest part of Fayoum, to the west of the rocky gorge of Illaoun, through which flows a natural side-channel of the Bahr Yousef, which probably, at some former geological period, was the principal current of the Nile. Fragments of a long dike, which in some places is not less that 30 feet high and 200 feet wide, may still be met with in the eastern part of Fayoum. It must once have constituted a semi-circular rampart, spreading round the outlet of the great basin of the Fayoum plains, and have penned back the water brought by the Bahr Yousef.

M. Linant has calculated that, during the hundred days of flood, this branch of the river, which represents on an average about the twenty-eighth part of the Nile, emptied into the basin a quantity of water equal to 466 cubic yards a second, and that the total mass of

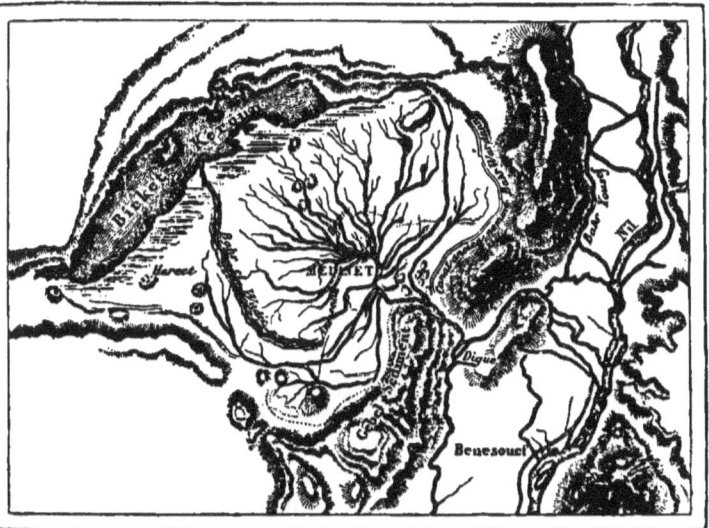

Fig. 126.—Map of Fayoum.

water contained in this gigantic reservoir, even after making allowance for evaporation, could not have been less than 3,694,000,000 cubic yards. This was sufficient to diminish very considerably the dangers resulting from the inundations of the Nile, and subsequently to afford all the water that was requisite for the irrigation of 444,000 acres. According to the statement of Herodotus, the surplus waters spread out to the west towards the Syrtes of Libya; that is to say, after crossing the lake, which is called Birket el Keyroun, it filled the bed of a channel now dried up, which carried the waters of the Nile into the western deserts. At the present day, Fayoum still possesses a magnificent system of irrigation, which may be compared to the ramifications of the arteries and blood-vessels of a living being : but forty-five centuries ago, Lake Mœris, which constantly changed its level according to the needs of agriculture, was like a heart from which the flow of life was shed out to nourish the great body of Egypt as far as distant Memphis. Nothing now remains of Mœris but the broken-down dikes, a few fragments of the two pyramids which were built up in its waters to the glory of Amenemha, and a thick layer of alluvium deposited on its basis by the troubled waters of the Bahr Yousef.*

Among European rivers, the Po is that which may be best compared to the Nile of the ancients, as regards the care with which its waters have been utilized for the fertilization of the soil. So far back as 1863, the Lombard agriculturists required for the watering

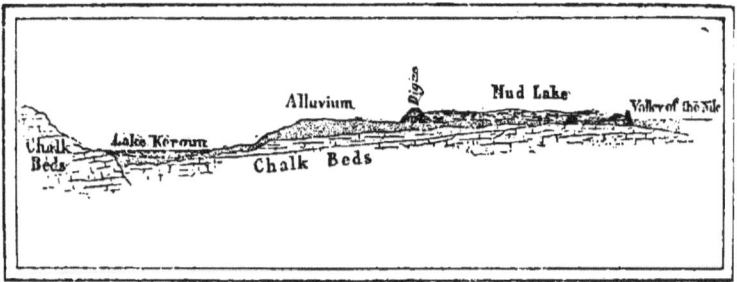

Fig. 127.—Section across Fayoum.

of their crops 59,000,000 of cubic yards of water a day, equal to 681 yards a second; that is a liquid mass equivalent to the average discharge of the Seine during the same period.† Since the above date, the great Cavour canal has been opened—a perfect artificial river —which requires for itself alone 144 cubic yards of water a second. Starting from Chivasso, below Turin, this river, which is not less than 55 yards wide at its commencement, spreads its fertilizing water on both sides in the already fertile plains of Lomellina; it receives, *en route*, numerous streams—the Elvo, the Sesia, the Agogna, the Terdoppio—and at Turbigo empties into the Tesino all that remains of its liquid mass. After having irrigated more than 494,000 acres. Next to the great canal of the Ganges, in Hindustan, it is the most

* Linant de Bellefonds, *Mémoire sur le Lac Mœris*.
† Elia Lombardini, *Politecnico*, January, 1863, quoted by Marsh.

important operation of this kind accomplished in modern times. There can be no doubt that the Po, once so dreaded on account of its sudden floods, will ultimately become, in conjunction with the other water-courses of Lombardy, a scientifically arranged system of agricultural canals.

Agriculturists should not only employ the water of torrents and streams for increasing their crops and nourishing the soil, but they should also make use of the sediment and débris of all kinds, which are washed away by the water from their up-stream banks. As an instance, let us take the Durance, a French river, which has been thoroughly surveyed and studied to ascertain the plan for utilizing its water and mud for the irrigation and manuring of the plains by the river-side. The eighteen channels, which are fed by this stream, can draw from it as much as 90 cubic yards a second; so that at any time when the whole of this liquid mass is being taken away at once, only 30 cubic yards remain in the bed of the Durance in low water seasons, or about a quarter of its regular discharge. According to the observations of M. Hervé-Mangon, which lasted from the 1st of November, 1859, to the 31st of October, 1860, the mass of mud brought down by the stream during the whole year represents a quantity of near 18,000,000 tons. Some idea may be formed of the enormous bulk of the mud which is washed away every year by the Durance from the upper portion of its basin, by picturing this mass in the form of a cube 242 yards on each side; if spread out uniformly on the ground this alluvium would cover in a year more than 108,000 acres, with a layer an inch thick, containing in a form of combination most suitable to the plants more azote than 100,000 tons of guano, and more carbon than 121,000 acres of forests (?). Unfortunately, as these canals are constructed with a view to irrigation only, nine-tenths of the mud is lost for manuring purposes; and the farmers purchase, at the cost of many thousand pounds a year, the very elements of fertilization which their stream washes down into the Mediterranean, although they might so easily avail themselves of them.

As every river possesses its own special peculiarities, the regulating works which the engineers have to undertake for the purpose of doing away with floods and distributing the water discharge, must be contrived in different ways, according to the form and capacity of the upper mountain hollows, the rapidity of the current, the suddenness of the floods, the porosity of the ground by the river-side, and the extent of the forests which clothe the hill-sides. The operation of regulating the discharge of a water-course is certainly very difficult to accomplish; in some river-basins it would require the labour of several generations; but suffice it to say, it is not impossible, and that it has already been successfully carried out in many parts of the globe. Although the greater part of the rivers, in both Europe and the civilised portion of America, have up to the present time remained free from man's guidance, and still occasionally devastate the cities and cultivated districts which lie upon their edges, there are at least a few of which the floods have been rendered harmless, thanks to the labour of the frightened inhabitants. Among the rivers which were once most dangerous, and are now almost entirely subdued, we may mention the Arno, which has been looked after for centuries by the skilful Tuscan engineers. At one time this river was most formidable, on account of its periodical inundations. From the year 1400 to 1761, no less than thirty-one disasters of this kind are recorded. Since 1761—the date when the improvements of the river were carried out—until 1835, there has not been a single serious flood.* The Po itself—the river which in flood-time hangs suspended, so to speak, over the surrounding plains—is now much less to be dreaded than heretofore, thanks to the irrigating channels which tap it, and also to the lateral embankments which border the whole of its lower course below Cremona.†

In this stream, as in a great many others, the surplus waters of the high floods come down too rapidly, and in masses too considerable, to afford any possibility of storing them up, or of turning them off in a lateral direction, without devastating the plains. It is necessary that the inhabitants should protect themselves by well-planned constructions against the threatening pressure of the water. The Egyptians dwelling in the delta built their cities on artificial hillocks above the level of the annual floods. The inhabitants of some parts of Holland, wishing to facilitate the "warping" of the fields, elevate their habitations above the ground, and the houses become so many islands in the midst of the floods. In recently colonised countries, where man's first care is to protect his habitation, all he does at first is to construct a circular embankment round the town or village. This was the procedure of the French colonists, after they had planted the first pile-work of New Orleans. The Americans, too, adopted this plan of protection for the Californian city, Sacramento, and the warehouses at Cairo, situated at the confluence of the Ohio and the Mississippi.

* Marsh, *Man and Nature*. † *Vide* p. 389.

EMBANKMENTS. 387

In like manner, the towns on the banks of the Loire are protected against the flood-wave by walls. Added to this, when the banks of a water-course are covered by cultivated fields, and the inundation would prove fatal to them — as in Louisiana, Lombardy, China, and many other countries in the temperate zone—

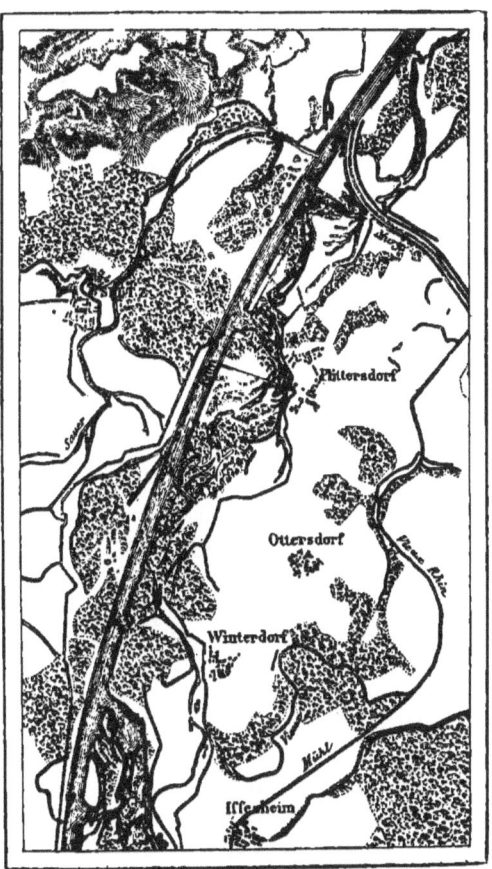

Fig. 128.—Dikes along the Rhine near Seltz.

it is necessary to raise longitudinal dikes on the edges of the streams which at flood-time are higher in their level than the surrounding plains. Thus shut in between their dikes, rivers are compelled to give up their wandering course, and to flow down to the sea through the channel which has been traced out for them. These longitudinal embankments, which, at any rate, are no ornament to nature, are sometimes a matter of absolute necessity; but if the constructors wish to prevent their dikes being broken through, and to avert the disasters which are the certain consequence of cracks, they must calculate beforehand the force of the liquid mass with which they will have to contend during extraordinary floods; and they must build their ramparts of materials sufficiently solid to resist without difficulty the lateral pressure of the water. They must likewise carefully protect their dikes against burrowing animals, for the embankments of the Po have several times been perforated by moles, and those of the Mississippi by musk-rats. It is necessary, also, to give the embankments a gentle bend, and to leave a sufficient width for the penned in river. The Loire, in front of Orleans, was once 3,827 yards wide, but has been reduced by the embankments to a bed of 306 yards. At Jargeau, it is only 273 yards wide at a place where it once had a lateral spread of 7,650 yards. In 1856 the Loire burst twenty-three breaches through these banks, which were said to be impenetrable;* as soon as the height of the flood rose in the river to more than $16\frac{1}{2}$ feet, cracks became inevitable. The losses occasioned by the breaking down of these too feeble ramparts, over which the flood-water rushed like a deluge, were so considerable that the question was often asked whether it would not be better to throw the dikes down entirely and to replace them with plantations of trees. The water, flowing without difficulty through the open barrier of the crowded trunks, would be distributed equally over the plains by the river-side; and would consequently never rise to the formidable height which it reached between the dikes. Added to this, its annual ravages would be in great part compensated for by the fertile alluvium which would be deposited by the sediment with which the water is charged. It has been calculated that if the vast basin of the Saone situated above the gorge of Pierre-Encise, were protected against the inundations of the river by means of dikes, confining the water to a bed 273 yards wide, the same as at Lyons, a liquid mass of 1,869,000,000 cubic yards, which, during inundations like those of 1840, now spreads over the plains, would then rush down upon the town in the space of a few days. On the other side of Lyons the Rhone affords a remarkable instance of the influence which the dimension of the bed exercise on the height of the flood. In 1856, in the wid

* Champion, *Inondations de la France.*

plain of Miribel, seven miles and a half above Lyons, the flood-water rose only nine feet and a half; but it rose to 20 feet—that is, more than double—in the narrow bed contained between Lyons and the Brotteaux.* In the valley of the Isere, the mean height of the flood-waters have undoubtedly risen since the construction of the side embankments, which were, in fact, placed too close to one another. This has been proved by the very exact observations of M. Dausse.

The embankments of the Po, more scientifically constructed than those of the Isere, were commenced many centuries ago, when the long night of the middle ages still darkened the rest of Europe. At a point below Cremona, where the continuous line of dikes commences, they are very wide apart, but the space through which the

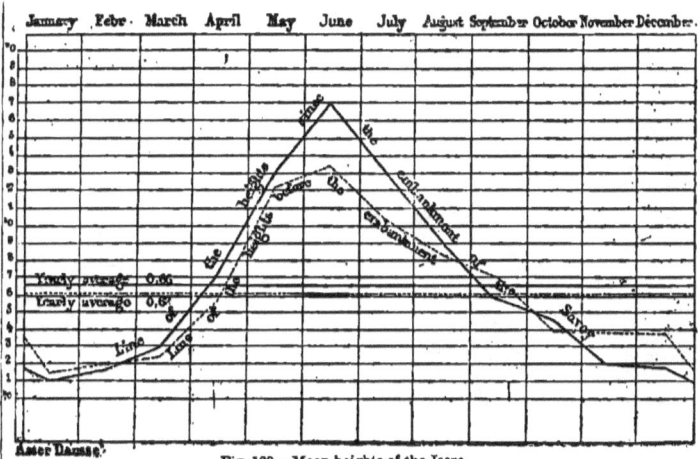

Fig. 129.—Mean heights of the Isere.

flood-waters can flow is gradually narrowed in down to the mouth of the river; from 6,564 yards it diminishes to 3,000, 2,000, and even 1,000 yards. Ultimately each of the branches of the delta is not more than 300 to 500 yards wide between the enclosing embankments. The fact is, that a great portion of the mass of water, finding between its upper dikes so considerable a space over which it is able to spread freely, remains stored in the plains above, and thus the flood-water tends constantly to diminish in a down-stream direction. The flat districts that lie between the dikes are called *golenas*. Each landholder may cultivate them and embank them as he likes; but on

* Gobin; L'Eveillé, *Commission Hydrométique de Lyon*, 1863.

the condition that his dikes shall be always nearly six feet lower than the principal embankments, so as not to offer any serious obstacles to high inundations. These *golenas*, therefore, with their dikes all round them, form so many settling reservoirs where the alluvium accumulates after each fresh flood, and their level is much higher than the plains outside the dikes.* Owing to the care with which these embankments are kept up by the syndicate of the riverside proprietors, cracks in them are very rare. Since 1705, the date at which a breach of more than 50 miles took place below Cremona, the reconstructed portion of this enormous rampart has not yielded at any point. Although lower down the river extraordinary inundations still occasionally break though the lateral embankments in some few places, any great disaster is in a measure prevented by the side

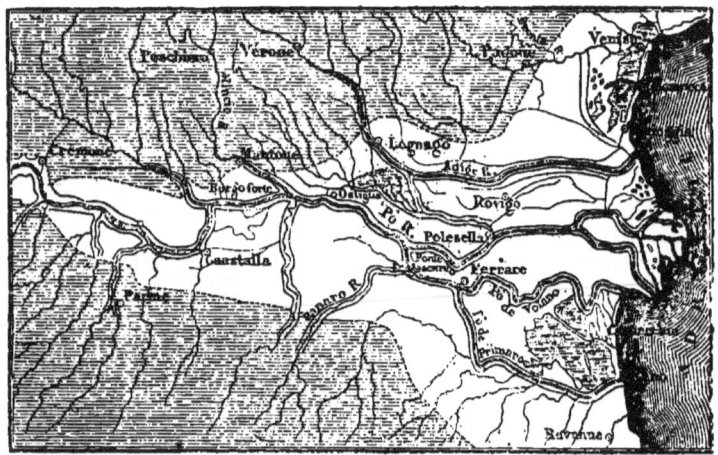

Fig. 130.—Dikes by the Po, from Cremona to the Sea.

channels, opened on both sides of the river, in the delta of the Po. Nevertheless, the system is not yet perfect. M. Lombardini thinks it very important that in the lower part of the river a considerable space should be left open to the flood-waters, so that the alluvium might be distributed on the plains on each side, instead of throwing out a promontory into the Adriatic Sea, and consequently raising the bed of the river.

Next to the embankments on the Dutch rivers, those of the Po form the most remarkable system of protection against inundations

* Elia Lombardini, *Dei Cangiamenti del Po*.

that has been devised in Europe; but they are inferior in importance to the embankments which run along a great portion of the Mississippi, and which, from their enormous size and length, form a source of admiration to every traveller. On the right bank of the river, from Cape Girardeau (Missouri) to the Pointe-à-la-Hache, situated below New Orleans, the embankments form a wall of 1,125 miles in length, only interrupted by the mouths of rivers and a few spots of rising ground. On the left bank, the base of the plateau, which the Mississippi here and there touches, has enabled the inhabitants to dispense with constructing any continuous dike; but they have been compelled to resort to embankments for protecting all the plains which extend from Memphis to Vicksburg, and from Bâton-Rouge to New Orleans. The ramparts that have been raised on the eastern bank are altogether more than 625 miles in length, and some of them are

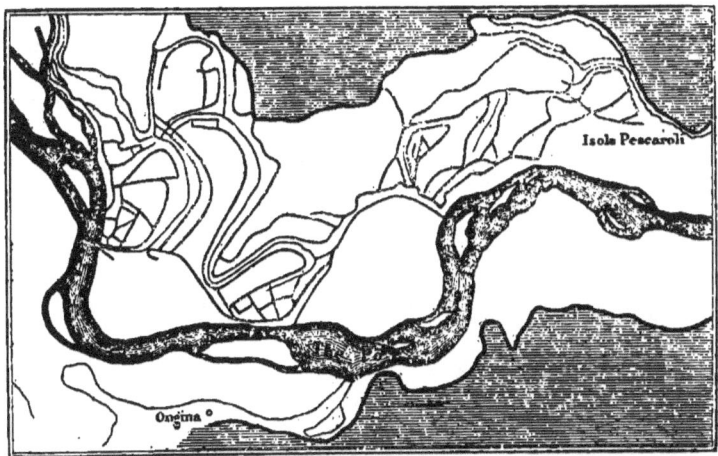

Fig. 131.—Golenas by the Po.

of very considerable dimensions; that which has been constructed at Yazoo-gate, in order to close a *bayou** of the Mississippi, is no less than 42 feet in height, 42 feet in width at the top, and 317 feet broad at the base. To these immense constructions we must add all the embankments formed along the tributaries of the Mississippi and the *bayous* of its delta; we must likewise take account of all the double and triple parallel dikes which have been raised in some of the spots which are most exposed to the action of the river. The whole of the

* *Vide* foot-note, p. 404.

embankments of the Mississippi must altogether reach a total length of at least 2,500 miles. It is true that in many of these imposing ramparts there is still much to be wished for in respect to solidity.

Every great flood on the Mississippi which has been recorded has formed one or more breaches in the embankments above New Orleans. The water then rushes like a cataract into the plains which extend

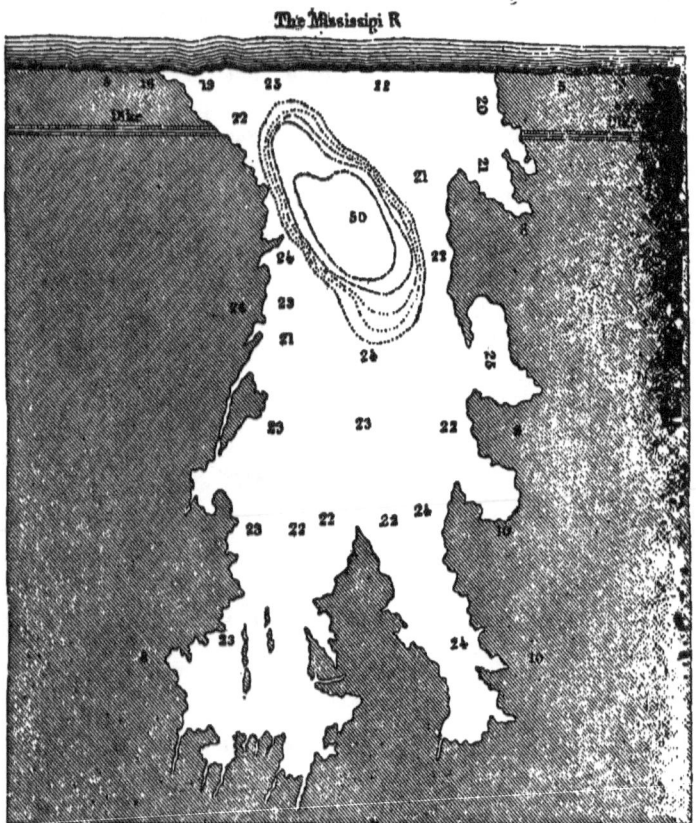

Fig. 132.—Gap formed near New Orleans.

below its level to a depth of 10, 12, and even 15 feet. It rapidly enlarges the opening by washing down the dikes for an extent of one or more miles, and then digging deep into the soil, hollows out for itself a new bed across the plantations. One of these temporary beds which the river made in 1850, 1859, and 1862, near the hamlet of Bonnet-Carré, discharged no less than 3,930 cubic yards of water a second; that is a sixth of the average liquid mass of the Mississippi. If the inhabitants had not succeeded in stopping it up on each occasion, the new river would, without doubt, have gradually become one of the branches of the delta of the Mississippi. In like manner the Hoang-Ho, having burst through its embankments, emptied itself into the sea, partly to the north and partly to the south of the peninsula of Chantoung, leaving a distance of 217 miles between its two mouths. The territory exposed to its ravages was not less in extent than the whole area of England. According to a tradition related by Ritter, which, however, is doubtless exaggerated, 200,000 individuals of the province of Honan were drowned during a civil war in consequence of the dikes being cut through.

CHAPTER LII.

THE MOUTHS OF RIVERS.—ESTUARIES.—LONG BANKS OF SAND.—DELTAS.—
NETWORK OF BRANCHES OF RIVERS IN ALLUVIAL PLAINS.

BELOW its confluence with its last tributary a river cannot fail to diminish in volume, on account of the evaporation of its water, and also of infiltration into the earth. There are, indeed, some streams which, as we have seen, gradually waste away without receiving any compensation from tributaries to make up for their liquid loss, and ultimately entirely dry up. Not only in the burning regions of the torrid zone, where rains are rare, but also in the great plains of the temperate zone, wherever the surface of the ground is too level to afford an incline for the running away of the water, we find many rivers flowing down from the mountains, and then, failing to make their way to the ocean or any inland sea, they disappear among the sands of the level country. Thus the Rio Dulce, the rivers Primo, Segundo, Quinto, and several other water-courses in the Argentine Republic, come to an end amid the *pampas* in a series of lagoons, which rise or fall, advance or retire, in the desert, according to the seasons of the year, or the quantity of water. Farther up-stream these rivers are navigable for boats, and sometimes cover the country round with their floods; but below their current becomes weakened, they break up into pools, and at last, becoming little more than liquid mud, they fail even in moistening the soil of the prairie. In a similar way the branch of the delta of the Rhine, which retains the name of the river, disappeared amid the sand-banks previously to 1806, the date at which a canal was dug through the dunes, and was protected against the sea by efficient flood-gates.

A river, however, can scarcely be considered to be worthy of its name, and can play no important part in history, if it fails to send down its water to the ocean in a constant and regular way. Only under these conditions is it accessible to ships, and in a position to connect the inland districts with those of the sea-coast. Just as a tree, the trunk of which, formed by the union of all its branches, brings into communication the atmosphere and the bowels of the earth; so the chief *trunk* of the river, in which all its affluents combine their liquid mass, links the sea to the mountains, and to the plains. By its ever-moving flow, by the junctions of its own current of fresh water with the salt waves of the rising tide, it brings together all parts of its basin, and gives life and energy to the earth, as the blood quickens the flesh which it moistens.

The oceanic portion of a river is characterised by the tides which twice every twenty-four hours change the direction of its current, and cause the water to flow back up-stream. In this small portion of its development, the action of a stream is completely modified; it is no longer a water-course, nor is it the ocean. It is, in fact, a common bed where the two elements meet and unite. The river-mouth is not only an entrance to a continent through which navigators may pass, it also opens an outlet to the sea-water, and enables it to ascend far inland, and to mingle with the liquid mass brought down by the river. That portion, therefore, of the channel where the junction takes place between the salt and fresh water constitutes a geographical division which is perfectly distinct from all the rest of the basin.

Most water-courses, however winding their course may have been, straighten as they approach the sea, and descend towards the shore by the shortest line possible, so as to form a right angle with the coast. This tendency may be partially explained by the fact that the steepest slope of the ground is generally inclined in this direction; but another cause, also, is the alternate action of the tide-wave, which takes place perpendicularly to the shore, the to-and-fro motion of which ultimately governs that of the river.

Added to this, a large number of rivers, when they reach the maritime portion of their course, spread out their banks very widely so as to form real gulfs, in which it would be impossible to trace out the precise limit which marks the river-mouth. When these bays are not original indentations of the coast, they owe their existence to the combined action of the river and the sea, which gradually cuts away the banks, and ultimately deposits them on some distant shore. Thus, fluviatile estuaries are generally found on those parts of the coast which are directly exposed to the force of the tides and storms. Estuaries are very numerous on the coasts of the open sea where the tide rises to a great height; but they are comparatively less frequent in land-locked seas, which preserve an almost unaltered level, such as the Mediterranean, the Baltic, and the Caribbean sea. Nevertheless, the shores of several inland seas—among others, the Euxine, so formidable for its winds—present river-estuaries similar to those on the oceanic coasts; the most remarkable are the *limans* of the Dniester and the Bug.

Almost all the rivers of Western Europe spread out into estuaries in the lower portion of their course. There are some among the number, as the Thames, the Severn, and other rivers of Great Britain, which are streams of no great importance above the gulfs at their outlets, and owe all their consequence to the powerful tide-waves of the Atlantic. In France, the Seine, the Loire, and the two combined rivers of the Garonne and the Dordogne, water basins which are better proportioned in their area to the dimensions of their estuaries; nevertheless, the quantity of fresh water sent down into these advanced bays of the ocean forms but a very small portion of the liquid mass which they contain. In the Gironde, which may be taken as a type of a marine estuary, the salt water generally ascends as far as Pauillac, 31 miles from the outlet; any one sailing on the river may readily notice the shifting line where the various liquid masses, some green and transparent, and others yellow and muddy, mingle with one another in long eddies.

At a point more than 10 miles from the sea-coast, the saltness of the Gironde is scarcely diminished by the admixture of fresh water. At one time, the low ground by the river-side was intersected by salt marshes, and the creek of Méchers, on the north bank, has been utilised for some years in the cultivation of oysters. The depth of the estuary is also very considerable. At Méchers, the Gironde, which at that place is 7½ miles wide, is from 50 to 100 feet deep even at low tide. At the outlet properly so called, the estuary contracts, and is only 3¼ miles wide; but in mid-channel, the sounding line finds no bottom at 100 feet. This enormous basin does not look like a river. If a spectator contemplates it, not from the point of a headland, but merely from the edge of the shore, at St. George or Royan, he cannot distinguish the whole extent of the opposite bank; all that is visible is a few clumps of pines, separated by the white line of the distant water, and these isolated clumps seem to form an archipelago; the Gironde appears like a sea dotted over with isles and islets. The colour and the appearance of the water are continually changing; it is as if several rivers, crossing one another in every direction, were flowing in one and the same bed. Sandbanks which show their white masses indistinctly under the green waves, the marine currents which meet and mingle with the turbid water of the ebbing tide, the gusts of wind which raise on the estuary a perfect network of winding ripples, the long trains of foam which incessantly shift their place; lastly, the submarine counter-currents which flow up to the surface and there spread out in sheets perfectly smooth — all these ever-changing phenomena are always modifying the magnificent spectacle afforded by the Bay of Gironde.

But what, after all, is this beautiful estuary of the French coast when compared with the grand outlets of some of the American rivers, such as the St. Lawrence, the current of the Amazons, and the Rio de la Plata. This last estuary, into which pour the gigantic Parana and Uruguay, more than 6 miles in width, is at the outlet no less than 155 miles across, and occupies a space of more than 15,400 square miles. Within a recent geological period, it stretched over a still wider area. At that time, the Parana had not filled up with its alluvium all the higher portion of the estuary, and probably, also, the surface of the pampas was covered by the seawater. Even in the present day, the now diminished gulf is nothing less than a real sea. Its bed, which prolongs in a gentle slope the surface of the Argentine plain, is hollowed out 66 to 100 feet below the level of the ocean. Currents and counter-currents, like those in the open sea, traverse the gulf in every direction. Furious winds, which seem to upheave the whole liquid mass, give rise to tempests which are more dreaded than those of the ocean, on account of the sandbanks and rocks which hem in the channels. The highest floods of the Uruguay and the Parana have no perceptible influence on the level of the Rio de la Plata, and seem lost like rivulets in the enormous estuary.

Although the winds and tide have such an effect in increasing the mouths of rivers, into which the waves enter in a direct line, their mode of action is very different when they are diffused along a sandy shore, which they meet at a very acute angle. In this case, the waves from the open sea, being driven obliquely against the coast, wash away from it large quantities of *débris*, which they deposit in front of the mouth of the adjacent river. Under the enormous pressure of the ocean, the current of the river bends and gradually doubles round in the same direction as the marine current, allowing a tongue of sand to form across its former bed. In the course of time, a narrow peninsula, having a sea-shore on one side and a river-bank on the other, divides the fresh water from the salt water for a distance of several miles, sometimes breaking up into islands, according to the various changes of the atmosphere, the current, and the tides. Thus, on the coast of New Granada, extending from the

Cape de la Vela to the foot of the snow-clad mountains of Santa Marta, all the river outlets are pushed towards the west by the current which runs along the shore towards the Gulf of Darien; mere embankments, ornamented here and there with green vine branches, and the violet corollas of a kind of bindweed, protect the still waters against the onset of the breakers.

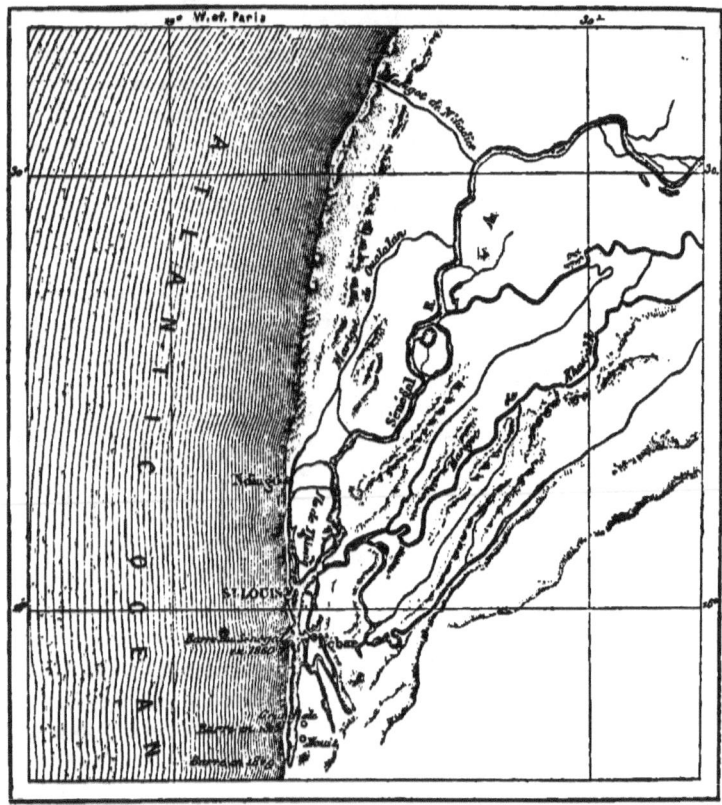

Fig. 133.—Belts of the Senegal.

The River Senegal exhibits one of the most remarkable examples of these belts, formed along the shore by the marine currents, and running across the outlet of a river. For a distance of more than 180 miles, the great water-course follows a direction perpendicular to the coast. In this way it reaches a point 15 miles from the sea, at which its course is arrested by a chain of dunes, and it is compelled to find an outlet through some other part of the sea-shore. At one time, the river, or at least one of its branches, continued in its direct path to the ocean, and on the spot where its former bed may still be traced, there is a narrow marshy flow, known under the name of the Marigot of N'diadier. Being thus driven in a south-west direction, the Senegal is compelled to approach the sea obliquely. Above St. Louis, the river is separated from the line of breakers by nothing but the narrow bank of the Guet-N'dar, on which the blacks have built their *faubourg*. Farther down the coast, the embankment of sand thrown up by the marine current running from the north continues for a considerable length, altering its position every year, owing to the double action of the river-floods and the sea-waves. At the present time, the mouth of the Senegal opens $2\frac{1}{2}$ miles south of St. Louis, and is ascending slowly towards the town. In 1849, it was 9 miles farther to the south; but, in 1825, it was near Gandiole, a little farther up the stream. This sandy rampart, which extends its graceful curve from north to south for more than 24 miles, is cut through by the current of the river, sometimes at one spot and sometimes at another; but it never fails to form again, owing to the action of the sea-waves. Until the operations of man have fixed the place of the mouth of the Senegal, it will continue to shift its position along the sandy dike.

In a similar way all the various streams which empty into the sea along the low coasts of the French *Landes* bend round towards the the south as soon as they reach a point at a short distance from the sea-shore. There is, in fact, a current produced by the swell which runs parallel to the shore of the *Landes*, a matter which is easily proved by noticing the drifting of any floating substance, or the bearing of shipwrecked vessels, which always point their sterns to the south. This current pushes before it masses of sand, which are mixed with the breakers, and thrown up upon the beach. The sandy points, which are constantly augmented by the additions brought by the waves, are thus elongated towards the south, and would ultimately reach the bases of the Pyrenean promontories if it were not for the tendency of the streams to rise, so as to increase their slopes, and thus to press with increasing weight on the sands which obstruct them. Formerly the waste channels, or "*courants*," of the Lakes of Soustons and St. Julian, flowed parallel to the sea for a length of several miles above their outlet, and fears were entertained that, in consequence of the lengthening of these streams, and the rise in their level, the lakes above would spread over the surrounding country. In order to avert this disaster, the inhabitants undertook to rectify the course of the waste channels, and thus to lower the level of the lacustral waters. This plan succeeded perfectly as regards the Soustons lake. Its level was sunk 10 feet, to the great advantage of the village near, which was enriched with a tract of fertile alluvium. The Lake of St. Julian was likewise lowered several feet by the alterations made in the "*courant*" of Contis; but the engineers met with considerable difficulty in mastering this watercourse, and in preventing it from flowing in a southerly direction parallel with the coast. They were several times compelled to lengthen the barrier which forced it to flow in a straight line down to the sea. As regards the more important stream of Mimizan, which serves as a waste-channel for several considerable lakes, an attempt was often made to dig out for it a regular bed in the direction of the coast, and to retain the flow of water in it; but the river would not be subdued, and, throwing down the barrier of piles and faggots which was opposed to it, continued to run towards the south and south-east. Miles of basket-work dikes, which were set up to guide its waters, now lie buried under the dunes.

During the course of the Middle Ages, and probably also in the previous historical era, the Lower Adour—which is now perpendicular to the coast—extended in a line parallel to the chain of dunes and the sea-shore for a length of about 12¼ miles. The river then fell into the sea at a short distance from the spot where the town of Cape Breton now stands. Towards the end of the fourteenth century a violent tempest obstructed this outlet, and the Adour, thrown back farther to the north, found no place of issue nearer than a point 22 miles from Bayonne; a village called Vieux Boucau (old mouth) marks the banks of the former river. At first sight it seems as if this ancient course of the Adour is to be explained in a similar way as the curves described towards the south by the streams of the *Landes;* but if this were the case, the current of the sea-swell in this part of the Gulf of Gascony ought to tend in a *northerly* direction. Now the action of the waves points, on the contrary, from north to south, as far as the mouth of the Bidassoa, and consequently the sandy points are lengthened with a southerly bearing. The belt of banks across the course of the Adour was turned towards the *north;* it is, therefore, necessary to look for the cause of this in the existence of a chain of dunes solidly based on the nearest Pyrenean rocks, which presented an insuperable barrier

CHANGES IN RIVER-MOUTHS.

to the river on the western side. In 1578 this chain was broken through at a point three miles and three-quarters below Bayonne, by means of a trench cut by Louis de Foix, the engineer, and still more by a formidable flood, which threatened to carry away the city. Since this date, the mouth of the Adour, yielding to the coast-current, constantly tends to bend round towards the south; and on this side the piers formed to maintain the river are carried the farthest. At the end of the seventeenth century—at a date when

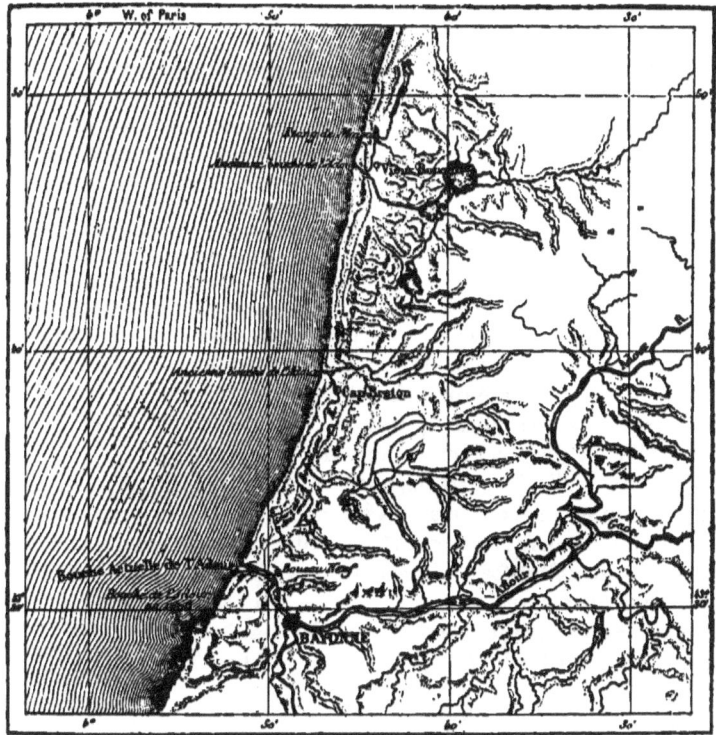

Fig. 134.—Old Course of the Adour.

these latter works had not been commenced—the river, bending gradually towards the south, emptied itself into the sea at the foot of the rocks of Chambre d'Amour, about two miles from Boucau-Neuf (new mouth). If the river had not been repelled on the right by the dikes constructed by Louis de Foix, it is very possible that it would again have turned towards the north.*

* Vionnois, *Annales des Ponts et Chaussées*, vol. xvi.

D D

One of the most wonderful phenomena on the face of the earth is the formation of those long banks of alluvium which affect a considerable number of streams, and, for a distance of hundreds of miles, protect a multitude of river outlets against the waves of the sea. A magnificent example of this formation exists on the coasts of Virginia and North Carolina. The rivers there, which flow on the surface of the ground, counterpoising the pressure of the ocean in the same way as the subterranean waters of Yucatan,* have formed out at sea an immense breakwater. This sandy dike—which is not less than 186 miles in length, bends round the continent in gracefully-winding curves, and encloses within its limits perfect seas—with their bays, archipelagoes, and currents; behind this the Tar, the Alligator, the Neuse, and several rivers run into the sea. An idea may be formed of the peculiarities presented by these long banks, common to several rivers, by comparing this dike with the altogether regularly formed littoral bank which lies in front of the River Cape-Fear, immediately to the south.

A third arrangement of the mouths of rivers is that which the ancient Greeks designated under the name of *delta* (Δ), on account of the triangular form so often assumed by the alluvial plain embraced between the branches of a river. This plain, which projects beyond the regular line of the coast, is nothing but a former estuary, which has gradually been filled up with mud and sediment of every kind. This alluvial plain cannot be formed to any great extent in places where the swell, the currents, and the tides are constantly disturbing the outlets of the rivers. It is necessary that the stream should be subject to conditions somewhat similar to those existing in still lakes, where deltas form without the least obstacle. These conditions are found in almost inland seas, with a scarcely perceptible current —such as the Mediterranean and the Baltic—which allow river-mouths to gradually fill up with mud. The alluvium which is brought down by the river is, certainly, soft, and has but little solidity; it is often roughly handled by the water at flood-times, and fails in preventing the liquid mass from forming forks, or even from dividing into numerous branches. But the sea, which assails these deposits, being constantly at about the same level, ultimately has the effect of consolidating them by dashing against them with its waves. On the contrary, when a river falls into a sea where the tides rise to a great height, and where the coast is alternately traversed by the rapid currents of the ebb and flow, no time is left for the deposit of

* *Vide* above, p. 286.

the river alluvium. This matter is first pushed back into the river by the flow of the tide, and then, being seized by the ebb, is carried out to great depths in the open sea. In this contest between the river and the ocean, the latter gets the advantage on account of the enormous mass of its waves, which, by their fluctuating movement of rising and falling, are incessantly scouring out the estuary through which the fresh water flows.

Amongst those rivers the deltas of which are incessantly gaining on the sea, we may mention, as belonging to the first class in this respect, the great affluents of the Mediterranean, the Danube, the Nile, the Po, and the Rhone; also, in the Caspian, the Terek, the Kouban, and the Volga. Other rivers possessing deltas fall into the sea at the extremity of some gulf well sheltered by a barrier of isles, and visited only by scanty tides. Of this kind are the Hoang-ho, the Yang-tse-kiang, and other water-courses, the alluvial shores of which continue to project more and more into the shallow Chinese Sea and the Gulf Pe-tchi-li. The delta of the Mississippi, which may serve as a type to all other formations of the same nature, pushes its way into an almost closed gulf, where the height of the regular tide never exceeds three feet. The only instance which can be mentioned of a great river-delta existing at the extremity of a gulf widely open to the ocean, is that of the Ganges and the Brahmahpootra. But it must not be forgotten that at the outlets of these rivers, the tide fluctuating between 1 foot and 16½ feet, never exceeds, on the average, 10 feet in height;* added to this, the delta, instead of pushing its way far into the sea, presents a flattened shape, and extends its low shores from east to west, giving a width of at least 186 miles. There is no doubt that in a more protracted sea, the delta of these two combined rivers of Hindostan, which bring down in their turbid waters so large a quantity of alluvium, would have thrown out a long promontory of delta exhibiting very different proportions.

In a cursory and rapid examination of a map, it would, however, be easy to fall into error as to the real character of certain river-outlets, and to look upon them as actual deltas, thrown out by the action of the river itself, instead of collections of soil deposited under the shelter of isles of marine formation. Thus, Holland, which is placed at the angle of the continent of Europe, appears at first sight to be the combined delta of the Scheldt, the Meuse, and the Rhine; but the outer shore is, in fact, an ancient coast cut through by the

* Beardmore, *Manual of Hydrology.*

waves of the ocean, and is composed of a vast semicircle of dunes, stretching from the mouths of the Scheldt, to those of the Ems and the Weser. Far from having gone beyond this original coast-belt, the greater part of the Dutch rivers have formed estuaries, and the wide sheets of the Bies-bosch, the Zuyder Zee, and the Dollart, constitute unquestionable testimony of the invasion of the sea-water. The alluvial tracts of Holland do not, therefore, present the character of a delta properly so called.

Deltas are not formed solely on the lower portion of a river's course; they also exist at all the points of the river where former lacustrine basins have been filled up by one or more several affluents. At these spots, the principal water-courses and its tributaries divide into several branches, radiating in a fan-like shape across the alluvial plain; sometimes, they even cross one another so as to form a complete net-work. About the middle of its course, the Mississippi receives two considerable affluents from the west, the Arkansas and the White River. The principal river and its two tributaries are united by a net-work of innumerable *bayous*,* which, at every inundation, change their course and their depth, falling alternately into one or the other of the three currents, according to the respective height of their waters. When the Mississippi is very high, it disgorges its surplus water into the system of *bayous*, and the latter empty into the Arkansas and the White River. During the low-water season, on the contrary, when the water poured by the Mississippi into the marshes above, has had sufficient time to flow from lagoon to lagoon down to the White River, the latter feeds the net-work of *bayous* which connects it with the Mississippi and the Arkansas. When the latter river is swollen more than usual after heavy rains in the western prairies, then the pressure of its water drives back that of the Mississippi, and, for a time, the Arkansas takes possession of the common delta. On the banks of the Amazon River all these phenomena take place with much more grandeur; at the mouth of the Japura especially, the principal current forms with its affluent an inextricable net-work of false rivers, which seem to flow indifferently in any direction, and, for a space of several thousand square miles, direct their surplus waters from marsh to marsh through the virgin forests. This system of *furos*, as they are called in South America, resembles those congestions in the human body, when the too great abundance of blood gives rise to a system of false arteries and veins.

* Derived from the French word *baie*. The Spaniards of La Plata give the name o *bahia* to natural channels of the same description.

CHAPTER LIII.

THE CHANNELS OF THE MISSISSIPPI.—"WORKING RIVERS."—SHIFTING OF THE POINT OF BIFURCATION.—RAISING OF THE RIVER-BED ABOVE THE DELTA. —ALTERATION IN THE SITUATION OF MOUTHS OF RIVERS.

IN a geographical point of view, it is important not to confound apparent deltas with the real deltas of alluvial earth. Thus, the basin of the Mississippi, in which there is opportunity for studying so many other hydrological phenomena, exhibits several instances of emissaries which must not be looked upon as branches of the delta. The Atchafalaya, in fact, is not a branch of the Mississippi, as it is not fed by the latter; it is, on the contrary, a continuation of the Red River, which sends down to the Atchafalaya a portion of its water directly, and another portion indirectly, by using for nearly a mile the bed of the Mississippi itself. The Plaquemine and Lafourche *bayous*, which, during floods, receive a small portion of the water of the Mississippi, are not regular fluviatile beds, like the branches of the Rhone, the Nile, and the Po; they are mere channels communicating between the inland lakes and marshes, and have become united to the Mississippi by an erosion of the banks of the river. It is, indeed, owing to the labour of man—that is, to the side embankments and the drainage of the marshes near it—that the Lafourche *bayou* has assumed the aspect of a river for so large a portion of its course, and now no longer disappears, as it once did, in a labyrinth of pools and marshes. The Manchac, or Iberville *bayou*, which used to reach the sea through the Amite River and the Lake Maurepas, is now completely obliterated by the alluvium and masses of entangled trees; but it has always been a mere flow of no great importance.* Thus, the delta proper only commences at the "Head of the Passes," and this sheath-like bed, through which the Mississippi rolls between two narrow banks of alluvium, one side of which is sea-shore and the other river-bank, is, geologically speaking, the sole bed of the river. Projecting from the continent like an arm, it pushes out for 62 miles into the sea, and spreads over the water the branches of its delta, like

* Humphreys and Abbot, *Report on the Mississippi River*, 1861.

the fingers of a gigantic hand. A Hindoo might well compare the extension of the mouths of the river to an immense flower opening over the ocean its serrated corolla.

These narrow embankments of mud, brought down into the open sea by the fresh water, present a striking spectacle. In several places these banks are only a few yards thick, and during storms the waves of the sea curl over the narrow belt of shore and mingle with the

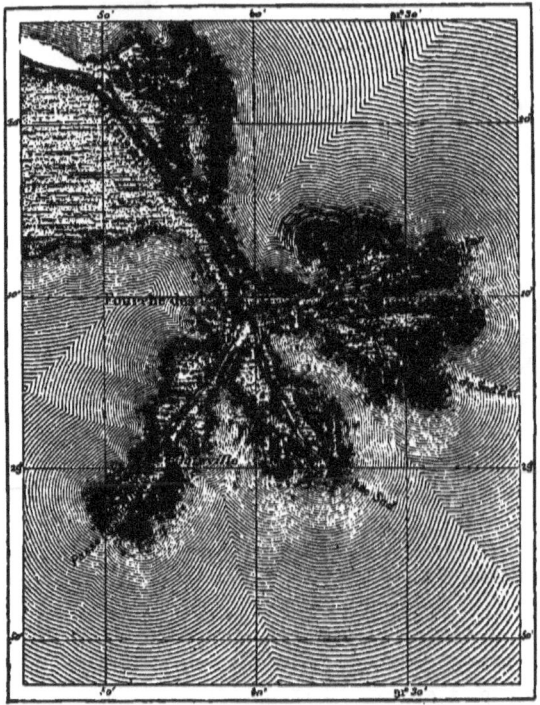

Fig. 135.—Mouths of the Mississippi.

river. The soil of the banks becomes perfectly spongy; it is not firm enough to allow even willows to take root, and the only vegetation is a species of tall reed (*Miegea macrosperma*), the fibrous roots of which give a little cohesion to the ooze, and prevent its being dissolved and washed away by the succession of tides. Farther down the reeds disappear, and the banks of mud form, are washed away, and form again, wandering, so to speak, between the river and the sea, at the will of the winds and tide. On the left bank of the south-west passage, which is used for the largest ships, the plank-built huts of a small pilots' village have been fixed as delicately as possible. These constructions are so light, and the ground that carries them is so unstable, that they have been compelled to anchor them like ships, fearing that a hurricane might blow them away.; still, the force of the wind often makes them drag on their anchors. Below, the banks of the Mississippi are reduced to a mere belt of reddish mud, cut through at intervals by wide cross streams; still farther down, even

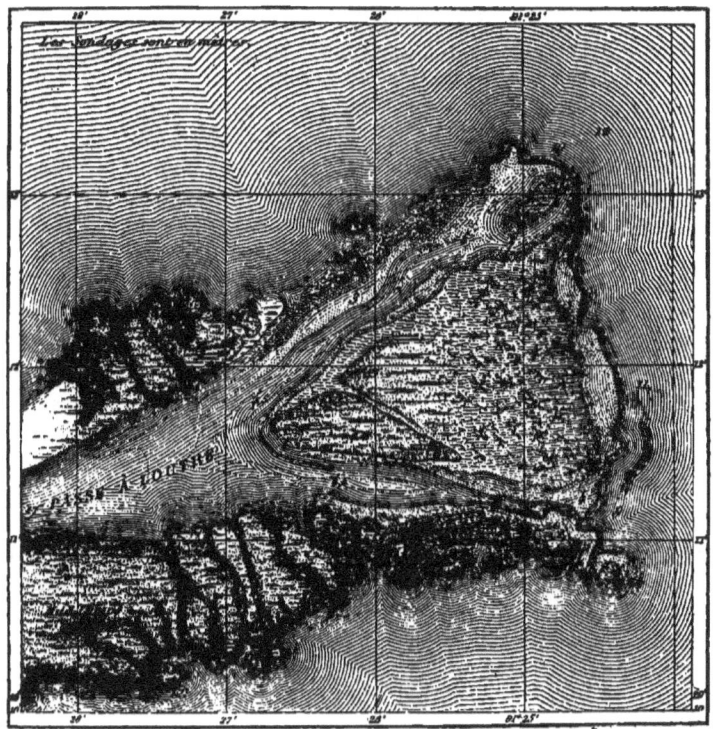

Fig. 136.—Channel of Loutre.

this narrow belt comes to an end, and the banks of the river are indicated by nothing but islets, which rise at increasing distances from one another, like the crests of submarine dunes. Soon the summits of these islets assume the appearance of a thin yellow palm floating on the surface of the water. Then all is mud; the land is so inundated with water that it resembles the sea, and the sea is so saturated with mud that it resembles the land. Finally, all trace of the banks disappears, and the thick water spreads freely over the ocean. After getting clear of the bar, the sheet of water which was the Mississippi preserves, during floods, the yellowish colour by which it can be distinguished for about twenty miles; but it loses in depth all that it gains in extent, and, gradually depositing the earthy matters which it holds in suspension, becomes ultimately perfectly mingled with the sea.

In calm weather, the union of the fresh and salt water presents an interesting spectacle, affording some similarity to the meeting of the tide and the river-current in an estuary. Gliding in layers of increasing thinness over the weightier masses of the ocean, the muddy water, on escaping from the mouths of the delta, swims like oil on the surface of the waves, and the sailors are able to collect it, without difficulty, by skimming the surface. Ships, as they pass, break through this light yellowish sheet and leave behind them a long track formed by the blue and transparent water of the sea. A contrast of the same nature is produced at the spot where the Gulf Stream causes the belt of the water of the Mississippi to swerve to the east; one might fancy that a straight line traced out by a ruler separated the two diversely coloured waters as far as the horizon. Finally, the sheet of fresh water, becoming very thin, is broken up into little turbid islets, surrounded with salt water. They are often full of vegetable *débris*—they are then edged with breakers in miniature, which give them a border of foam.* The sounding-line let down to the bottom of the sea off the mouth of the river finds the mud of the Mississippi as far as the coral banks of the Florida coast. The accompanying plates show the difference in the depth of the sea between the axis of the Mississippi and those portions of the gulf which are situated immediately to the south.

The fluviatile tracts of alluvium, which are constantly forming before our eyes, may be classed among the most important geological phenomena in the history of the globe. Owing to the quantity of mud which the masses of running water bring down to their outlets, the shore-line is incessantly changing, and continents are increased in area. Carl Ritter has given the name of "working rivers" (*fleuves travailleurs*) to those water-courses which deposit a large quantity of alluvium in deltas and push their shores farther and farther into the midst of the sea. Every river, indeed, takes its share in

* Kohl, *Zeitschrift für allgemeine Erdkunde*, September, 1864.

ALLUVIUM OF RIVERS.

this labour; but in great deltas the earth quite visibly encroaches upon the ocean. At the mouths of several rivers the lifetime of a man would be a period long enough for the salt bay to be converted into a plain, and the floating sea-weed beds to become a magnificent forest.

The deltas themselves, the vast plains which, as Herodotus says, are the "gifts of rivers," bear witness to the geological importance

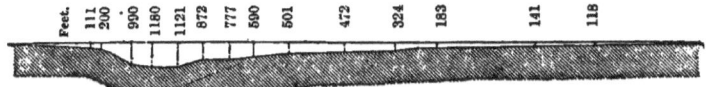

Fig. 137.—Depths of the Gulf of Mexico in the axis of the Mississippi Current.

of running waters in the formation of continents. But the investigations which have been made up to the present day enable us to estimate the progressive course of these alluvial formations in but a small number of rivers. In fact, the problem which has to be resolved is a very complex one. In the first place, it would be indispensable to prepare at intervals exact charts of the sea-coast and the depths of the sea in the vicinity; next, it would be requisite to strictly apportion the quantity of sediment brought down in each

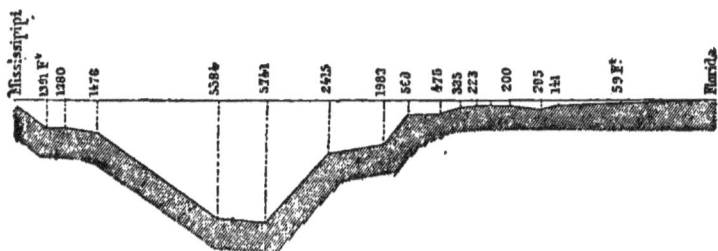

Fig. 138.—Depths of the Gulf of Mexico South of the Mississippi Current.

season of the year by the water of the river, and to ascertain the amount of alluvium which is lost along the coast. Lastly, in the beds of the delta itself, it would be necessary to distinguish between the *débris* washed away from the adjacent coast and the matter which is brought down by the river; for when a muddy point is formed, the currents along the shore always drive upon it a constantly increasing bank of sand. Some day, doubtless, more exact observations will enable us to trace out the journey of the alluvium down the river that carries it along; we shall ascertain the average time that elapses before the rock rolled down by the torrent is broken up into pebbles, and then in succession reduced to gravel, sand, and impalpable mud; we shall learn the number of resting-places that the *débris* avail themselves of in bend after bend from the river's source to the sea. Perhaps, even by the mere observation of the alluvial layers, we shall be able to discover the age of the bed, as we ascertain the age of a tree by its concentric rings. We must, however, confess that this class of geographical observations is scarcely inaugurated, and that it would require an enormous staff of savants, which does not at present exist. We are, therefore, compelled to form rather rough estimations as to the results of the labour of rivers; this is the case as regards the Hoang-ho, which is probably more loaded with alluvium than any river in the Old World.

This river owes its name, Hoang (yellow), to its muddy sediment, which, far out at sea, soils the purity of the sea-water, and is carried by the currents as far as the coasts of the Corea. The delta which it has formed during the present period extends over at least 96,000 square miles, and constitutes one of the most important provinces in China. The tracts of alluvium have joined to the mainland the mountainous mass of Chantung, which once stood alone in the midst of the sea. Fresh islets have slowly risen from the bed of the sea, and the detritus is deposited in quantities so great that, according to a calculation made by Staunton at the end of the last century, they would be sufficient in the course of sixty-six days to form an isle a square mile in extent and 118 feet in depth. According to the calculations of the same author, the whole of the Yellow Sea is destined to disappear entirely in about 24,000 years; but this period should be at least doubled, for the waters of this sea are much deeper than Staunton stated.

The English authors who have written on the subject of the lower regions of the Sunderbunds—that prodigious mass of alluvium brought down by the Ganges and the Brahmapootra, the terrible "son of Brahmah"—affords us but uncertain information as to the lengthening of the mouths of the rivers. According to Rennell, the Ganges alone sends down in its water from 5 to 6 cubic yards of mud a second; nevertheless, the line of shore extending from the mouth of the Hoogly to the estuary of Huringota, which consequently limits the Gangetic portion of the delta, appears to have been subject to but very slight modifications during historic times. The promontories and the islets of the eastern portion of the delta encroach much more rapidly on the sea; for on this side the waters

DELTA OF THE GANGES

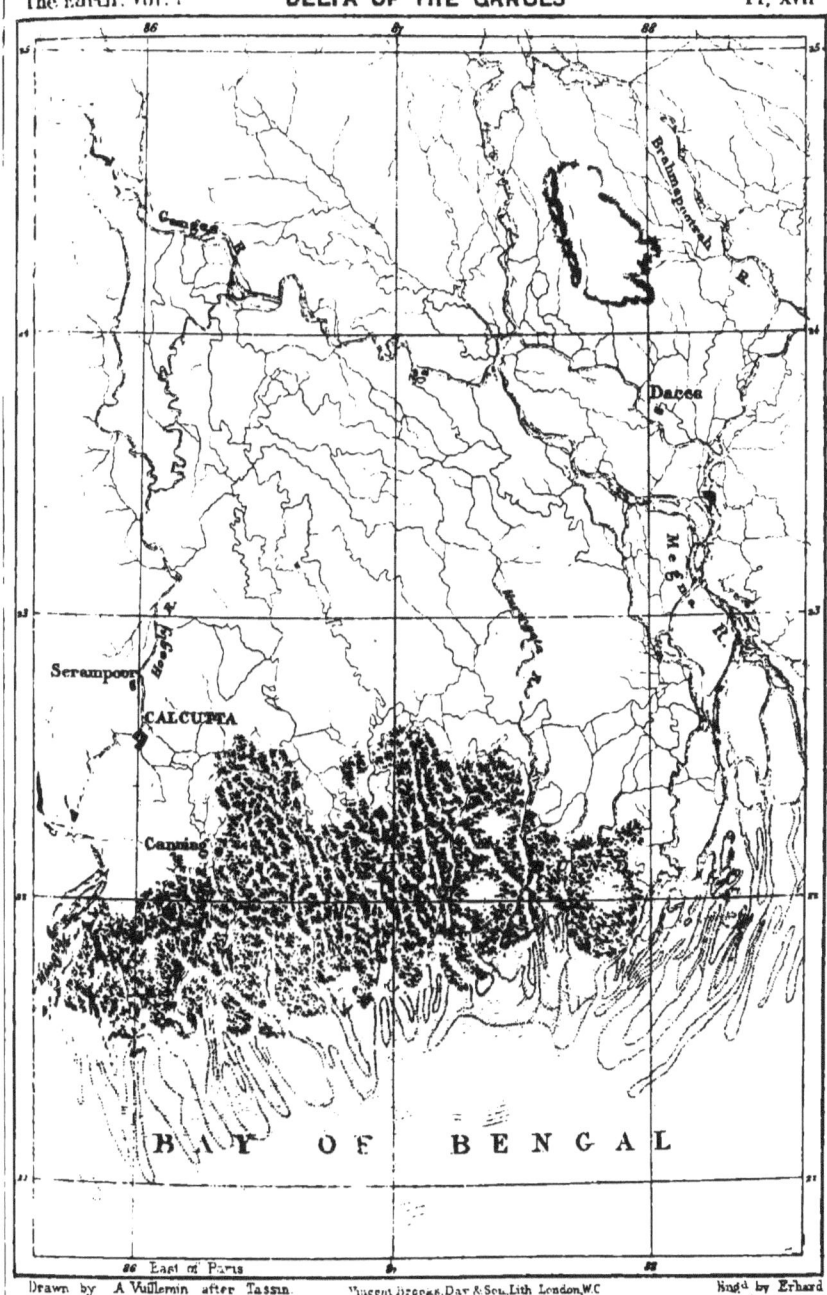

of the Brahmapootra, which on the average are charged with twice as much mud as those of the Ganges, pour into the Bay of Bengal.* A great quantity of the alluvium which is brought down by the two rivers is lost in the immense depths of the marine depression, which lies about 31 miles from the mouth of the Ganges, which is called the "Great Swatch."

The Nile—that typical river which was the subject of study to the Egyptian hierophants thousands of years ago—which spreads out the graceful delta formed of its own alluvium, is incomparably better known as regards its lower course than any river of Asia. This great water-course, which may be compared in the length of its bed to the

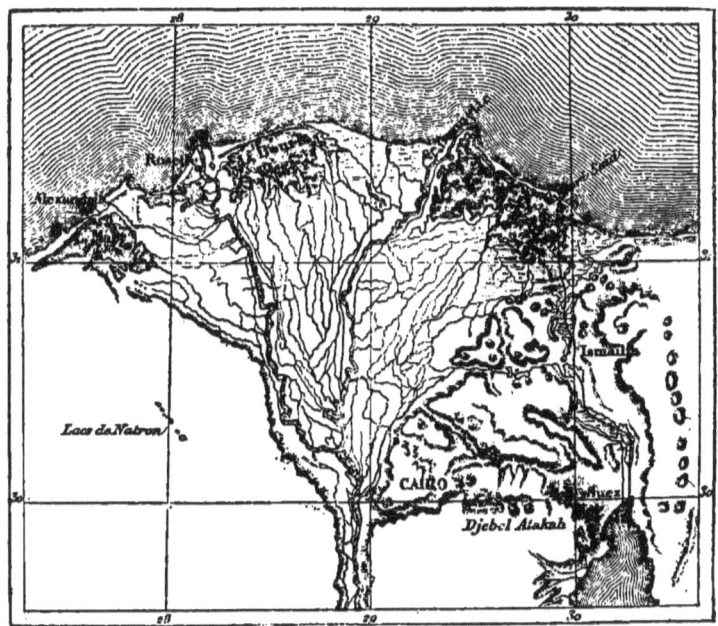

Fig. 139.—Delta of the Nile.

Mississippi and the Amazon, scarcely surpasses rivers of the third class, such as the Rhone or the Po, in the importance of its liquid mass, and is much inferior to them in the quantity of its alluvium. It has been calculated that if all the mud brought down by the mouths of the Nile was thrown up uniformly on the coast, the latter

* Ferguson, *Zeitschrift für Erdkunde*, 1864.

would advance about 13 feet a year. The low points of alluvium which are deposited near the Rosetta and Damietta mouths increase on the average—the one 34 acres, and the other 39 acres, every year, which gives only 3 feet of annual progress for the front of the delta, the convexity of which is 186 miles in length. If the advance of the alluvial deposits was not more rapid during past ages than it is at present, it must have taken the Nile no less than 74,253 years to deposit, grain by grain, the triangular plain of the delta, comprising an area of 8,610 square miles.*

The fact is, that the Nile leaves the greater part of its alluvium on the plains by the river-side; added to this, the extension of the water over the two banks, and the diminution of the current which results, necessarily cause the fall of a certain quantity of sediment on the bottom of the river-bed. The French *savants* of the Egyptian expedition found that the rise in the bottom averaged 4·960 inches a century. This gradual elevation of the bed doubtless corresponds with a similar change in the level of the two banks of the river. By measuring the bed of alluvium in which the statue of Rameses II. is buried at Memphis, Mr. Horner came to the conclusion that during the last 3,215 years the soil of Egypt had risen 3·043 inches in each century. It is probable that in future the soil will be raised more and more rapidly every year, owing to the "warpings" which are incessantly carried on by the agricultural inhabitants on each side of the river. Now that a vast system of skilful cultivation has appropriated the banks of the Nile, and that steam-pumps are drawing off the water of the river in every season, the liquid discharge and the mass of sediment must diminish at the mouth; and if this impoverishment of the Nile continues to go on in the same proportion, we might perhaps calculate the future date when the Nile, being exhausted by the irrigation canals, will no longer send down to the Mediterranean either a drop of water or a grain of sand.

It may be readily understood, that the best known river-delta must be looked for in Europe, and in that country of Europe which, for so many centuries, has devoted itself most earnestly to all questions relating to hydraulics and irrigation. The delta we speak of, is that of the Po. Owing to the testimony afforded by history, the monuments left by the ancients, and the operations of the engineers of the Middle Ages, we are enabled to follow with the mind's-eye the progress made by the alluvium of the river during the last twenty centuries. In some spots, especially round the lagoon of

* Elia Lombardini, *Essai sur l'Hydrologie du Nil.*

Comacchio, there are secondary deltas, the encroachments of which may be measured with mathematical exactitude, for these tracts are, so to speak, of human creation, and have been altogether deposited since the opening of artificial channels and sluices.

Notwithstanding the shortness of its course, the Po is one of the most remarkable "working rivers" in the whole world. The gradual subsidence of the shores of the Adriatic, which is estimated by Donati at 6 feet at least, since the foundation of Venice, does not prevent the river encroaching without intermission on the domain of the sea. Ravenna, which once, like another Venice, stood in the

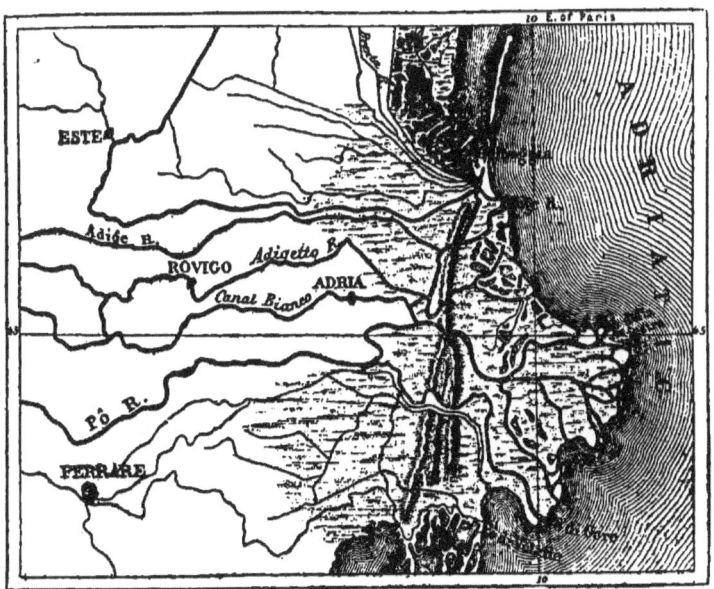

Fig. 140.—Mouths of the Po.

midst of lagoons, its outer rampart being bathed by the Adriatic, is now situated far from the gulf in a plain filled with the alluvium of the Po. We also know that the town of Adria, the ancient emporium of the Adriatic, to which, indeed, it gave its name, is now 21 miles from the extreme point of the shore. This is a proof that in two thousand years the annual average progress of the delta has been 55 feet; but at the present day the advance of the alluvial tracts is much more rapid. The patient investigations of M. Lombardini have established the fact, that the river brings down every year, 15,015,600 cubic yards of mud and ooze,* that is about 1.781 cubic yards a second, and enlarges the shore of its delta 76 yards. A chain of dunes, now left inland by the encroachments of the alluvial deposit, still points out the direction of the former sea-coast. The enormous amount of increase in the deposit at its mouth, which is thus accomplished by a river of the third class, is readily explained by the embankments, which compel the Po to carry down to the sea the whole of its alluvium, whilst the Nile and the Ganges, during each period of flood, spread over a great area of land, the level of which they raise by their deposits.

The Rhone is the most active amongst the French rivers in the

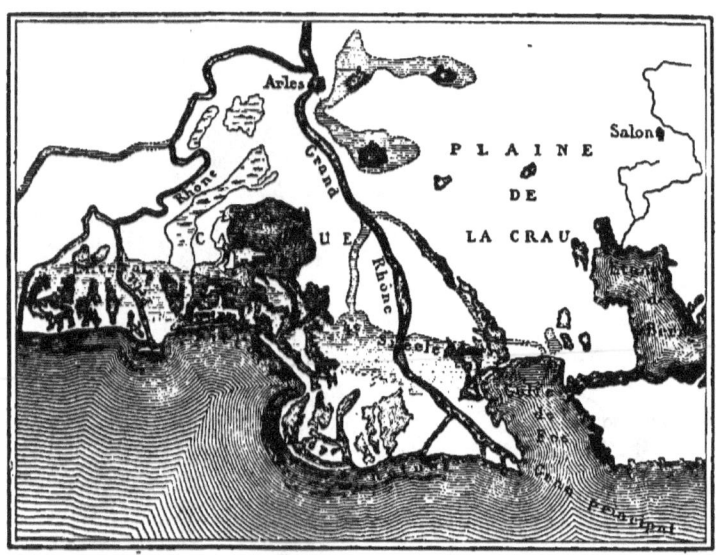

Fig. 141.—Delta of the Rhone in the Fourth Century, and at the Present Time.

formation of a delta. The promontory deposited by its current in the open sea projects much more decidedly beyond the regular line of coast than the delta of the Nile, and advances every year with a rapidity which may almost be compared to that of the Po. In the fourth century the town of Arles was only 16 miles from the sea, whilst at the present day it is 29 miles removed from it. The

* According to M. Ch. Hartley, the Danube, which discharges five times as much water as the Po, brings down to the sea only 46,500,000 cubic yards a year.

advance of the alluvium has, therefore, been 13 miles during the space of fourteen centuries, or about 52 feet a year.* The annual average elongation of the shores of the principal branch of the river is, therefore, about 164 feet. But this does not prove that the *débris* brought down by the river are increased threefold, in consequence of the embankment of the land by the river-side; for the Rhone has frequently shifted the position of its outlets, by opening them alternately on both sides of the banks of mud caused by its own deposits. In this way, the increase of the delta takes place at several points in succession; on one side the alluvium encroaches rapidly, and in other places it remains almost stationary.

In the Rhone, as in the Po, an endeavour has been made to estimate, by means of the annual discharge, the quantity of matter deposited by the river. This mass is about 22,000,000 cubic yards every year. It certainly is a fact that, by direct measurements of the increase of the delta, and by soundings made on the bar, M. Reybert found that the total quantity of matter brought down from 1841 to 1858, amounted to 419,000,000 of cubic yards, which would be equivalent to an annual increase of 25,000,000 a year; but this difference may be explained by taking into account the enormous quantity of *infusoria*, and small shell-fish which exist in all the newly-formed soft banks. Some specimens of the mud taken from the mouth of the Rhone contain, as M. Delesse has ascertained, as much as 30 per cent. of carbonate of lime, proceeding no doubt from the remains of the shell-fish. It also appears that the proportion of the alluvium of the Rhone which is brought down to the sea, and afterwards carried away by the currents to distant shores, is very slight. Almost all the mud is absorbed in the construction of the delta, and forms the *teys* or muddy islets which make their appearance on each side of the mouth. The soil which is thus brought down by the river is generally very fertile. The mud of the Rhone is no less productive than that of the Nile, and sanitary and irrigatory operations would soon render La Camargue another Egypt. In this respect, France has much to learn from the ancient land of the Pharaohs.

The delta of the Mississippi advances even more rapidly than that of the Po. Among all the questions in respect to the great river of the New World, the yearly prolongation of its alluvium has most of all excited the curiosity of science. How many yards does the Mississippi advance into the sea during the course of each year?

* E. Desjardins, *Aperçu Historique sur les Embouchures du Rhône.*

How many square miles does it add to the mainland in a century? How many thousands of years must it have been at work in forming its delta, and depositing its enormous burden of alluvium? Many geologists have, each in their turn, endeavoured to answer these questions, by basing on data, which are sometimes only hypothetical, the very different results at which they have arrived. Thus, M. Elie de Beaumont, who, at that time, had not the necessary elements at his disposal, estimated the progress of the delta at 382 yards a year. M. Thomassy, comparing the ancient French charts with the American surveys, has felt warranted in fixing the annual conquest effected by the Mississippi at about 110 yards. Messrs. Humphreys and Abbot, looking upon the old chart as being too incorrect to serve as the base of a serious calculation, are satisfied with comparing the charts of Talcott and of the Coast Survey, and judge the annual prolongation of the delta to be 86 yards. M. Ellet, one of the most conscientious investigators of the action of the river, reduces the probable elongation of the delta to 22 feet, so as to make allowance for the erosion exercised by the sea. Lastly, M. Kohl, whose hypotheses it is very difficult to understand, even if you have the maps before you, maintains that the delta of the Mississippi remains nearly stationary.* It must be confessed, that the differences on the point would be serious enough to render doubt "the best pillow for the wise man," if it were not that the calculations of M. Thomassy and the learned explorers, Humphreys and Abbot, undoubtedly surpass all the others in scientific value. The average advance of the delta during the two last centuries must, therefore, be estimated at from 86 to 110 yards.

This rapid progress in the alluvium is, perhaps, very much owing to the cutting down of the forests, which has rendered the soil of the banks much more movable.† To this cause for the growth of the delta, must be added the construction of high embankments on the banks of the Mississippi and its tributaries; for, as only a small portion of the mud is able to settle at the sides, a much more considerable mass is carried down to the mouth; nevertheless, the delta is not increased in proportion. The more the points of alluvium gain on the water, the deeper is the spot in the gulf in which the matter (estimated at seven cubic yards a second) is deposited. At the lower extremity of the delta of the Mississippi, the thickness of the bed of sediment is not less than 98 feet; and soundings have

* *Zeitschrift für Erdkunde*, September, 1862.
† Marcou, *Bulletin de la Société de Géographie*, July, 1865.

shown that the river will soon reach the edge of the deep abyss through which the Gulf Stream flows. At 11 miles from the south-west channel, the bottom of the sea is 885 feet from the surface, and this depth rapidly increases to more than 5,000 feet. Being, of course, unable to fill up these gulfs where the rapid currents would carry the alluvium into the open sea, the Mississippi must be content with obstructing the lateral bays, or with extending towards the east, in the direction of Florida. Some day, the delta of this river will be bounded on the southern side by a rapid slope, like that which is formed by the Rhone in the Lake of Geneva, and by the Congo in the Gulf of Guinea. At the mouth of this latter water-course, the sounding-lead falls rapidly from 30 to 1,600 or 2,000 feet.

When the river-outlets are left to themselves, the spot in the river where the bifurcation takes place, gradually shifts its position in a down-stream direction in proportion as the mouths advance

Fig. 142.—Height of the Layers in the Delta.

towards the sea. In fact, the current striking against the upper point of the delta, must necessarily wash away the two banks of the island which it has itself formed by the deposit of its alluvium. A remarkable instance of this alteration in the place of bifurcation of the river-outlet may be noticed in the Egyptian delta. At the time of Herodotus, Memphis was the spot where the Nile divided into two branches; it now forms its fork at Cairo, more than 18 miles from the spot where it took place 2,400 years ago. The upper point of the delta will henceforth remain stationary, owing to the barriers constructed just at the beginning of the two principal branches of the river.

The elongation of the delta has a proportionate and constant tendency to raise the bed of the river above the mouth. The calm and immense river which empties itself into the sea, obeys the very same laws as the boisterous torrent pouring into a lake. In proportion as it pushes its branches further into the sea, it must form a slope considerable enough to insure the discharge of the mass of water. This slope can only be produced by the gradual raising of the river-bed. It is evident that this rise will be the more rapid, the better the shores are protected by embankments against inundations;

for the alluvium must, in this case, all descend to the sea and lengthen the extreme points of the delta.

The results produced on the action of rivers by lateral embankments have, however, been singularly exaggerated. Pessimists have often pointed to the example of the Po as a proof of the rapid heightening of the river-level which is brought about by the construction of embankments; "but this oft-repeated assertion is not based on any real fact. Cuvier was entirely mistaken in stating, according to a communication from M. de Prony, that the surface of the water of the Po is now higher than the roofs of the houses in Ferrara."* This, unfortunately, is one of those accredited errors which it is difficult to dispel, on account of the great names which countenance them. Elia Lombardini has proved, by strict measurements, that the mean level of the Po exceeds in but very few spots, the level of the ground in the adjacent country. In 1830, at the time of one of the highest floods of the century, the surface of the Po was scarcely ten feet above the level of the pavement in front of the palace at Ferrara. The mean height of the water over the whole course of the river is considerably below that of the neighbouring plains. To make up for this, the streams of the Reno, the Adige, and the Brenta, which empty into the delta of the Po, have certain portions of their beds higher than the adjacent country. The fact is, that having so lately left the mountain gorges, they still retain their characteristics as torrents, and, like all mountain streams, raise a bank of *débris* below the ravines of erosion.† The exceptional height of the Adige, the Reno, and the Brenta must not, therefore, be attributed to the dikes which border the lower portion of the course of these streams, but to the impetuosity of the water above. The calculations of MM. Humphreys and Abbot prove that the mouths of the Mississippi must project 24 miles farther into the sea for the river to rise only one foot under the ramparts of Fort St. Philip, 31 miles above the south-west channel.

If, however, rivers which are subject to high floods, such as the Nile, the Po, and the Mississippi, are, during inundations, higher in level than the plains by the river-side, this fact is owing to the lining of alluvium which is gradually formed on the banks. During the period of flood, the waters which pour over the banks are retarded by a thousand various obstacles—trunks of trees, bunches of plants, mounds, palisades, buildings—and consequently they

---

* *Discours sur les Révolutions du Globe.*
† *Vide* above, p. 338.

deposit on the ground much of the sediment which they contain; before they leave the banks and flow far and wide into the plains they are comparatively purified. The effect of this is a gradual elevation of the banks and the ground near them to a level somewhat above that of the country generally. Above New Orleans, the natural inclination of the soil is very marked, from the shores of the Mississippi to the marshes in the interior, the difference in level is not less than 13 to 16 feet, and at some points even this considerable slope is exceeded. The banks of the islands scattered about in the lower courses of rivers are likewise raised by inundations to a point above the level of the surrounding country. The Lower Parana,* the Volga,† and a number of other large water-courses present, near their mouths, multitudes of islands, the raised banks of which circle round pools or marshes.

The elevation of the lower course of a river above the surface of the surrounding plains, explains in the most simple way the con-

Fig. 143.—Section of the Mississippi at Plaquemine.

tinual shifting of the outlets of the delta. As soon as a breach is made in the lining of the bank, a considerable portion of the running water immediately escapes through this opening and descends to the sea over a new bed which it hollows out for itself across the low-lying tracts, marshes, and lagoons; these are natural crevices, similar to those which occur in embankments raised by man's labour. Thus, when the economy of a river has not been modified by human agency, its outlets are of a changing character and move across the delta, depositing their sediment in the lagoons, so as gradually to elevate the soil, and to bring it everywhere to the level of the high floods. Every delta becomes modified, even during the historical period, in the number, direction, and importance of its branches. Of the seven famous mouths of the Nile, five have now ceased to exist except during floods; the two which still remain open—those of Rosetta and Damietta—appear, according to Herodotus' statement, to have been dug out by the labour of man. During the last 3,000 years, the branches of the Lower Hoang-ho have under-

* Martin de Moussy, *Confédération Argentine*.
† De Baer, *Kaspische Studien*.

gone similar modifications in their course, which are more remarkable on account of the immense extent of ground over which they have constantly wandered.* Still more strangely, the Amou-Daria in Tartary, which now falls into the sea of Aral, was in former days a tributary of another sea, and flowed into the Caspian; the traces of its abandoned bed may still be seen here and there in the desert.

In consequence of the incessant modifications to which the lower portions of rivers are subject, it often occurs that two water-courses

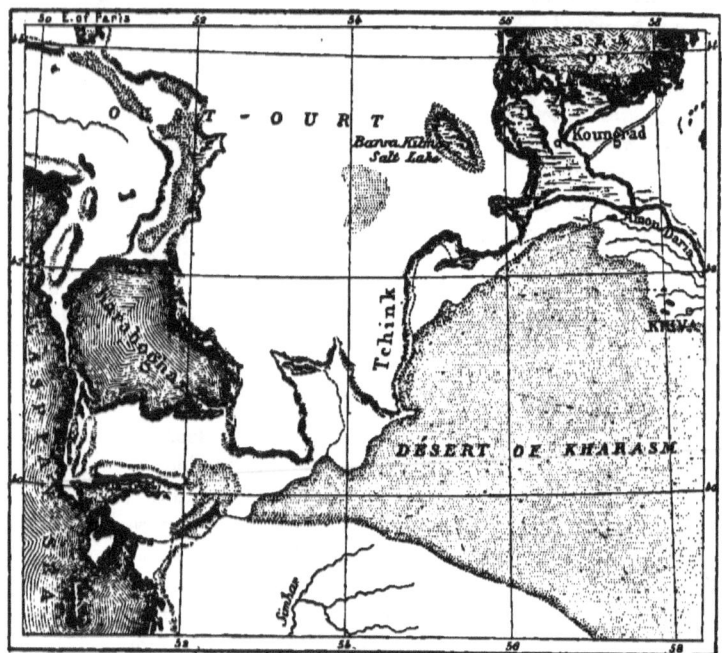

Fig. 144.—Ancient Course of the Amou-Daria.

which were once perfectly distinct and independent of one another, become united in their deltas and principal outlets. We may mention the instance of the Shat-el-Arab. In like manner, the Adige and the Po, which communicate with one another by lateral branches, have a tendency to join one another completely in a common bed, and nothing but extensive operations has prevented, up to the present time, the perfect junction of these two rivers. The Mississippi, so remarkable in all other respects, presents, according to Ellet,

* *Vide* above, p. 410.

the phenomenon of three former rivers united in one. At one time the river Ouachita ran down to the sea through the Atchafalaya, which is now an overflow-channel of the Mississippi, but was then a distinct river. The Red River, too, flowed in the valley of the Têche, where it has left numerous traces of its passage. The opposite windings of the Red River and the Mississippi gradually approached one another and then united; the Ouachita-Atchafalaya has been, as it were, cut into two parts, one of which, the northern portion, is become an affluent, and the other, the southern portion, an effluent of the Mississippi. Similar phenomena are observed in the delta common to the Ganges and the Brahmapootra. There seems to be a real conflict between the branches of these two rivers; they first come together and are then mutually repelled; they sever one another and fill each other up.*

Thus, whilst in some cases distinct rivers unite, others, on the contrary, which were once combined, are now separate, and take contrary directions. As a striking instance of this double series of hydrological phenomena, we may mention the two rivers of Cilicia, once called the Sarus and the Pyramus, now known as the Seihoun and the Djihoun. These streams, which project their alluvial deposits more than six miles beyond the outline of the former coast, fall into the sea sometimes through two distinct mouths, sometimes through one outlet common to the two rivers. Since the days of Xenophon, the two streams which then flowed in beds some distance from one another, have united three times, and three times have again separated. In the space of twenty-three centuries, says M. Langlois, six complete revolutions have taken place in succession in the course of action of the Sarus and the Pyramus.

* Ferguson, *Zeitschrift für Erdkunde*, 1864.

CHAPTER LIV.

BARS OF RIVERS.—OPERATIONS UNDERTAKEN FOR DEEPENING THE MOUTHS OF RIVERS.

NOTHING is more variable than the channels at the mouth of a river. Thus—only to mention the Mississippi,—this river has now five channels, the south-west, the south, the south-east, the north, and the Loutre, which is a ramification of the one preceding. Sometimes one and sometimes another of these outlets becomes the real mouth of the river, and the stream takes to them and abandons them in turn. The fact is, that the Mississippi, having considerably elongated its principal outlet by the alluvium it has brought down, is compelled to seek some bed which is shorter and consequently more inclined, in order to pour down its mass of water; when this fresh outlet is likewise pushed out too far into the sea to afford the requisite slope, the river turns either to the right or left to clear for itself a third place of issue. At the time of the first attempts at colonisation in Louisiana, the south-east channel was the principal one; but this gradually became obstructed, and the north-east mouth was next the most important. The mass of water in this channel diminished every year; and in 1853 there was not more than 8 feet of water on the bar, and small coasters were the only vessels which ventured over it. Since 1843, the south-west channel has become the real mouth of the river, through which almost all large ships try to enter. In 1853 there were 16½ feet of water; but constant labour was necessary to maintain even this depth, for the quantity of water constantly tends to diminish, whilst in the Loutre Channel it is gradually increasing. Some hydrographers think that this latter mouth will ultimately become the true Mississippi; it already has 13 feet of water on the bar; and, in order to avoid a considerable circuit, nearly all the steamers running between Cuba, Mobile, and New Orleans, now attempt to pass through it.

However much they may shift their course, still most rivers are obstructed by a bank of sand or mud, to which mariners have given the name of "bar." These banks of alluvium are for the most part deposited in the form of a crescent, off the mouth of the river, and, turning their convex sides towards the open sea, mark the precise spot of the line of breakers which rise in rough weather. They may be deposited in different modes, according to the quantity and impetus of the river-water, the mass of sediment which the latter holds in suspension, the configuration of the coast, and the general direction of the winds and currents out at sea. There are, however, a few hydrological problems which have given rise to lively discussions among geographers and engineers, for which, too, many various or contradictory solutions have been propounded. The fact is, that the question is altogether a complex one, and presents itself under a new aspect at the mouth of each particular river. It certainly is the case, that, as regards all rivers, the collision of the two liquid masses flowing in contrary directions is the primary cause of the formation

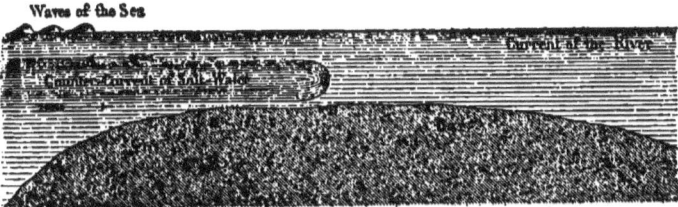

Fig. 145.—Longitudinal Section of the Bar of the Mississippi.

of bars; but the materials which are employed and the progress of the work vary singularly.

At first sight, the origin of a bar seems a matter easily to be understood, especially in the case of rivers with waters much charged with mud. It is thought that the current of fresh water, being suddenly arrested in its career by the sea water, immediately lets drop on the bottom the matter which it held in suspension, and thus gradually forms the kind of sill which rises between the bed of the river and the ocean. This, however, is not the exact mode of formation. The flow of fresh water, being but little retarded, continues its movement above the salt water coming in a contrary direction. The sediment which is let fall by the current of the river is taken hold of by the counter-current and borne up stream. At the same time the heavier alluvium, which makes its way to the sea by gliding over the bottom of the river-bed, is arrested in its progress and is mingled with the sand and the innumerable organic remains driven in by the waves. Thus an increasing cushion of mud is formed in front of the rising tide flowing to meet the river, and in this way the heaps of *débris* which constitute the bar are gradually accumulated. This obstacle, being produced by the shock of two opposing currents, shifts coincidently with the scene of the conflict. During floods, the impetus of the mass of fresh water becomes sufficiently strong to remove the whole bar and to carry it further in advance; but, on the other hand, when the water of the river is low, the tide resumes the preponderance, and the bar is again driven back. The barrier shifts its place, sometimes in one direction sometimes in another, and is incessantly seeking to preserve its equilibrium between the two opposed forces which impel it.

The bars of the delta of the Mississippi may be quoted as an instance of this mode of formation. Over the bar which obstructs the entry of the principal channel, and the most practicable of all those on the coast of the Gulf, there is an average depth of $16\frac{1}{2}$ feet. The alluvium of the bed, being kept in constant motion by the waves and the current of the river, is in an almost liquid state. Vessels have been known to cross the bar without any other assistance than their sails, although their hulls were, for more than half a mile buried in the mud to a depth of 6 feet. Notwithstanding the soft nature of the ooze, vessels may still incur considerable danger in crossing the bar. Those that do not avail themselves of a steam-tug are sometimes taken athwart by the wind and driven irretrievably upon the banks. It is often impossible to get them off again; the motion of the keel stirs up and sends into the current the smaller particles of the mud, but the heavy sand remains, and ultimately becomes cemented round the bottom of the ship.

There are some bars which are almost entirely the work of the sea; these are banks of sand or shingle which the waves throw up across the outlet of a river, thus continuing the line of shore. Barriers of this kind form in front of water-courses running into a sea agitated by violent storms, or raised every day by a very strong tide. The flow coming from outside ascends far into the river-mouth, and forces the current meeting it to deposit its heavier alluvium at some considerable distance above the bar properly so called. The earthy particles held in suspension by the current of the river cannot be precipitated on account of the continual agitation which is kept up at the entry by the breakers and the swell; they remain mixed with the masses of water, and are driven up-stream by the flow, or carried out to sea by the ebbing tide. Even the fine sand which the waves throw up on the bar is not allowed to remain there for long; it is again stirred up by the water which brought it, and it finds a resting-place only in those spots where the motion of the waves ceases. The heavy sand, the shingle, and the stones which the waves drive before them without carrying them along in eddies are the only materials which constitute the bar. Like the banks of mud in river-deltas, this line of *débris* is incessantly shifting its place, sometimes up-stream and sometimes down-stream, seeking the exact line where an equilibrium exists between the ebb and the flow. When the river is flooded, the force of the water running down carries the bar farther out to sea; on the contrary, when the river is low, the tide gains the ascendency, and pushes the sand up into the mouth of the stream.

In France, these phenomena have been best studied at the formid-able bar of the Adour. Thanks to the submarine charts which the engineers prepare twice every month, we may trace out, so to speak, by the eye, all the fluctuations of the bank of *débris*, and all the causes of its movements can readily be taken into account. In this bar, however, there is every evidence to show that the materials forming it are brought up by the waves. The soundings that have been made in the bed of the Adour, as far up as $15\frac{1}{2}$ miles above Bayonne, uniformly show a bottom of mud or fine sand; but it is ascertained that the bank at the mouth is composed of heavier sand and shingle, proceeding, no doubt, from the cliffs of the Spanish coast.*

The bars of rivers, which have always been an obstacle and a source of danger, are at the present time more troublesome to deep naviga-tion than they have ever before been. It certainly is the case that, thanks to the steam tugs, vessels with a light draught of water are able to follow the direct channel, and can cross the difficult part in the space of a few minutes; but, nowadays, commerce is no longer contented to employ the small vessels of former times; it requires ships of heavy burden carrying large quantities of merchandise, and drawing a considerable depth of water. Many a river-port, once the resort of whole navies, is now abandoned on account of the bar which cuts it off from the ocean, and is frequented only by coasting vessels; commercial vitality has gradually left it. Thus the deepening of their river-mouths is become a most important question in some sea-coast towns. If they could only succeed in doing away with the bar, these towns would increase suddenly in wealth, population, and importance. If the bank of sand must remain fixed across the outlet of the river, the city is on its way to certain ruin. Every engineer

* Vionnois, *Annales des Ponts et Chaussées*, 3rd series, vol. xvi.

recommends his own special plan as being adapted to avert the danger; each promises that he will correct those river-outlets which Vauban characterised as "incorrigible." But only too frequently, operations are undertaken without taking account of the numerous causes which determine the formation and fluctuations of the bar. Amongst all the immense works which have been carried out at the mouths of rivers, many have become useless, or even absolutely injurious to navigation. Millions and millions of money have been thus cast into the ocean and purely wasted.

The most simple means, and that, indeed, which is always resorted to in the first place, is *dredging;* but this plan is evidently merely provisional, and in the present state of science can scarcely be considered as a remedy of a lasting character. Moving an obstacle is not doing away with it; besides, the flotilla of dredgers which can be employed on a bar in removing the alluvium is always insufficient in number. Even if they were constantly at work, there would be little or no result, for the inexhaustible ocean would take up the task of providing the alluvium and raising the bar; the obstacle would be merely shifted in place.

Instead of moving the mud, the more simple plan has often been tried of sending it into the current by keeping the water in a constant state of agitation. For a length of time it has been a recognised fact, that, after the passage of several ships, there is an increased depth of water on bars composed of mud and fine sand: the particles stirred up by the keels are carried away by the current. This phenomenon may readily be produced by artificial means. More than a century back a French company applied this remedy in the principal channel of the Mississippi, by causing heavy iron harrows to be dragged over the shifting bed of the river. Recently, in 1852, the same plan was applied, and the Federal Government employed on the bar a certain number of steamboats, which kept the mud on the bottom incessantly in motion by means of drags or harrows, and thus prevented its precipitation. According to a popular tradition, mentioned by M. Engelhardt, a Turkish Pacha formed the same idea as the American engineers. He obliged every vessel which left the Danube to drag astern, while crossing the bar, a harrow attached to a heavy chain. In both cases the agitation of the water seems to have produced a favourable result. The Pacha succeeded in maintaining a channel of about 13 feet deep through the Soulina bar, where formerly it was not above half this depth. By the same plan the American engineers obtained nearly 20 feet of water in the south-west channel. In a similar way, in order to force the torrents to deepen their beds, the inhabitants of the Piedmontese Alps used to plough up the tracts of pebbles which were brought down by the floods.*

Unfortunately, this simple method of improving the condition of the bar produces no lasting result, and the work always has to be begun over again; for the bank forms again whenever the drags allow a moment's respite to the sediment held in suspension by the water. Besides, when the water is low in the river, the operations must be suspended, or the sea would drive back all the sediment in an up-stream direction, and thus contribute to the silting up of the river-bed. It must also be remarked that measures of this kind are only practicable in rivers where the bar is composed of small particles, and is not subject to all the fury of the wind and the billows. At the mouth of the Adour, for instance, what immense harrows it would be necessary to use to move the beds of shingle driven up by the storms!

A system of moles and jetties is, therefore, the plan that has generally been resorted to by engineers in the improvement of the mouths of rivers. This plan is somewhat similar to that of the embankments which have been employed, with various degrees of success, on the middle courses of rivers; but the marine dikes have not always produced favourable results, and a great many experiments which seemed to offer good chances of success have entirely failed. Among the undertakings of this kind which have been the most costly and the most useless, we may mention those which have been carried out in the delta of the Rhone. It was hoped that, by confining the mass of water in a narrower channel, and compelling it to run into the sea through a single outlet, a current would be produced which would be strong enough to clear out the passage to a considerable depth, and thus to allow ships of a deep draught of water to enter the river. In 1852, Surell, the engineer, closed up the various *graus*† through which the water of the Rhone found lateral outlets, and lengthened the two banks of the principal mouth by means of dikes converging one upon another, thus doubling the force of the current. The water of the river did, in fact, accomplish the work of erosion, and cleared out the channel; but fresh alluvium being incessantly brought down by the flow of the Rhone, and thrown

* Chabrol, *Statistique du Département de Montenotte.*
† From the Latin *gradus*, a step, passage. In the south of France this name is given to the outlets of rivers, which connect the shore lakes with the sea and the mountain defiles.

up by the waves of the sea, a new bar formed across the mouth outside. Before the operations of embanking were begun, the average depth of the channel was 5 feet 10 inches; at the present time it is the same, after having varied from 13 feet to $6\frac{1}{2}$ feet, and 3 feet 8 inches, according to the quantity of water sent down by the current.*

A similar undertaking, attempted, in 1857, in the south-west channel of the Mississippi, had not a more favourable result, a curvilineal jetty, 1,849 yards in length, having been carried away by a tempest. However, the commissioner appointed by the Federal Government to study the course of action of the delta of the Mississippi, recommended the reconstruction of convergent jetties, as being a plan which was likely to keep the channel clear. Looking forward to the constant increase of the alluvium of the river in the now contracted channel, they advised besides, as an indispensable matter, that the jetties should be lengthened about 245 yards every year, leaving it open to abandon a channel and choose a fresh one when the dikes of the first outlet should have attained any immoderate length.† These operations would be enormous in their character; but if, as was hoped, they would result in maintaining a depth of more than 19 feet in the channel, a tax might be imposed every year on the immense commerce of the Lower Mississippi which would be amply sufficient to defray the costs of construction.

The Adour, which does not carry down to its mouth such large quantities of alluvium as the Rhone and the Mississippi, is one of those few rivers where engineers have obtained, at least temporarily, favourable results. Besides, it must be confessed that the works undertaken for the improvement of the bar have lasted for a good many years; for it was in 1694 that Ferry, the engineer, constructed at the southern point a jetty which was to fix and deepen the channel, and since this date there has been no cessation in this interminable labour. The lateral dikes have been continued on to the sea, either in rock-work or by means of piles, and tending in various directions, which were adopted, in despair, after each successive failure. Finally, from 1855 to 1860, the jetties of the two points were slightly bent round to the north, and continued, to an equal distance into the sea, to a point 550 yards from the shore. The northern jetty, resting upon a base of rocks hidden under the water, is constructed with openings along all its length, and allows the current to flow between the piles.

* Minard, *Des Embouchures des Rivières Navigables.*
† Humphreys and Abbott, *Report on the Mississippi River.*

The southern jetty, which was to prevent the mouth from bending round to the south, is solid for a length of 220 yards, and keeps the bar tending in a direction leading to the coast. Since these operations were finished, the condition of the river-mouth has been subject to much change. In February, 1862, when the level of the water was very low, a bank of sand formed exactly across the mouth, and compelled the water of the Adour to escape laterally through the openings between the dikes. Nevertheless, the general state of the channel exhibits a considerable improvement. The channel has taken a fixed direction towards the west, and no longer spreads out to the south; the bar has been driven farther out to sea, but the average depth of water on it is greater. Before the improvements were begun, at low-water it was covered with only 5 feet to 5 feet 10 inches of water; the depth is now as much as $10\frac{1}{2}$ feet. This is an immense result, which, according to the calculations of an author who is sufficiently sceptical in matters of this kind,* represents a clear annual gain of forty thousand pounds for the commerce of Bayonne. If, in the future, the bank of shingle should attain to the same height as before, it would be necessary to again lengthen the jetties of the Adour for several hundred yards further into the gulf, and, by means of the experience already acquired, the engineering operations would again produce a considerable temporary improvement. According to M. Bouquet de la Grye, whose opinion is a very valuable one, it would be a very useful measure if the barriers were curved round in a south-westerly direction, so that the waves from the offing should not beat directly in front of the current and ebbing tide. But the chances in favour of any decisive amelioration of the bar must be much increased in proportion as the jetties are lengthened. When these artificial promontories, like headlands, are surrounded by a deep sea, the *débris* which the currents and waves deposit at their base will be no longer incessantly agitated by the breakers, and thus raised into the form of a bar; they will accumulate but slowly, and a long series of years, or even centuries, must elapse ere they could seriously modify the submarine features.†

The operations undertaken at Soulina—one of the mouths of the Danube—appear to have met with great success, but through an entirely special cause. Mr. Charles Hartley, the skilful constructor of the Soulina jetties, has taken care to push them out more than a hundred yards into the sea, as far as a point where a current generally passes along the shore tending from north to south. This

* Minard, *Des Embouchures des Rivières Navigables.*
† Mongol-Bey, *Percement de l'Isthme de Suez.*

current catches hold of all the alluvium which glides down over the bottom of the bed of the river, and thus hinders the formation of a fresh bar.* The average depth of the channel—which was only

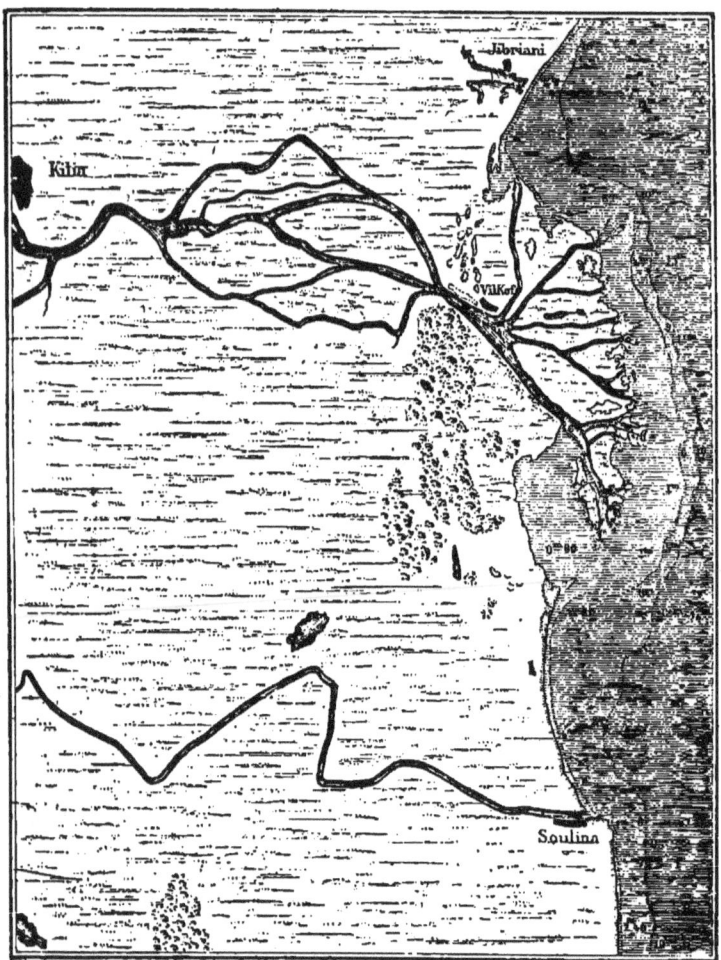

Fig. 146.—Mouths of the Danube. Arms of Kilia and Soulina.

9 feet before the commencement of the works—is now not less than 16½ feet, since the dikes have been constructed. It is cer-

* *Zeitschrift für Allgemeine Erdkunde.*

tainly a fact, that the gradual encroachments of the whole delta of the Danube will have the effect of pushing the current itself farther out to sea, and sooner or later, a second mound of sand will obstruct the mouth of the Soulina. According to an approximate calculation, based, however, on plenty of hypotheses, the works finished in 1860 will not become completely useless until the year 1916.*

There are other mouths of rivers, especially those of the Oder, in Prussia, and the Meuse in Holland, which have been permanently improved by engineering operations; we are not, therefore, entirely warranted in repeating the words of Vauban, and characterising all

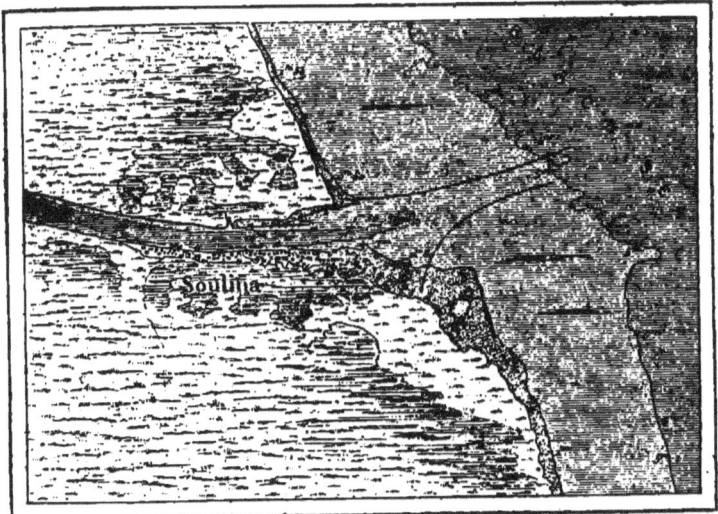

Fig. 147.—Jetties of Soulina.

the bars of navigable rivers as "incorrigible." The results obtained on the Clyde, in Scotland, may especially be classed among the most important triumphs of engineering art. The water of that river was once so very shallow that ships of a deep draught of water were compelled to stop 15½ miles below Glasgow, and the merchandise had to be reshipped in barges: at the present time great three-masters easily come up close to the quays. Besides, in a great many cases it might, perhaps, be possible to divert the mouth of a river in places where it was not possible to force a passage, and thus, by indirect means, to obtain the depth of water necessary for large ships.

* Engelhardt, Etudes sur les Embouchures du Danube.

A deep canal, protected against alluvium by a system of sluices, would then replace the natural channel. This plan, according to M. Desjardins, is that which was adopted by the ancients in the case of the Tiber and the western branch of the Nile, which was diverted towards Alexandria. This plan, too, has been proposed by several American engineers, for insuring to New Orleans a magnificent port worthy of the river which bathes its quays. Moreover, a work of this kind is being carried out at the mouth of the Rhone, by the digging out of the canal of St. Louis. This navigable channel is $2\frac{1}{2}$ miles long, and 66 yards wide, and is intended to connect the river with the Gulf of Fos—so called from a former navigable canal, dug out by Marius (*fossæ marianæ*). Vessels drawing 24 feet of water will thus be able to ascend as far as the port of Arles, which at the present day has been almost abandoned by commerce, on account of the deficiency of water in the channels of the Rhone. If the excavation of the canal of St. Louis meets with success—which scarcely seems a doubtful point—this great undertaking will serve as a model for the subjection of other mouths of rivers, which, up to the present time, have continued rebellious to the operations of engineers.

CHAPTER LV.

ALTERATION IN THE POSITION OF WATER-COURSES IN CONSEQUENCE OF THE ROTATION OF THE EARTH.—MASSES OF WATER BROUGHT DOWN TO THE SEA BY RIVERS.—GENERAL CONSIDERATIONS.

THE sudden changes in river-beds produced by the rupture of their dikes, as well as the movements, and even obliterations of their mouths, constitute, generally speaking, catastrophes of a serious character; and it may readily be conceived that the imagination of man has looked upon these incidents as among the most important facts in the history of rivers. Yet this is not the case. However great may be the influence of these sudden modifications in the action of water-courses, they are but phenomena of a secondary class in comparison with those more durable changes which the rotation of the earth brings into the economy of every river and the general physiognomy of its basin. The fact is, that the water of rivers, like that of the ocean and the aërial waves, is subject to the influence of all the great astronomical laws. Rivers, as well as the winds, have a natural tendency to shift their course, so as to effect an arc of revolution round the planet.

In fact, the running water which the earth carries round in its diurnal movement is affected differently from the solid bodies which lie upon the ground. Whilst the latter, just as the mere inequalities in the terrestrial surface, describe their daily orbit round the central axis, the fluid particles which glide over the rotundity of the globe traverse in succession various latitudes, and their movement consequently varies. The speed of rotation being completely nullified at the mathematical points which act as poles, and increasing gradually as far as the equatorial regions, where it exceeds 1,470 feet a second, everything movable which tends from one of the poles to the equator must necessarily remain in the rear of the increasingly rapid terrestrial movement which carries it round, and must consequently deviate towards the west—that is, to the right hand in the northern hemisphere, and to the left in the southern. In like manner, any movable body which takes its course from the equator to one of the poles, exceeds—owing to its acquired speed—the angular movement of the globe, and inevitably deviates to the east; that is, to the right in the northern hemisphere, and to the left in the southern. These facts have been rendered perceptible by the celebrated experiments of M. Foucault on the pendulum of the Pantheon; and they can easily be verified by every one, by causing the rotation of two suspended globes, and by allowing some coloured liquid to glide over their surfaces.* The trade winds and all atmospheric currents obey this law of deviation, as well as the Gulf Stream and the other flows of the ocean. Even balls rushing from the mouth of a cannon are subject to this law; and sometimes the locomotives on our railroads, when they run off the lines. This law applies equally to all water-courses, and—provided that the configuration of the ground allows it, and that the oscillations of the terrestrial surface do not hinder it—it causes running water to deviate regularly to the right in the northern hemisphere, and to the left in the southern. With regard to those rivers which flow in a line parallel to the equator, there is no force which compels them to eat away either one or the other of their banks; but they are retarded in their course if they flow to the east, and are, on the contrary, accelerated if they run towards the west.

This is the law which, for some time past, several geographers have pointed out; which, however, M. de Baer has had the honour of completely bringing to light. The only difficulty is to make a choice among the numerous rivers, which may be mentioned as examples of water-courses modifying their course in the direction pre-supposed by this theory. South of the equator there are the affluents of the gigantic Rio de la Plata, which, after having watered on the west the extent of the *pampas*, are incessantly wearing away their left banks. In the northern hemisphere there is the Euphrates, which endeavours to pour itself bodily into the bed of the Hindiah, to the right of its own course;† there is also the Ganges, which abandons the town of Gour, in the midst of the jungles, and shifts in its delta four or five miles to the west.‡ There is the Indus, wearing away the stony hills of its western bank, so as to move its delta for more than 600 miles in that direction. There is the Nile, leaving its ancient bed in the Libyan desert, in order to carry its waters by the side of the Arabian chain of mountains. In like manner, in Europe, the Gironde,

---

* A. Herschel, *Intellectual Observer*, November, 1865.
† *Mittheilungen von Petermann*, vol. xi., 1862.
‡ Ferguson, *Zeitschrift für Erdkunde*, April, 1864.

the Loire, and the Elbe wear away the escarpments of their right bank; and the Vistula deepens its eastern mouth at the expense of that to the left. The Rhine, in the plains of Alsace, is gradually increasing its distance from the base of the Vosges, and is approaching the mountains of the Black Forest; and so long as its course was not fixed by the continuous rampart of its embankments, it constantly gained on the territory of Baden, and bent round to the west of the hills, along the foot of which it had previously flowed.* A still more remarkable fact is exemplified by the Danube, which passes in succession through a series of defiles, and always develops its windings towards the right below each gate of rocks through which it has to pass. The river shifts its place under the influence of the movement of terrestrial rotation, in the same way that a cord fastened at certain points would bend under the influence of a current.† Thus, when entering the plains of Hungary, which were once a vast lake, the Danube, instead of crossing diagonally the level tract bathed by its waters, bends suddenly to the south, and then to the east, so as to take the course of the great central depression round the high ground on its right.

In European and Asiatic Russia the normal displacement of rivers affords an especial opportunity for most interesting studies. In these countries, in fact, all those conditions are united which are most favourable to the gradual encroachment of the rivers on their right banks; they have a very considerable length of course, and the liquid masses are powerful enough to readily clear away any obstacles; there are enormous floods which periodically increase the force of erosion in the currents, and the cliffs are composed of friable rocks; lastly, the sharp curvature of the globe is the cause of a rapid change in the speed of rotation in the various latitudes. Two centuries ago, the principal mouth of the Volga flowed directly to the east of Astrakhan; since that time the great current has successively hollowed out for itself fresh beds, tending more and more to the right; and at the present day the branch navigated by vessels turns to the south-south-west. Above the delta the river has everywhere shifted its bed towards the west, and opposite Tchernoï-Iar, the Achtouba, the former bed of the Volga, now lies 12½ miles from the principal current. The twenty-three towns which have been built on the western bank, also called the upper bank on account of its high cliffs, have been almost all demolished in detail, house by house,

* Bourlot, *Variations de Latitude et de Climat*.
† Von Süss, *Der Boden der Stadt Wien*.

and street by street; and being thus undermined on one side, they have been compelled to advance on the other into steppe-land. On the east, the plains once washed by the river are scarcely raised above the average level of the water: during inundations they are converted into

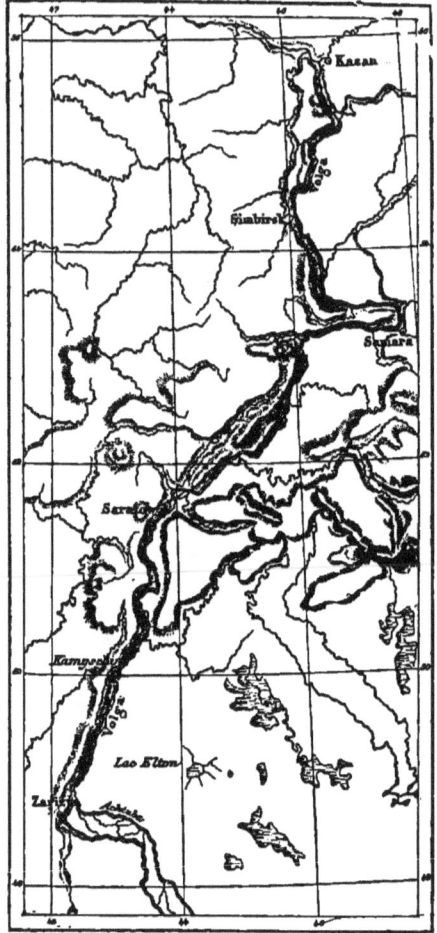

Fig. 148.—Middle Course of the Volga.

perfect seas; therefore the people have not been able to build more than three towns on the eastern bank. One of these towns—Kasan—was once situated at the confluence of the Kasanka and the Volga, but

DIRECTION OF RIVER-COURSES. 437

is now two miles from the latter river; it has, so to speak, travelled to the east. In Siberia the water-courses move to the right still more rapidly. The modern towns of Yakutsk, Tobolsk, Semipalatinsk, and Narym have already been partially rebuilt. Along these watercourses, the right bank, which is undermined by the current, is almost invariably higher than the left bank bordering on the *toundras*, which once served as the bed of the river. This is a fact of such a general nature, that map-designers admit it as an axiom, and never fail to draw the right bank as being the highest, and the most escarped.

A large number of rapids and cataracts—among others, the magnificent falls of Trolhäta—also afford examples of a continuous displacement produced by the rotation of the globe. Similar phenomena are likewise observed in those river-like arms of the sea which are formed by the sea-water passing through a narrow channel; thus the force of the current is exercised mostly on the right-hand

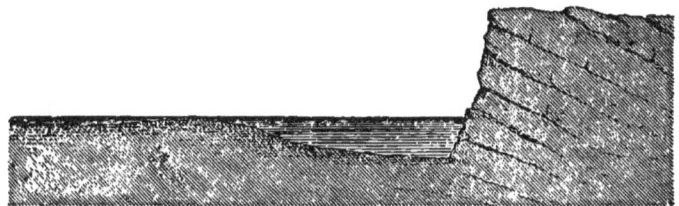

Fig. 149.—Left and Right River-banks.

side in the Straits of Kertch and the Bosphorus, and the greatest amount of erosion takes place on this side. The law is of general effect, and applies to all the rivers which flow on the surface of the earth. The great rushes of water in former geological periods have likewise in their flow worn away the ground on the right-hand side. In the north of the Pyrenees, the *gaves*, which radiate so remarkably round the plateaux of Lourdes and Lannemezan, all flow through valleys of erosion, commanded on the east by high cliffs worn away at their base, and on the west by long slopes of easy access, on which the *débris* are deposited.*

Among the important rivers which, in consequence of local circumstances, seem to contradict this law of the displacement of running water, we may mention the Mississippi and the Rhone. Instead of gaining on its right bank, and eroding the base of the heights which rise on the west, the great American river impinges in fifteen places

* Leymerie.

against the cliffs of the eastern plateau, and, throughout the whole of its course, constantly tends towards the left. As soon as it enters the marshy plain of its delta at a point below Bâton Rouge, it flows almost in a straight line towards the south-east, to form the remarkable peninsula of mud through which it falls into the sea. It must be remarked, that the direction taken by the water of the current of the Mississippi is exactly the same as that of all the rivers which run

Fig. 150.—Radiation of the " Gaves," North Pyrenees.

into the Gulf of Mexico—the Rio Grande and its ffluents, the Rio Pecos, the Nueces, the Colorado of Texas, the Brazos, the Trinity, the Neches: these rivers, which uniformly tend towards the south-east, are parallel to the ridges of the Rocky Mountains. If it is a fact, as many geologists seem to have established, that the western chains of North America are undergoing a movement of upheaval, whilst the Carolinas, Georgia, and other neighbouring regions are gradually subsiding, it might very well be the case that the lower course of the Mississippi and all the Texan rivers tends to the east, in consequence of the slow movements of elevation and depression to which the North American continent is subjected.*

With regard to the Rhone, the mouth of which likewise flows in a south-east direction, instead of tending to the right, as it once did, and following the vast bed now adopted by the smaller Rhone, it is possible that its course may have been modified during historic periods by the impetuosity of the *Mistral*. However strange an assertion of this kind may appear at first sight, it perhaps merits the attention of geographers. In fact, it seems a matter beyond all doubt that, in consequence of the gradual cutting down of the woods of the Cevennes and the central plateau of France, the *Mistral* has continued to increase in violence since the ages of the Roman occupation. If this be so, this turbulent wind must necessarily impel the waters of the Rhone towards the left bank in the direction they are taking at the present day. The aërial current beating down from the Cevennes on the marshes of the Camargue must necessarily have pressed upon the current of the river, and marked out for it the line which it had to follow in hollowing out for itself a fresh bed. Everything in nature takes its share in effecting modifications; every feature in the planet owes its form to the breath of the winds, the currents of the waters, and the movements of the soil, quite as much as to the motion of the globe in space.

Rivers, taken as a whole, being merely the arterial system of continents, renewing the liquid mass of the seas, whence the waters return again to the interior of the land in the form of clouds and rain, it is important to know, at least approximately, the quantity of river-water which is flowing on the surface of the globe. For many years back, various hypotheses have been regarded on this point; but any very precise data are still wanting, and nothing but observations taken for a series of years will render it possible to arrive at any accurate knowledge of this hydrological fact, so important in the economy of the globe. Buffon supposed that the mass of water emptied out by the whole of the rivers running into the sea would represent, in 812 years, a quantity of water equal to that of the ocean ; but the data on which he based his supposition are not of sufficient authority to render it of much use to discuss his opinion at the present day. Among the most important calculations which have been recently made, taking as their starting-point the quantity of rain falling annually on the earth, we must mention that of

* *Vide* the chapter on "Upheavals and Depressions."

Metcalfe. He estimates the total mass of water brought down by the rivers at 176,000,000,000 of cubic yards of water every day. Keith Johnston considers that the daily average discharge of the rivers of the earth is 229,000,000,000 of cubic yards, or more than 2,620,000 cubic yards a second.

This estimate is certainly much too high, for, by adopting another method, more in conformity to the rules of direct observation, and consequently more scientific—that is, by adding up the masses of water rolled down by the rivers which have been already gauged in various parts of the world by engineers and geographers—we find that the total discharge of a collection of river-basins comprehending an area of 4,246,000 square miles does not exceed a little more than 72,000 cubic yards a second. Now, these basins, which are those of the principal rivers of Western Europe, including the Danube,* as well as those of the Nile, the Chat-el-Arab, the Ganges, the Hoang-ho, the Mississippi, and the Atrato, form a tenth part of the terrestrial

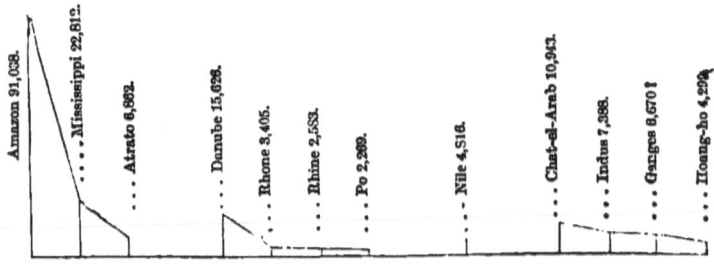

Fig. 151.—Diagram showing the Comparative Discharge of Rivers; in Cubic Yards.

surface, the waters of which flow down to the ocean. If, therefore, the proportion of water which runs from the surface of the ground into the sea were everywhere the same, the liquid mass of fresh water combining with the salt waves would not be more than 850,000 cubic yards a second. We must, however, take into account the enormous quantity of water discharged by certain rivers in the tropical zone, and especially the Amazon, the delivery of which is probably 100,000 to 130,000 cubic yards.† If, therefore, we add a third to the total river-discharge obtained by the previous calculations, we shall have for the whole mass a maximum of over

* The discharge attributed to the Danube is probably too great; according to M. Hartley, the real discharge is only 11,123 cubic yards a second.
† At the defile of Obydos, the section of water which flows every second to the sea is, according to Spix and Martius, 23,113 cubic yards at low-water time; in floods, says Avé-Lallemant, it is 319,176 yards at the same spot.

1,100,000 cubic yards a second. This is a quantity which represents an average fall of about 11 inches of rain over the entire surface of each basin, an average much larger than that of most of the rivers which have been studied up to the present time. If we admit that the average depth of the seas is 5,400 yards, the quantity of water which flows down the surface rivers of the Continent would not equal that which fills up the abysses of the ocean until after a lapse of fifty millions of years.

This is evidently nothing but a provisional calculation, which will be gradually rectified as the facts relative to the hydrology of the globe become better known. When the mean discharge of all visible water-courses is accurately gauged, when the force of subterranean streams has been disclosed by the investigations of meteorologists as to the fall and evaporation of rain, then it will be more easy to calculate, within a few millions, the total mass of liquid which is annually poured into the sea by the rivers of the continents. No doubt, within a period not very distant, the measures which have been adopted with so much precision as regards the Mississippi, the Po, the Rhone, and the Nile, will be applied with equal care to other river-mouths.

The investigations which have been simultaneously made as to the proportion of sediment which exists in suspension in rivers will enable us also to resolve the often-discussed question as to the actual importance of the alluvium of rivers. Without mentioning here the streams which are literally liquid mud, or sometimes even avalanches of mud, there are some rivers like the Missouri, the waters of which are so charged with sediment, that the drift-wood, being completely penetrated with muddy particles, is ultimately entirely submerged, and covers the bottom of the river.* There are, on the contrary, other rivers, such as the St. Lawrence, which send down to the ocean water which is generally pure and transparent. During floods the Durance holds in suspension as much as 21 thousandths of mud;† the Garonne sometimes contains 10 thousandths;‡ the Rhine 6 thousandths only.§ It will be readily understood that the quantity of alluvium held in suspension must necessarily vary in different rivers, according to the more or less compact nature of the soils through which they pass; thus, observations made on any particular watercourse have nothing more than a local value. The estimations made

* *Continental Monthly.* June, 1864.
† Payen. The average for the whole of the year is not more than one-thousandth.
‡ Baumgarten.
§ Payen.

by various geographers as to the average quantity of alluvium contained in running water differ prodigiously one from another. In the last century, Eustache Manfredi, who, taking account of the enormous deposits produced by the Po, exaggerated the work of this kind accomplished by other rivers, and estimated the average proportion of muddy matter at $\frac{1}{150}$ of the liquid mass of rivers. But in this estimate he doubtless included the sand and mud which are impelled by the current along the bottom of the bed, the bulk of which is probably twice as large as that of the floating matter. Hartsoeker, in his *Traité de Physique*, admitted that the proportion of alluvium was $\frac{1}{100}$; whilst another author of the same epoch, the writer of the *Recherches Philosophiques sur les Américains*, was led by his observations and calculations to fix the amounts of *débris* existing in the water of rivers at only $\frac{1}{51,000}$.*

The differences between the various estimates are naturally quite as great when an endeavour is made to reckon approximately the time that it would take for the alluvium emptied out at the mouths of rivers to raise the level of the ocean to a given point. Manfredi supposes that the detritus carried down to the sea would be sufficient to raise its bed a yard in 3,300 years. Tyler thinks himself warranted, by his calculations as to the alluvium of the Mississippi, in asserting that the deposits of rivers would elevate the level of the ocean only 2 inches in 10,000 years, or about a yard in 180,000 years. These are estimates of a very different character; but when one reflects on the greatness of the sea, and on the littleness of rivers compared with the immense reservoir, even the last-named estimate seems too high. If we admit that the average proportion of the earthy matter carried down into the sea is about $\frac{1}{1000}$ of the entire liquid mass of rivers, and if we adopt, as the total discharge of running water, the approximate quantity to which the critical examination of the known facts of fluviatile hydrography has led us, we shall find that the mass of alluvium deposited every second at the mouths of rivers would be equivalent in bulk to 436 cubic yards, or every year a body of matter equal to 4,000 square miles in area, and a yard thick. This, however, would be an almost infinitesimal quantity in comparison with the enormous abysses of the ocean.

Yet the earth belongs to all time, and during the course of ages any geological work must ultimately be accomplished. These rivers,

* This question is treated on in detail in Von Hoff's work, *Veränderungen der Erdoberfläche*, vol i.

almost imperceptible, so to speak, in comparison with the ocean, are gradually eating away mountains and plateaux, and filling up the abysses of the sea with their accumulated *débris*. These deposits have the effect of raising the average level of the waters of the ocean, and of causing them to cover low shores. There is, therefore, a double cause operating in the modification of the relief and outline of continental masses. If the only force in action on the surface of the globe was that of running waters, the elevated parts of the earth would be constantly becoming lower, the sea would incessantly encroach on the coasts, and, sooner or later, the planet would become an immense globe, covered with a thin sheet of water. Owing, however, to the geological movements of the earth's strata, a transformation of this kind is not to be dreaded; but still, from the action of the water of rivers, continents and seas are undergoing changes of the very highest geographical importance. The Baltic Sea has already become something between an inland sea and a chain of fresh-water lakes. The liquid mass poured into it by rivers continues always the same, whilst the area and depth of its basin are constantly diminishing. In the long course of ages, its water will ultimately become perfectly fresh, and the straits of the Sound will be only the European St. Lawrence.

Some day, Bory de Saint-Vincent tells us, the Mediterranean itself will become nothing more than a chain of lakes, and then a gigantic river. The Sea of Azof is already being gradually converted into a stream, as its shores are getting nearer and nearer together, whilst its bed remains perceptibly the same.* The tracts of water which extend from the mouth of the Don to the straits of the Dardanelles might be compared to the Lakes Superior, Huron, and Michigan; the isles of the Archipelago will some day overlook a labyrinth of lagoons similar to those which border the Baltic Sea; the Gulf of Venice will be only an elongation of the valley of the Po; and the two great basins of the Mediterranean, separated by the Siculo-African bar, will become two lakes of increasingly contracted dimensions, the waters of which will feed the greatest river in the world. Then the Dnieper, the Danube, and the Po will be but mere tributaries. Perhaps even the Nile, which is now of no great size at its mouth, may lose all its water by means of evaporation before it reaches the Mediterranean Sea, and will become nothing but a water-course of an entirely continental character, such as the Chary, the Houach, and the Jordan.

* *Zeitschrift für Erdkunde*, May, 1862.

Certainly it would be difficult to exaggerate the importance of the part played by rivers in the history both of the earth and mankind. They distribute uniformly the snow and rain which fall at the various points of their basins, and fertilise the whole territories by their innumerable ramifications. They powder up the rocks of the mountains, and spread the matter which results in fertile alluvium over the plains, forming also new tracts of land at their mouths. They equalise climates. Rivers coming from the south warm with their vapour northern districts, whilst rivers flowing in a contrary direction moderate the heat in more southern latitudes. Added to this, watercourses, those powerful workers, do not limit themselves to carrying down water, alluvium, and climate; they also roll down in their flow the history and life of nations. The course of the river's current is the path down which descended the canoe of the savage warrior, and is now the highway for the fleets of commerce bearing peace and comfort. Steam has converted rivers into roads, which can be traversed both in a downward and upward direction, and a floating population is constantly pervading their surface. Far from forming a barrier to nations, rivers are the means of mobilising them: they are continents set in motion. Aided by rivers, the mountaineers of the Alps and Pyrenees make their way to the Atlantic and the Mediterranean, whilst the inhabitants of the sea-coast ascend to the elevated districts in the interior of the continent.

In the present day water-courses no longer assume, in the history of civilisation, the high importance they once possessed, for now they are not the only ways of communication between nations. No river can now be all that the Nile was to the Egyptians, at once their father and their god, the cause from which sprung both a race of husbandmen and also the harvests which they gathered on the river-mud, warmed by the rays of the sun. Another Ganges, with its sacred waves, will never again flow over the surface of the earth, for man is no longer the slave of nature. He can now develop artificial roads, which are shorter and more speedy than the roads formed by nature; and this second, and even more vital nature, which he has created by the labour of his own hands, supersedes his adoration of that first nature which he has succeeded in regulating. Nevertheless, rivers will be more important as servants than they have ever been as gods. They bear upon their waters ships, and the products with which they are freighted, and serve as arteries to vast organisms of mountains, valleys, and plains, which are sprinkled over with thousands of towns, and millions of inhabitants. They vivify the earth by their motion, carve it out afresh by their erosions, and add to it by their ever-increasing deltas. Some day, when the hand of man will be enabled to guide rivers and to trace out for them their beds, he will employ these potent workmen to carve out a nature in harmony with his own will; water-courses will wear away the hills, fill up lakes, and throw out promontories into the sea in obedience to his orders: their eternal and mighty vitality will become the complement of ours.

CHAPTER LVI.

LAKES.—FORMATION OF LAKES.—THEIR INCREASE AND DIMINUTION.—THEIR FORM AND THEIR DEPTH.—LAKES LYING IN SUCCESSIVE GRADATIONS OF ELEVATION.

COLLECTIONS of water—ponds, pools, lakes, or inland seas—are formed in every depression of the ground which receives a larger quantity of liquid, either from rivers or directly from the clouds, than it can get rid of through its affluents, or transfer to the atmosphere in the form of vapour. Hence arises that infinite variety of lacustrine sheets of water which gives so much grace or grandeur to landscapes, and exercises such a considerable influence on the action of rivers, on climates, on the productions of the soil, and consequently on the development of mankind.

The liquid mass contained in any basin on the surface of the earth does not increase to an indefinite extent, even when considerable quantities of water are constantly being poured into it by its tributaries. Either the basin completely fills up, and the overflow is emptied out through the lowest depression in its rim, or the lacustrine sheet, gradually enlarging in area, ultimately presents a surface sufficiently extensive for evaporation to establish an equipoise to the supply of water.

Perfect equality between the mass of water received and that which escapes does not, however, exist in any lake, and consequently the level never ceases to fluctuate; sometimes it rises and sometimes it sinks, according to the various seasons and years. After heavy falls of rain, or at the time of the melting of the snow, some pools are changed into perfect lakes, in the same way as, during long periods of drought, some lacustrine basins entirely dry up. The great phenomena of the vitality of the globe—such as the upheavals and sinkings of the ground, the growth of mountain ridges, the encroachments or retirement of the shores of continents, the alternations of the winds and rains, land-slips, and the rupture of natural dikes—all have the effect of either giving rise to and increasing, or of doing away with or diminishing, the masses of water which are collected in the interior of continents. Like everything else which exists on the surface of the globe, lakes have their periods of increase and decrease, and even within the limited period during which man has begun to record the annals of his planet, numbers of fresh lakes have made their appearance, whilst many others have entirely dried up, or have considerably diminished in extent.

In mountainous regions it is a well-known fact that the fall of rocks and the advance of glaciers have often caused the formation of considerable lakes. In like manner, some of the large lakes of the Landes have appeared since the Middle Ages, owing to the cutting down of the trees upon the dunes, and the shifting of the latter towards the east.* On the other hand, instances of lakes which have disappeared owing to natural causes, without being subjected to any human labour in the process of their exhaustion, are likewise very numerous.

Thus the plain of Oisans, in the Alps of Dauphiny, having been suddenly closed up in 1181 by a downfall of rocks which came from the sides of the Voudène, the waters of the Romanche, the Olle, and the Vénéon accumulated above the obstacle, and spread out into a lake of $6\frac{1}{2}$ miles in length. Villages, vast plains, and whole forests were swallowed up under a liquid sheet of an average depth of 33 feet, and the local employment gradually became that of fishing. The lake existed for thirty-eight years, and then the barrier of *débris* suddenly yielded under the pressure of the water, and the body of liquid rushed like a deluge over Grenoble and all the towns and plains on the banks of the Isère. At the commencement of the fourteenth century the former lake, which had received the name of the lake of Saint-Laurent, was completely dried up.

The formation of lakes of this kind above some dam of rubbish, and their sudden disappearance when these dams are broken down, may, however, be considered as accidental phenomena, and not dependent directly upon climate. In this latter respect, the changes in level which are exhibited by some great lacustral sheets, such as the lakes of Titicaca and Van, are much more remarkable facts. Travellers assert that the area of the immense Bolivian lake has always been diminishing since the commencement of the historical period. Its water once bathed the walls of Tia-Huanacu, one of the principal cities of the Incas; but this locality is now situated $12\frac{1}{2}$ miles from the lake, and more than 130 feet above the level of its

* *Vide* in vol. ii. the chapter on "Dunes."

water. This would be a remarkable proof of the increase of dryness on the high plateaux of Bolivia.* On the other hand, the height of the lake of Van continues to increase—a fact which is confirmed by travellers every year. The inhabitants on its shores are frequently obliged to turn the sea-shore roads further inland; ancient villages have been swallowed up, and in some spots the ruins buried by the water are still visible. Finally, the town of Erdjisch, which was once separated from the lake by a great plain, is nowadays invaded by the water, and the city of Van itself, which was once far from the shore, is now quite close to it. A legend, which explains in its own way the constant swelling of the water, relates that some capricious nomads having obstructed an outflow of the lake, afterwards made useless efforts to re-establish the former outlet; but, since this date, the irritated lake has never left off covering a fresh extent of plain every year.†

As a simple process of reasoning must point out, lakes are most numerous and most extensive in those countries where rain falls in considerable quantities, and the surface of which, although but slightly undulated, is nevertheless formed of compact rocks which do not allow the water to flow away into the depths below, and retain it as if in natural basins. Of this kind are the regions of North America, in which lies the fresh-water Mediterranean crossed by the St. Lawrence, the Winnipeg, Winnipegoos, Bear and Slave Lakes, and several other sheets of water of less extent. In these districts there is certainly less rain than in the tropical zone, and even than in most of the countries of the temperate zone; for the depth of rain and snow water does not attain to more than three feet in a year. But the granitic soil retains in the shallower depressions the moisture which falls from the atmosphere; evaporation does not take place actively, and the slopes towards the different seas are not sufficiently inclined for the numerous rivers to be able to pour down to the ocean all the surplus waters.

The island of Newfoundland is also in great part granitic, and is likewise covered by lakes maintained by the constant humidity which prevails in those parts of the sea. In like manner, in Europe, the eastern valleys of the Scandinavian mountains and the plains of Sweden exhibit a perfect labyrinth of lakes, some of which are very small whilst others stretch away and are lost on the distant horizon, save where they are dotted over with archipelagoes, rocks, and islets, like

* Pentland; Bollaert, *Antiquities*.
† Otto Blau, *Mittheilungen von Petermann*, vol. vii., 1863.

the Lake of Mälar, which contains no less than 1,260 islands. On the other side of the Gulf of Bothnia, the granite plains of Finland are sprinkled still thicker with lakes than those of Scandinavia itself, so that the whole country may be considered as an immense sheet of water intersected by innumerable isthmuses crossing one another in every direction.

Labyrinths of lakes of an altogether similar character are also

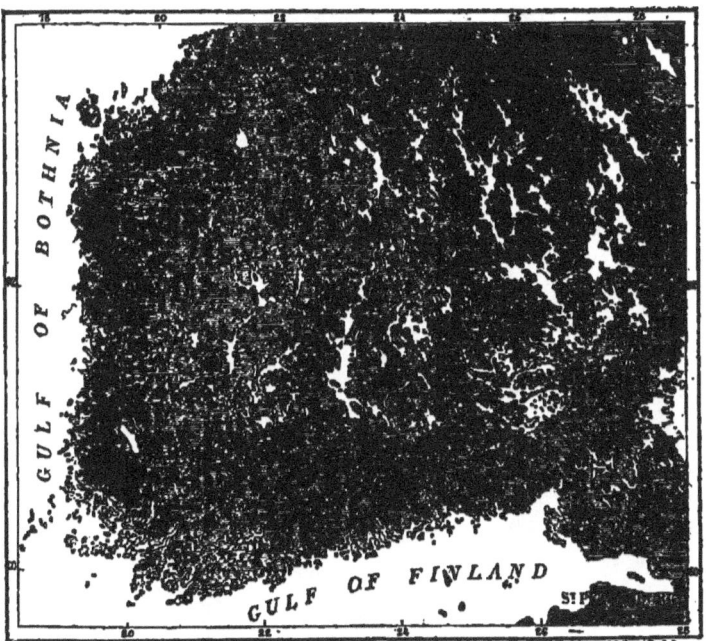

Fig. 152.—Lakes of Finland.

found in countries where the soil, although not rocky, lies on a clayey or ochreous subsoil, which is entirely impervious to water. Thus, there used to exist in the French *landes* a great number of pools which the bed of *alios* retained on the surface;* these are at the present time mostly dried up. In the same way Sologne, Brenne, and some other solitudes in central France, were dotted over with shallow pieces of water. La Dombes, a plateau of about 300 feet in height, which extends to the north-east of Lyons, between the Rhone,

* *Vide* above, p. 84.

G G

the Saone, the Veyle, and the Ain, is also covered with a multitude of pools, occupying altogether an area of more than 47,000 acres. It is a fact, that in this part of France man has unfortunately lent his aid to the work of nature. Most of the pools of the Dombes are of artificial origin, and their being laid dry would cost even less than

Fig. 153.—The Dombes.

their construction. They serve as fish-ponds for the wretched inhabitants of the neighbouring villages, and then, being emptied and cultivated for cereals, they are again filled up and stocked with fish.

The form of a lake always bears some relation to the general relief

of the ground in the depression of which its water is contained; its outline and the profile of its bed harmonise perfectly with the continental architecture. The water of alluvial, oozy soil, is spread out in vast marshes, in which it is difficult to point out the precise spot where the dry ground ends and the water begins. The liquid sheets of low plains, deserts, and level plateaux present generally more sharply defined outlines; but their depth is but slight in comparison to their extent, and the least fluctuation in level considerably modifies the line of their banks. The lakes of more undulating regions are in general tolerably deep in proportion to their extent, and present bays and promontories of a more varied and picturesque

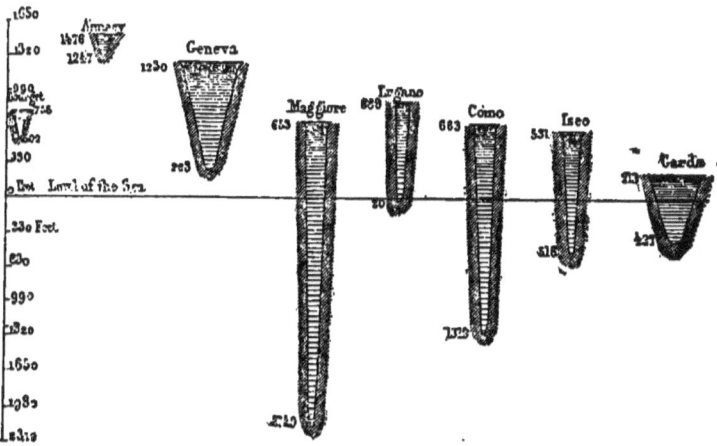

Fig. 154.—Altitudes and Depths of Lakes in Italy and Savoy.

character than the sheets of water in the plains. But the place where lakes exhibit all their beauty is round the bases of lofty mountains. There, torrents run down into them, falling over in rapids and cascades; green glens slope down to their very margin; the spurs of the mountain plunge straight down into their waters, and the shores between the headlands are traced out in gracefully curved bays. By the harmony and variety of lines presented by their outline, these lakes seem almost a necessary feature of the landscape, and their horizontal surface, by the contrast which it affords, gives a more noble appearance to the surrounding mountains.

Lakes, like seas, are in general all the deeper as the cliffs which overhang them are the more steeply escarped; indeed the cavities which are filled up by the water seem to correspond in their dimensions to the height of the upheaved masses. Thus, to bring forward no other instances than those of the Alpine lakes, the deepest of these lakes are found at the southern base of the Alps, which, on this side, present their steepest slopes. Lake Maggiore, the level of, which is 652 feet above the Adriatic, is no less than 2,800 feet in depth; the Lake Como is 1,981 feet deep in the lowest part of its

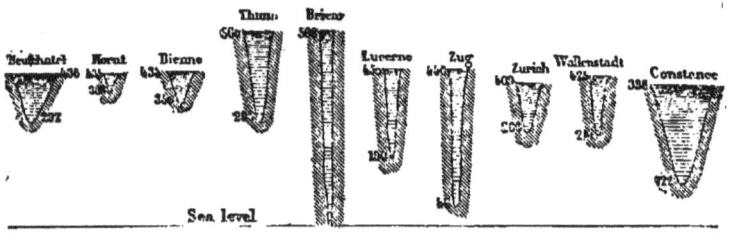

Fig. 155.—Altitudes and Depths of Lakes in North Switzerland.

basin. The lakes of Garda and Iseo are not so deep, but still deep enough to descend far below the level of the sea. If we could suppose the whole body of the Alps cut down to the level of the sea, the abysses of the water in the lakes of Maggiore, Como, Garda, and Iseo would still be respectively 2,149, 1,318, 518, and 426 feet in depth; whilst, on the other side of the Alps, the only lake which has a bed below the level of the sea-water is, perhaps, that of Brienz, if it be true, as Saussure asserts, that it is 1,968 feet in depth.*
The two annexed plates represent the respective altitudes of the

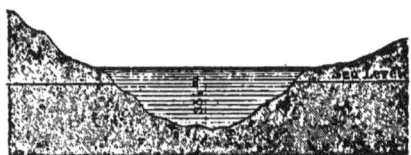

Fig. 156.—Lake Maggiore.

principal lakes of the Central Alps in comparison with the level of the sea. The results depicted in these plates have, however, unfortunately only an approximate value; for in the Alps, which are, nevertheless, visited and studied by so many scientific men, accurate sounding operations have not yet been made in some of the most important lakes. In each of these plates the depth has been

* Desor, *Schweitzer Seen*, in the *Album de Combe-Varin*.

exaggerated a hundred-fold in comparison with the breadth. In order that a clear idea may be formed of the shape of the Alpine lakes, it is necessary to annex here the actual outline of the depression of Lake Maggiore, the deepest of all the lacustral basins in the Alps, and of the Lake of Neuchâtel, the principal sheet of water in the Jura.

Glancing at a map of the Alps, it is impossible to avoid remarking at first sight that the lakes are distributed in a certain order as regards the great groups of mountains. Thus the Maritime Alps, those of Viso, Provence, and Dauphiny, and also the Mont-Blanc group, have but a very small number of lakes, and even these better deserve the name of ponds. On the east of Switzerland the various ranges of the Alps, which extend as far as Turkey, are likewise almost devoid of lakes, except in Southern Bavaria and the districts of Salzburg, where several masses of water fill up some narrow valleys which open nearly uniformly from south to north between parallel chains of mountains. The noble lakes which form the glory

Fig. 157.—Lake of Neuchâtel.

of the Alps are all situated round the central group (of which the Saint-Gothard occupies the middle), and in the valleys and plains which, under various names, form the western limit of the parallel ridges of the Jura.

These lakes, which evidently owe their origin to the star-like form of the chains which radiate round the Saint-Gothard, and to the intersection of the Alpine system by that of the Jura, have in general elongated basins, tending either from south-west to north-east, or, perpendicularly to this direction, from south-east to north-west. The waters of the valleys of the Jura—for instance, the Lakes of Joux and Saint-Point—lie in the former direction, likewise the great bodies of water situated at the base of the limestone mountains—the lakes of Neuchâtel, Bienne, and Morat. The Alpine lakes of Brienz, Sarnen, and those of Engadine also lie in this direction; and even the lakes on the Italian side, Maggiore and Garde, are nearly parallel to the lacustral basins and mountainous ridges of the Jura. On the other hand the great Alpine lakes of Constance, Zurich, Sempach, Zug, and Thun, all stretch in a contrary direction to those above named. With regard to the two magnificent inland seas of Switzerland, the lakes of Geneva and Lucerne, they owe their admirable shape to a combination of the two types. The Leman is a lake of the Jura in its lower part, and an Alpine lake in its upper part; towards the middle the two sheets meet and cross one another. In the lake of Lucerne the two basins cross one another at right angles, and thus give to the whole body of water the shape of a cross.

It must likewise be remarked that the largest lakes are found on the courses of the most plentiful rivers, which goes to prove that the same geological laws have presided in the formation of the valleys and in hollowing out the lacustral basins. The lake of Constance, the largest of all, receives the Rhine, the largest river in Switzerland. The Leman is crossed by the Rhone; the Aar flows into the two lakes of Brienz and Thun; the Reuss enters the lake of Lucerne, the Linth that of Zurich. An arrangement of this

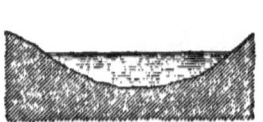 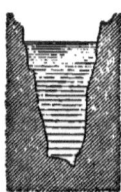 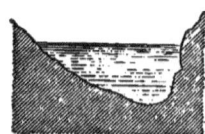

Fig. 158.—Valley. Fig. 159.—Cluse. Fig. 160.—Combe.

kind can hardly be fortuitous, but must depend on the general structure of the great groups of the Alps.

In those mountains which possess an architecture of almost perfect regularity, as, for instance, those of the Jura, it is easy to classify the various lakes according to the form of the depression which is filled by their waters. Thus the lacustral sheets which spread out in a valley between two parallel ridges of mountains generally exhibit outlines which are scarcely at all broken and disposed in a regular oval; the slopes of the bed are gently inclined towards the central part, the average depth of which is not, however, very great. Here and there, and especially at the two ends, the banks are marshy, and it is difficult to determine the exact spot at which the firm ground begins. Among these valley lakes we may mention those of Joux, Saint-Point, and Bourget.

In a similar way the *combes* and *cluses* which serve as reservoirs to the lakes confer on the water which they contain special charac-

FORMS OF LAKES. 455

teristics very different from those of the lacustrine basins in valleys.

Fig. 161.—Stages of Lakes in the Valley of Oo.

Thus the lakes in *cluses*, lying crosswise to a chain of limestone mountains, are generally narrow, and the escarpments of the high cliffs which command them descend to a great depth below the surface of the water. With regard to the lakes in *combes*, the amphitheatre-like reservoirs which contain them give to the surroundings of each basin a magnificent aspect of grandeur and majesty. The water in them is deeper than that of the valley lakes, but not so deep as that of the lakes in *cluses*; and it is just in the lower portion of the lacustrine cavity that the section of water presents the greatest thickness.

It is, however, but seldom that—in the Swiss Alps and other mountainous countries with a deeply indented vertical outline—we do not find lakes which present characteristics of all the different types. In some parts of their basins they are lakes of the valley,

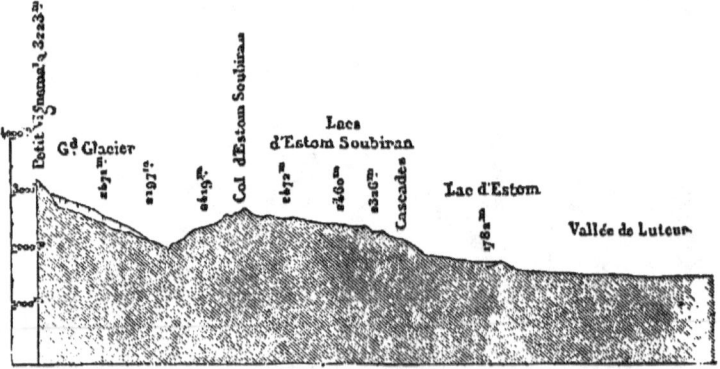

Fig. 162.—Lake-stages of Estom Soubiran and Estom.

and in others they are lakes like those contained in *cluses* or *combes*. For this reason what a diversity of appearance they present in their shores, what picturesque beauty in the windings of their bays and the succession of their headlands!

Between mountain ridges which are not arranged in long parallel lines like those of the Jura, valley lakes are not mere oval sheets of water; they extend in long windings like the Lakes Maggiore, Como, and Lugano; or in those valleys in which wide basins and contractions alternate in succession, the lakes spread out and become narrow alternately. Numerous instances of this may be seen in the Scandinavian mountains. In a general way, however, these lakes are cut up into several pieces of water lying in gradation, one above the other, as if on enormous steps, and are connected by narrow defiles,

POSITIONS OF LAKES. 457

down which the water of a torrent pours in cascades. These lakes,

Fig. 163.—Lakes of Nors Elf.

lying on graduated levels, are found in the high valleys of almost every mountainous country. In Switzerland, the three lakes of Lungern, Sarnen, and Alpnach, which are traversed by the river Aar, may be mentioned; in the Pyrenees, the mountain lakes of Oo and La Têt, and the lakes of the valleys of Couplan, Aygues-Cluses, and Estom Soubiran, belong to the same class of lacustrine basins. In the Carpathians they form those charming little pools of water to which the name of "*meeraugen*" (eyes of the sea) has been given; lastly, in Scandinavia, lakes situated on graduated levels may be reckoned by hundreds.

There, all the rivers, almost without exception, are, from their source to their mouth, nothing but chains of lakes connected with one another by rapids and cascades. They are, in fact, water-courses in process of formation, which have not as yet hollowed out for themselves regular beds, but flow in all the natural depressions of the soil through narrow channels which have been opened since the ground itself has risen above the level of the sea. The land of Scandinavia, having risen only at a recent epoch by a gradual movement of emergence, which at the present time is still continuing, the rivers have not yet had time either to fill up with *débris* the lakes that they meet with in their course, or to pierce wide valleys through the rocks.*

* *Vide* the chapter on "Upheavals and Depressions."

CHAPTER LVII.

VARIOUS PHENOMENA IN LAKES.— COLOUR OF THEIR WATERS.—SEICHES.—
CURRENTS AND TIDES.—FORMATION OF ICE IN LAKES.

LAKES are not only distinguished from each other by their shape and the depth of their basin, they also vary in the appearance of their water, and even in this respect the diversity of the matters held in suspension or solution in the liquid mass is not always sufficient to explain the remarkable contrast presented by adjacent sheets of water. The colour and transparency of the liquid differ astonishingly in most mountain lakes. Some are of an emerald green, others of a sapphire blue, a few even have a milky shade. There are some, indeed, the water of which is transparent, that have a brown or yellowish colour. In every case, whatever may be the natural hue of each of these lakes, they incessantly vary on account of the reflection of the rays of the sun, the clouds, or the colour of the sky and the refraction of the light. One lake, the water of which not far from the bank is of a yellowish green, owing to the rocky bottom just visible below the undulations of the surface, is of a deep blue above the invisible abysses of its central portion. Another lake presents a well-defined difference of colour between the tranquil water of its basin and that which is brought in by the rapid current of the river which crosses it. In other places, again, the eddies light up the surface with reflections of a bronzed or greenish hue; even the particles of sand or ooze, as well as the chemical substances dissolved in the water, must necessarily, however infinitesimal their tenuity may be, tinge the liquid sheet with various shades. Vegetable mould gives to lakes a colour more or less shaded with red or brown ; clay gives them a yellowish tinge. As to the *débris* of rocks and pebbles, these, according to Tyndall, are the agents which confer on the Lake of Geneva and other mountain lakes their lovely azure colour. The most wonderfully transparent water, which, too, is the most devoid of all impurity, is in general a sea-green hue. It is said that objects are sometimes visible in it at a depth of 80 and even 100 feet.

All long and narrow lakes, over which atmospheric variations often take effect in a sudden and violent manner, frequently exhibit abrupt oscillations of level, which can only be explained by a difference in the pressure of the air. Such are the *seiches* of the Lake of Geneva and the *Ruhssen* of the Lake of Constance, which are noticed sometimes at one point sometimes at another. In these purely local swellings of the water, the latter may rise all at once some inches or even a yard above the level of the surrounding surface. The outbreak of subterranean tributaries cannot be taken as an explanation of the cause of this sudden rise, for it takes place at the foot of mountains of a compact formation, which certainly do not conceal any considerable streams in the depths of their rocks. Added to this, on the surface of many lakes and inland seas the phenomena of *seiches* have been observed around islets and mere rocks.

Schulten has proved that the *seiches* of the Baltic, which are in every respect similar to those of the Lake of Geneva, are in direct connection with the height of the barometrical column. When the pressure of the air diminishes the water begins to swell, and when the barometer again rises up the surface of the sea sinks, only the movements of the water are always a few minutes earlier than those of the instrument, on account of the greater mobility of the aqueous particles. Now, as the total variation between the different heights of the barometrical column at the level of the sea corresponds to a variation of about a yard in a column of water, it follows that the most considerable *seiches* cannot exceed this height. This has, in fact, been verified by observers in the Baltic as well as in the Lake of Geneva, and in the great lakes of North America. In the midst of the open sea *seiches* would likewise be produced, especially during hurricanes; but the liquid mass being at full liberty, and able to spread out freely all round the rising of the wave, the phenomena is there more difficult to notice than in narrower lakes.* It is probable that the phenomenon known by the Sicilians by the name of *marubia* (from *mare ebriaco*, "drunken sea") is also a swelling of the water accompanied by the barometrical depression. It is observed on all the coasts of Sicily, but especially off Mazzara, at the precise spot where the Mediterranean, contracting into the form of a strait, is severed into two basins by a submarine ledge which approaches the surface. Daubeny considers that these movements of water are a sign of some volcanic vibration of the soil; yet the description which he himself gives of the movements seems to indicate that they are *seiches* similar

* Anton von Etzel, *Die Ostsee*.

to those in the Lake of Geneva and the Baltic. When the *marubia* occurs the air is calm and the horizon misty; suddenly the water, stirred up in short waves, raises its level about 23 inches, and then, after an interval of from half an hour to two hours, the south wind begins to blow, and a heavy storm rises.

Lakes, moreover, which are, indeed, inland seas of fresh or salt water, must exhibit phenomena similar to those of the ocean. Lacustrine sheets of water have also their tempests, their swells, their breakers, and their bores; and certain bays of Lakes Superior, Ladoga, and Baikal, are not less dangerous than the Black Sea and the Bay of Biscay. The waves raised by the wind in the more confined areas of lakes are neither so high nor so rapid as those of the sea, because they have not so vast a field on which they can spread out, and because they do not move over a sufficiently great depth of water. They are short, compact, and "chopping," and from this very fact they are more formidable to any ship against which they incessantly dash. Added to this, the water of most lakes being fresh, and in consequence lighter than that of the ocean, it is also more readily stirred up, and the wind has scarcely commenced to blow before the surface of the lake is roughened with foaming billows.

With regard to the currents, it is evident that in lakes they cannot be developed with the same regularity as in great seas which lie open and exposed from the poles to the equator; but currents are nevertheless produced in every spot where any perceptible difference in temperature exists between two adjacent regions on the surface of the sea. There is of necessity a flow of cold water towards the sides of the lake whenever the superficial liquid layers, heated by one cause or another, are comparatively lighter and suffer a greater loss from evaporation. Besides these lateral currents, which are sometimes difficult to be certain about, there are also, in lakes as in the sea, interchanging currents flowing between the upper sheet of water and the masses underneath. Moreover, all the rivers which cross an open lake or fall into a closed-up lake, like the Rhone, the Rhine, the Reuss, and the Jordan, determine the formation of local currents, from each side of which the water of the basin flows back in a contrary direction. Lastly, lacustrine basins also have their tides, although these phenomena are generally nearly imperceptible, and are only discovered by a long and attentive series of observations of the oscillations of the level. In Lake Michigan the height of the tide reaches to about 3 inches.

One of the most curious phenomena in the lakes of the northern temperate, and polar zones, is that of the formation of ice. In winter, when the sheet of water is perfectly still, needles of ice, radiating one from the other at angles of from 60 to 120 degrees, appear on the surface; then joining their network together, they soon form a level sheet of ice. On the contrary, when the water is violently agitated by a storm, the first needles of ice being incessantly bruised and rubbed against each other, agglomerate in discs rounded by the friction, and the whole of the congealed mass ultimately presents an uneven surface like that of rivers with a rapid and violent current. The ice of lakes is generally much more regular and transparent than that of water-courses, in which the process of crystallisation is nearly always being disturbed. When a prism of this pure ice is exposed to the influence of a ray of the sun concentrated upon it by a lens, a multitude of little corollas, with six sepals arranged round a glittering point, suddenly appear in the thickness of the prism. This is one of the most charming sights which the beauties of nature can present to the eyes of an observer.*

When the whole extent of the covering of ice is solidified over the water, it does not remain immovable until the thaw; on the contrary, it is constantly agitated by various movements, according to the state of the atmosphere and the phenomena which are going on in the liquid mass beneath. If the temperature diminish, the lower side of the frozen crust is immediately increased by a fresh layer of ice more expanded than the water; the sheet must therefore necessarily rise and form a somewhat curved surface. If the cold become less intense, the solid mass consequently grows thinner, and forms hollows in some places. When the level of the lakes rises, owing to any larger quantity of water being poured into it by its affluents, the arch of ice is upheaved unequally by the liquid sheets which flow beneath it. If the supply of water diminish, and the level of the lake consequently sinks, the solid cover simultaneously gives way, owing to its own weight, and splits up so as to follow the downward movement of the water. Lastly, the long undulations which are produced in the liquid mass by shocks received on the surface, the large quantity of air which makes its way under the sheet of ice either in considerable bodies or isolated bubbles, even the gas incessantly being evolved by the respiration of the fish—all combine in producing the same result; that is, the upheaval of the ice. The comparatively thin crust which separates the hidden water from the great atmospheric ocean is constantly being drawn some-

* Tyndall, *Glaciers of the Alps*.

times in one direction, sometimes in another. Enormous crevices, generally tending in the direction of the greatest length of the lake, open suddenly with a terrible crash; the roaring of the air which penetrates under the icy layer, or which escapes from it, is mingled with the crackling of the breaking crystals; we have simultaneously noises like the rolling of thunder and the rattling of musketry. On that part of the Lake of Constance which is called the Untersee, M. Deiche has noticed cracks in the ice which were six miles long and 13 to 16 feet wide.

In the great lakes of North America, and also in those of Siberia, especially in Lake Baikal, the phenomenon of the formation of ice takes place in the most magnificent way. During three months of winter, the mighty Baikal, the inland sea in which seals live and coral-stems grow as in the ocean, is covered by a field of ice, presenting in some places a thickness of 6 to 9 feet. The vast sheet of water, extending over an area of more than 1,400 square miles, and surrounded by mountains as high as the Alps, and glittering with glaciers, is nothing but a solid mass, on which caravans of travellers venture without fear. Sometimes, when the ice begins to form, a sudden tempest reduces it to fragments, which, under the pressure of fresh pieces of ice, brought by the waves and currents, are piled up one on the other, intermingling in a kind of chaos which calls to mind the *séracs* of the Alpine glaciers. Subsequently, when the water is entirely covered with its heavy shell, the latter is occasionally rent asunder, and shrill whistlings, dull cracking noises, prolonged thunder-like rumblings, mingled with innumerable partial crepitations, are heard whilst the ice is bending and breaking. The water springs out from the fissure in vertical sheets, and falling down again on the surface, forms risings on each side of the crack, which is sometimes more than a yard wide. Sometimes, a fragment of the broken layer of ice sinks below the general level; another piece, being pressed on in every direction by the frozen masses, curves perceptibly in the middle. All these movements of the solid crust produce long undulations in the water beneath. Travellers, borne along rapidly in their sledges over the ice of the lake, feel distinctly the shock of the waves breaking against the lower side of the trembling floor beneath them. On the sides of the cliffs which border on the lake, may be noticed heaps of solidified flakes, sometimes resembling a cascade; this is the foam which is jetted out at the time of the violent rupture of the ice, and has hardened upon the rocks before it had time to fall.* In a general way, Lake Baikal freezes so rapidly that, according to the statement of the natives, the ice begins by adhering to the bottom of the lake, from which it afterwards becomes detached with a terrible noise and rises to the surface.† But this fact, which could not take place unless the temperature of the deep water was much lower than that of the surface which is traversed by freezing winds, has not yet been scientifically verified. It is, on the contrary, very probable that the water on the bottom remains constantly liquid. At the temperature of 39° F. the aqueous particles acquire their greatest density, and consequently their heaviest specific gravity. In obedience to the law of gravity, the layers which are at 39° F. of temperature are those which must lie upon the bottom of the lake, and therefore ice can only be formed on the surface. The direct observations which have been made as to the temperature of the Swiss lakes confirm this theory. In the Lake of Geneva, the effects of meteorological variations are not felt below a depth of 236 feet, and deeper still the constant temperature is 42° F. In the Lake of Constance, the temperature is lower; there it is only 39° F., and in the Lake of Lausanne, 39° 12′ F.; this comparatively slight excess of heat is probably owing to the natural warmth of the ground.‡ Added to this, in the environs of Boston, where all the small lakes are regularly worked during winter, and furnish for the demands of commerce more than 200,000 tons of ice a year, the solid layer of ice has never been noticed to form in the first place at the bottom of the basin.

* Russell-Killough, *Seize Mille Lieues.*
† Carl Ritter, *Erdkunde.*
‡ Buff, *Physik der Erde.*

CHAPTER LVIII.

LAKES ACTING AS REGULATORS OF THE RIVERS WHICH PASS THROUGH THEM.—
FRESH-WATER AND SALT-WATER LAKES.—THE CASPIAN SEA.

THOSE lakes which receive a superabundant quantity of water—and these constitute the most numerous class—give rise to a river which carries off the surplus of the liquid mass poured into the basin by the upper affluents. These lacustrine reservoirs may then be considered as expansions to some extent of the fluviatile valley; in this point of view, the Lake of Geneva would be the Rhone, and become a hundred times wider and deeper. The Lake of Constance would be an immense hollow of the Rhine, containing in its reservoir nearly a hundred times as much water as all the rest of the river. In like manner, the great inland lakes of North America—Superior, Michigan, Huron, Erie, and Ontario—form the first part of the course of the St. Lawrence, a river of such slight importance in comparison to the vast basins which feed it.

The large basins in which the water of a river is spread out, before it again takes its course down to the ocean, regulate the discharge of their out-flows all the more efficiently the more extensive the area over which they extend. Very considerable inundations in a stream produce comparatively but a slight rise in the level of a lake, because the water has to be diffused over the whole surface of the basin, and loses in depth all that it gains in breadth. During the season when the ice is melting,—that is, in spring and summer,—the Lake of Geneva rises on the average 6 feet above the low-water of winter, and consequently contains a surplus mass of 1,572,000,000 cubic yards of water. The gauges used at Geneva establish the fact that the discharge of the Rhone at its issue from the lake is at its maximum 53 cubic yards; now, as the various affluents of the lake supply more than 1,400 cubic yards during their highest floods, it is evident that the Lake of Geneva acts as a complete regulator. It keeps back at least one-half of the inundation-water, which it subsequently empties down gradually when its tributaries have retired to their usual level. It is certain that, owing to the regulating action exercised over the discharge of the river, the plains on the banks of the middle course of the Rhone, from Geneva to Lyons, are comparatively protected against floods. The equilibrium in the action of the river would be still more complete if a dam were constructed at Geneva, with flood-gates to regulate at will the discharge of the water.*

Lakes which are crossed by rivers must be, almost without exception, fresh-water basins, as the saline particles which are carried into the basin by one or more affluents are conveyed out of it with the surplus water. Still, lakes of no great area, which are mostly fed by salt springs, discharge brackish water through their out-flows. As regards lakes without any outlet, it is evident that the saline particles brought into them by tributaries cannot make their escape, and must consequently be deposited on the edges, or must more and more saturate the liquid mass. Except they are fed by affluents entirely devoid of saline matter, lakes which are without any communication with the sea must, therefore, more or less resemble the ocean in the composition of their waters. Almost all the lakes without outlet are filled with water more or less saline. It must, however, be understood, that the proportion of salt varies in all inland basins, and the transition is most gradual between the condition of water called fresh, and that of brackish or salt water.

The largest inland sea devoid of any outlet—the Caspian—is the remains of that great central sea which once extended from the Euxine to the Frozen Ocean. It is probable that the slow upheaval of Siberia and Tartary has gradually separated the Caspian from the Gulf of Obi and the Sea of Aral, and that, subsequently, the rupture of the Bosphorus, by lowering the level of the water of the Black Sea, has laid dry the Ponto-Caspian isthmus, which is now traversed by the waters of the Manytch. Be that as it may, it is certain that by remaining isolated in the middle of the land, the Caspian has lost by evaporation a larger quantity of water than is supplied to it by its tributary rivers; for it has gradually diminished in extent, and its level has sunk more than 80 feet below that of the Black Sea. If the Caspian was again to fill up the whole concavity of its basin to a height corresponding to that of the adjacent open seas, it would inundate the whole plain of the Volga below Saratov, and would cover the surface of the steppes for an area of several hundred thousand square miles.

The Caspian Sea is divided into three distinct parts. The northern portion—the bottom of which continues the almost imperceptible

* L. L. and E. Vallée, *Du Barrage de Genève.*

LAKES WITH NO OUTLETS.

slope of the steppe—is a vast marsh, which is nowhere more than 48

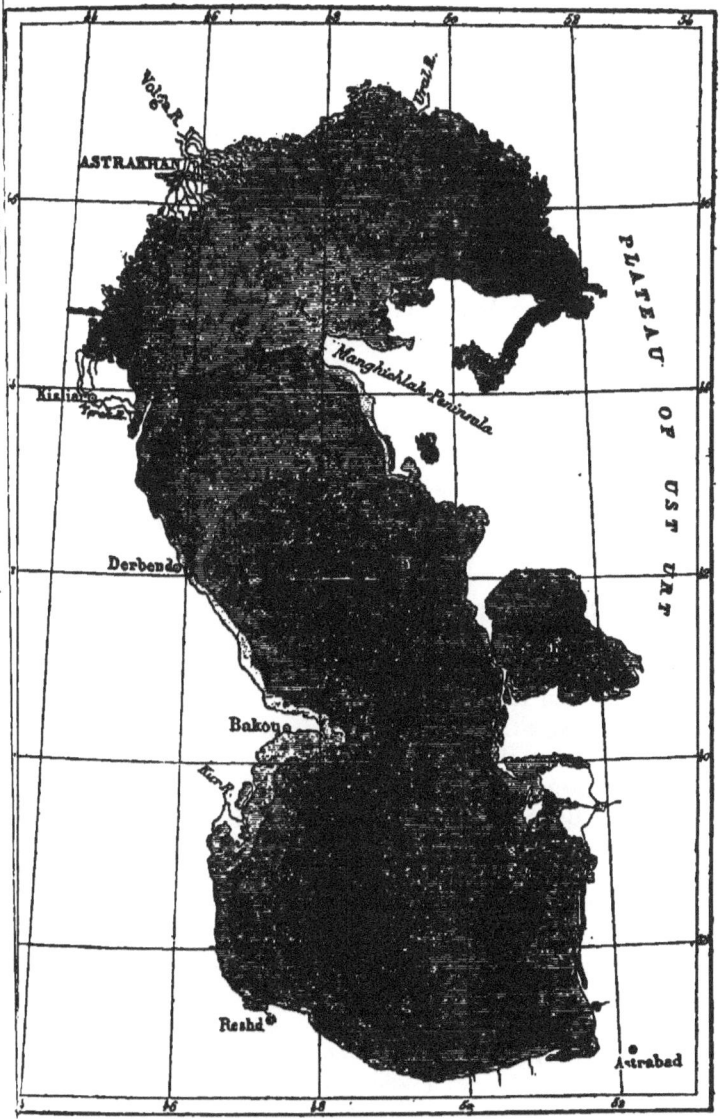

Fig. 164.—Caspian Sea.

50 feet deep, which, too, several rivers are constantly engaged in

filling up with their alluvium. In the southern part of this sea of steppes, lies the central basin of the Caspian, which is bounded on the south by the promontory of Apcheron, a prolongation of the Caucasus. The southern basin, mostly surrounded by high mountains, the escarpments of which extend beneath the water, is also the deepest. In some spots soundings have been made of 1,772 and 2,953 feet.

The saltness of the water is very unequal in different parts of the Caspian. On the north, the Terek, the Oural, and especially the Volga, bring down to the sea an enormous liquid mass, so much so that the total saltness is only from 15 to 16 ten-thousandths, and at many of the post stations, where there is a deficiency of springs, they drink the sea-water without either dislike or danger. The central and southern basins, on the contrary, contain water which is completely salt. It is proved by the experiments of M. de Baer that the average saltness is about nine-thousandths; this is a degree of saltness about one-third of that of the waters of the Atlantic Ocean.

Is the saturation of the Caspian diminishing during the course of ages, or is it, on the contrary, in process of increase? At first sight one is tempted to admit the fact of the increase of the saltness as an evident matter, since the soil of the surrounding steppes is gradually yielding up to the sea the salt which it contains. The rain and snow-water, when penetrating through the surface layer of sand, carry with them the saline particles, and concentrate them in the clayey subsoil. In every place where the ground is hollowed out by the ravines, so numerous on the steppes, the saline clay is washed away by the water, and carries the matter with which it is charged into the Caspian Sea, either directly or through the bed of a river. It appears, then, that the waters of the Caspian ought to present an increasingly large proportion of salt.

Yet M. de Baer, who has devoted more study to this inland sea than any other *savant*, does not believe in any increase in the degree of saltness in the waters of the Caspian; and, in his idea, if the proportion of salt be undergoing any change at all, it is diminution. In fact, in the plains abandoned by the sea banks of shells are here and there to be met with, which are identically similar to those of the shell-fish which now inhabit the Caspian. The dimensions of these shells, being always proportional to the quantity of salt contained in the water, ought to indicate the degree of saltness of the former sea, and thus give a point of comparison. Now the shells which are picked up in the vicinity of the Lake of Elton, more than 200 miles from the present sea-shore, are as large as those of the molluscs which now inhabit the open Caspian, at a point 60 miles from the mouth of the Volga. Near Astrakhan, where the sea-water, being mingled with that of the river, must be comparatively fresh, the shells left by the retirement of the sea indicate a degree of saltness equivalent to that of the water in the central basin. Moreover, in the environs of Baku, on the sides of the hills which overlook the water, amid the rocks, shells of molluscs are found which are much larger than those of the same species now swimming in the sea some yards lower down. This fact alone is sufficient to afford considerable probability to M. de Baer's hypothesis as to the decrease of saltness in the waters of the Caspian. The Black Sea, however, with which the great inland sea of Russia formerly communicated, contains proportionately twice the amount of salt.

How can this decrease be possible? How is it that the salt brought down by the rivers and rivulets is able to escape from the vast basin which has received it, and to separate itself from the sea-water with which it is mingled? Nothing can be more simple; by the regular movements of its waves, the Caspian—the same as all other seas—throws up banks of sand in front of the shallow bays of its shores, and thus converts gulfs and creeks into lagoons, into which the sea-water runs only through a narrow channel. Evaporation, which is very active in these regions bordering on the burning desert, is constantly tending to sink the level of these basins, whilst the sea-water charged with salt flows in without intermission to maintain the equilibrium; in this way are formed perfect magazines of salt, which are incessantly being increased. When, after heavy storms or a long continuation of dry weather, the channel which communicates between the sea and the lagoon ultimately becomes dried up, the sheet of water, now completely isolated, diminishes rapidly in area, or is even completely absorbed by the atmosphere; nothing being left of it but a layer of salt of variable thickness, which is formed at the expense of the sea. Thus it is that the lagoons recover from the Caspian the salt which the rivers of the steppes carry down to it. The only question is to know if equality exists between the in-comings and the out-goings, or if, in conformity to M. de Baer's theory, the loss of salt is more considerable than the gain. A long series of accurate observations could alone solve this problem.

The formation of these saline reservoirs may be studied all round the circumference of the Caspian Sea. A former bay, situated not far from Novo-Petrosk, on the eastern coast, is nowadays divided into a large number of basins, which present every degree of saline concentration. One basin still occasionally receives water from the sea, and has deposited on its banks only a very thin layer of salt. A second, likewise full of water, has its bottom hidden by a thick crust of rose-coloured crystals like a pavement of marble. A third exhibits a compact mass of salt, in which glitter here and there pools of water, situated more than a yard below the level of the sea. Lastly, another has lost by means of evaporation all the water which once filled it, and the strata of salt which carpet its bed are partly covered by sand. The same facts are found existing farther to the south, in the environs of the Bay of Alexander, and also quite at the extremity of the northern basin, at the point where the arm of the sea lies, which is known under the name of Karasu (black water). The saltness of the Karasu exceeds that of the Gulf of Suez, the saltest of all the seas which communicate with the ocean; in this part of the Caspian the proportion of marine salt rises to nearly 4 hundredths, and all the salts combined form 57 thousandths of the water; animal life, therefore, must there be almost if not entirely suppressed.

Among the thousands of bays and lagoons in which the salts of the Caspian are stored up none is more remarkable than the Karaboghaz, a kind of inland sea which probably connected the Hyrcanian Sea with the Lake of Aral, and into which perhaps the Oxus emptied itself when this river was still a tributary of the Caspian. This vast gulf communicates with the sea by a narrow mouth, which, in its most contracted part, is from 150 to 160 yards wide; the bar will not allow vessels to enter which draw more than 5 feet of water. A current coming from the open sea is always running through the strait with a speed of three knots an hour. The west winds accelerate it, and the winds which blow in an opposite direction retard it, but it never flows with less rapidity than a knot and a half. All the navigators of the Caspian, and all the Turkoman nomads who wander on its shores, have been struck with the unswerving, inexorable advance of this river of salt water, rolling over the shoals towards a gulf which even recently none had ever ventured to navigate. In the view of the natives this inland sea could be nothing but an abyss, a "black gulf," as is expressed by the name Karaboghaz, into which the waters of the Caspian dive down in order to flow through subterranean channels into the Persian Gulf or the Black Sea. It is perhaps to some vague rumours as to the existence of the Karaboghaz, that we must attribute the statements of Aristotle about the strange gulfs in the Euxine, in which the waters of the Hyrcanian

Sea bubble up after having flowed hundreds of miles through the realms of Pluto.

The existence of this current, which conveys the salt waves of the Caspian into the vast gulf of Karaboghaz, is nowadays most satisfactorily explained. In this basin, exposed as it is to every wind and the most intense summer heat, the evaporation is considerable; the water is, therefore, constantly diminishing, and the deficit can only be supplied by a continual fresh flow. Investigations, which can be readily made in the narrow and shallow channel of the Karaboghaz, have failed to ascertain the existence of a sub-marine countercurrent conveying back to the Caspian the salter water of the gulf. It is, therefore, very probable that it is the atmosphere only which absorbs the water brought by the Caspian current; but though deprived of its water by evaporation, the immense marsh retains the salt; the saline matter is more concentrated in it, and the water is more and more saturated with it every day. Already, it is said, no animal can live in it; seals which used to frequent it are no longer found there, and even its banks are devoid of all vegetation. Layers of salt begin to be deposited on the mud at the bottom, and the sounding line when scarcely out of the water is covered with saline crystals. M. de Baer has made the attempt to calculate approximately the quantity of salt of which the Caspian is every day deprived for the benefit of the "black gulf." Taking only the lowest estimates of the degree of saltness of the Caspian water, the width and depth of the channel, and the speed of the current, he has proved that the Karaboghaz receives daily 350,000 tons of salt, that is, as much as is consumed in the whole Russian empire during a period of six months. If, in consequence of violent storms, or the slow action of the sea, the bar should close up between the Caspian and the Karaboghaz, the latter would quickly diminish in extent; its banks would be converted into immense fields of salt, and the sheet of water which might remain in the centre of the basin would become only a marsh. Perhaps, indeed, it would disappear altogether, like that sea which used to lie between Lake Elton and the River Oural, the former existence of which is made known only by a depression in the ground of about 79 feet below the level of the Caspian, and 151 feet below that of the Black Sea. Like a tree letting fall its fruit upon the ground, the Russian Mediterranean detaches from its bosom the bays and gulfs on its coasts, and scatters them over the steppe in the form of lakes and pools.

The comparative observations which have been made as to the

average level of the Caspian Sea are not yet numerous enough to warrant us in admitting, with certain geographers, as a proved fact that there is a constant diminution of the water in this inland sea. We are likewise ignorant what foundation there may be for the opinion of some of the inhabitants of the coasts, mentioned by Humboldt in his *Asie Centrale*, according to which, the Caspian Sea experiences a succession of rises and falls every twenty-five to forty-five years. It seems, however, probable that the oscillations in its level are of no great importance, and that the quantity of water removed by evaporation is on the average exactly replaced by the liquid mass accruing from rivers and rain. An equilibrium is nearly established between the supply and the loss.

There is one point which is certain, that at the epoch when the Caspian Sea was separated from the Euxine, its level sank in a comparatively rapid way on account of the excess of the evaporation. A proof of this fact may be seen on the sides of the rocks which were once washed by the waves of the Caspian. At the height of 65 to 80 feet above the present level of the water, these former shoal-rocks have been furrowed out into tooth-shaped points and needles; lower down, on the contrary, the rocks bear no trace of the erosive action of the water, evidently because the level of the sea sank too rapidly to allow the waves sufficient time to attack successfully the cliff walls.

The innumerable indentations which cut into the shore between the mouths of the Kouma and those of the Oural, and principally south of the Volga, constitute another striking instance of the rapidity with which the level of the Caspian must have sunk after the sill of the Isthmus of Manytch emerged from the water. For a space of more than 248 miles the shore is gashed with very long and narrow channels, twelve, twenty, and even thirty miles in length, and throws out into the sea a multitude of peninsulas, which are prolonged for a great distance into the water by isles likewise disposed in parallel ranges and separated by long channels. These tongues of land form a kind of chain, which is interrupted here and there by the sea-water, and sink by successive falls from isle to islet, and from islet to marsh. The thousands of channels which separate these narrow embankments of land are an immense labyrinth, unexplored even by fishermen; it requires a map of the most detailed character to give any idea of this strange swarm of isles, islets, channels, and bays.

The *bugors* or chains of hillocks which run between the parallel bays, and farther inland, are connected with the level ground of the steppes, are in general very narrow, their length varying from a hundred yards to three or even four miles. They usually rise to the unpretending elevation of 26 to 30 feet; but some attain double this height. Seen from a balloon, the *ensemble* of the *bugors* would resemble a tract of marshy land turned up by a gigantic ploughshare. Immediately to the west of the Volga, the *limans*, or furrows which separate the *bugors*, are always changed into rivers. During the inundation of the river, the current pours into these channels the overflow of its waters charged with mud; then, after the flood is over, the sea again penetrates them, and there is thus produced in these channels a constant backward and forward motion between the sea and the Volga. Farther to the south, the narrow valleys of the *limans*, not being so often filled up by the flood-waters, do not in general present a continuous sheet of water, but only a chain of lakes, separated from each other by sandy isthmuses.

If we compare the whole of these ranges of hillocks to a border of fringe attached to the continent, we shall observe that these fringes spread out somewhat like a fan, on one side towards the north, on the other towards the south. They are all like the extremities of radii diverging from a common centre which would lie in the depression of Manytch, on the ledge which separates the slopes of the two seas. How can this arrangement be explained, except by the fact of the rapid sinking of the level of the Caspian waters hollowing out in the soft soil the narrow furrows which so astonish us? Thus, on the muddy banks of a reservoir when the sluice-gate is opened, small *limans* are formed, separated by *bugors* in miniature. A very remarkable fact, which again tends to confirm the result of M. de Baer's investigations, is that all the *bugors* of the Caspian shore are stratified, and the superimposed beds assume the form of concentric arches. The strata of the strongest clay are, as it were, the nuclei round which are deposited the earth that is more mingled with sand. This distribution of the strata is owing to the action of the currents of water which gave to the *bugors* their present appearance. It may, in fact, be readily understood that the strata of clay and sand, being undermined laterally by the water running down the channels, bent over on both sides towards the currents which washed their bases; hence arise these stratifications in the form of an arch.

CHAPTER LIX.

THE DEAD SEA.—THE SALT LAKES OF ASIA MINOR AND THE RUSSIAN STEPPES.—
THE GREAT SALT LAKE.—THE MELR'IR.

ALTHOUGH the Caspian is the largest of all the inland seas, the Asphaltite Lake is in some respects the most curious on account of its position in a deep fissure of the earth, many hundreds of feet below the level of the Mediterranean. Since Schubert discovered, at the beginning of the century, this single instance of a similar depression, it has been ascertained by exact measurements, that over an area of nearly 186 miles the whole valley ascending towards the base of Lebanon and lying parallel to the sea-shore of Palestine is lower than the ocean. Below the small Lake of Houleh the River Jordan, which traverses the valley, flows into a cavity which deepens by quickly recurring steps below the ideal sea-line. The level of Lake Asphaltites, in which the waters of the river are lost, is 1,280 feet lower than that of the sea. The greatest depth reached by the sounding-line exceeds 984 feet, and is, therefore, 2,270 feet below the level of the Mediterranean. Thus the depression into which the Jordan falls is deeper than the whole extent of the Adriatic and several other marine basins in communication with the ocean. Lake Asphaltites, however, does not merit the name of sea from its depth and its intense saltness only; it also possesses its principal current, flowing from north to south, and continuing the course of the Jordan, and its counter-currents flowing on both sides parallel to the shore.* The surface of the Dead Sea exceeds 460 square miles; but, as is proved by the horizontal layers of gypseous marl and the beds of salt deposited in stages on the slopes of the surrounding mountains, the level of the lake was formerly much higher than it is at present; † and probably the water filled all the elongated space comprehended between the foot of Lebanon and the entrance to Arabah at the north of the Red Sea. The drying up of the ancient sea of the Sahara, and the consequent diminution of rains and

* Vignes, *Voyage d'Exploration à la Mer Morte.*
† Lartet, *Bulletin de la Société Géologique de France,* vol. xxii.

The Earth. Vol. 1   THE BUGORS OF THE CASPIAN SEA   Pl. XIX.

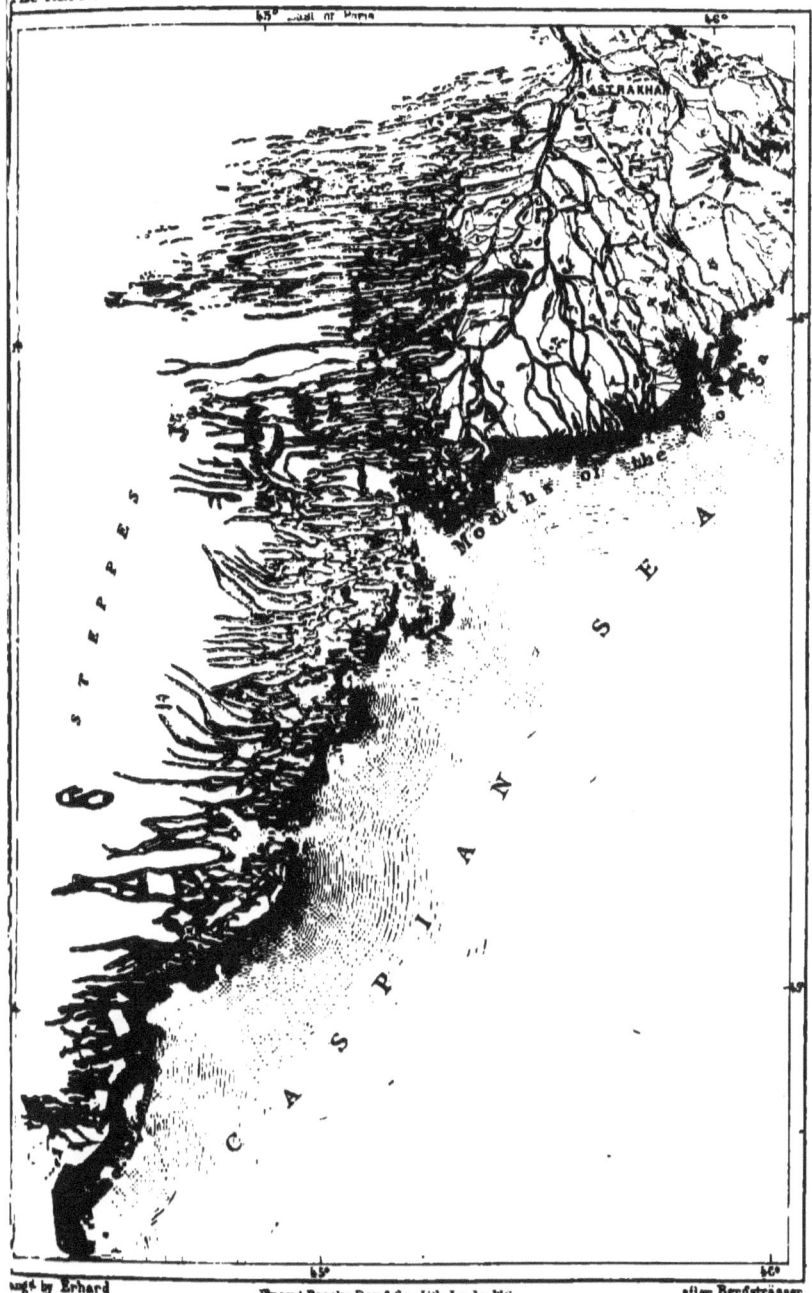

engd by Erhard   Vincent Brooks, Day & Son, Lith. London, W.C   after Bergsträsser.
CHAPMAN & HALL, LONDON.

increase of evaporation is, perhaps, the cause which gradually

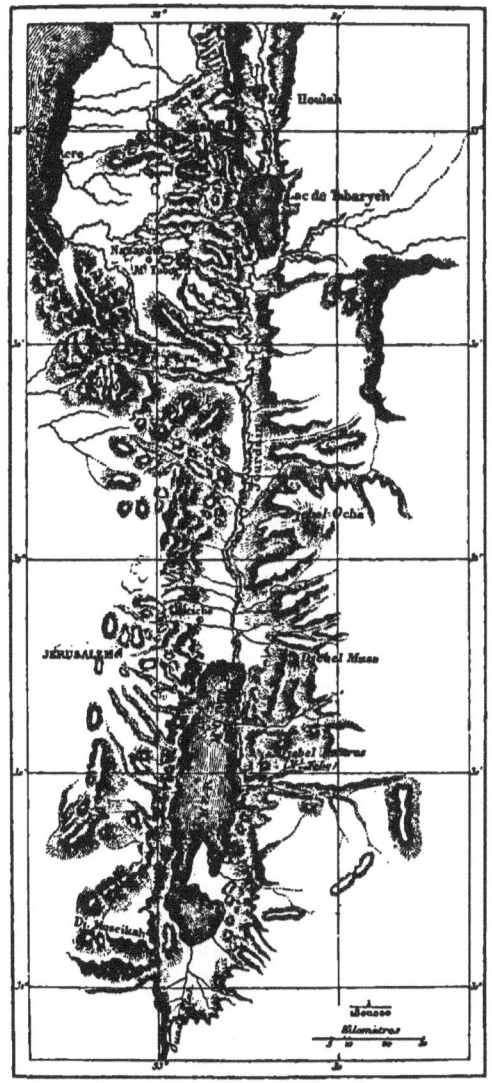

Fig. 165.—The Dead Sea and the Jordan.

lowered, century after century, this ancient sea, called so appropriately to this day, the "Dead Sea."

In fact, the landscape thoroughly presents an aspect of death. The rocks are bare; nearly every spot on the shores is sterile; the waters themselves nourish with difficulty but a few living beings of the lowest order; the fish, crustaceans, and insects brought down by the Jordan and the surrounding mountain-torrents, immediately die; aquatic plants are unable to grow. Off the mouth of a rivulet, the Wady-Mojeb, small fish are carried as far as a point in the lake where the density of the water is 1·115, but beyond this spot they inevitably perish. The only animals that have been found in the mud at the bottom are some species of foraminifera, classified by Ehrenberg, the micrographer. This almost complete absence of living organisms was formerly attributed to the enormous proportion of sea-salt which is found in the water of the Dead Sea. This pro-

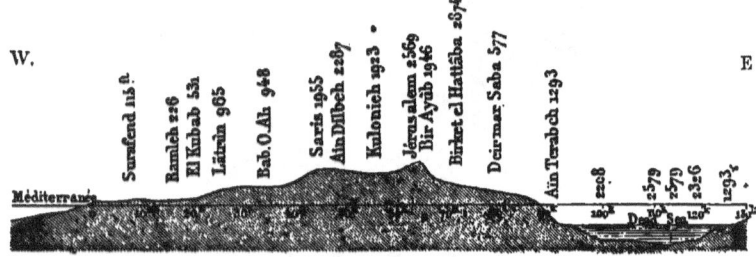

Fig. 166.—Section of Palestine from West to East.

portion is, in fact, very considerable, for it is twice as great as that in the Mediterranean; but there is, on the border of the lake, a small pond, the water of which is not less salt than that of the Dead Sea, and yet a large quantity of small fish live in it, which are immediately killed by an immersion for a few moments in Lake Asphaltites.* It is, then, probably chloride of magnesium and bromine which render the waters of this inland sea so completely destructive to animal life.

Chemical analyses have shown that the matters contained in the Dead Sea differ greatly from those of sea-water, not only in proportion but also in number. Thus, chloride of magnesium is found in this lake in much greater abundance than sea-salt itself; the proportion of bromine is also most extraordinary, as it varies from less than 15 to more than 67 thousandth parts of the water. On the other hand, iodine, a substance the presence of which is so characteristic of

* Lartet, *Bulletin de la Société Géologique de France*, vol. xxiii.

the waters of the ocean, appears to be completely wanting in the water of the Dead Sea; neither are phosphorus, silver, cæsium, rubidium, nor lithium found in its water. It must be concluded that the Lake Asphaltites has never, since its formation, constituted a part of the sea, and that it is not, as was long supposed, an ancient prolongation of the Red Sea, separated from the rest of this gulf by the upheaving of the entrance of Arabah. Ehrenberg, however, had already come to this conclusion by ascertaining that not one of the foraminifera found in the mud of the Dead Sea belongs to any species discovered in the Red Sea. M. Lartet thinks that the chemical substances contained in the water of Lake Asphaltites proceed from thermal springs spouting out on the shores, and especially from the bed of the lake. One fact which tends to confirm this hypothesis is, that the quantity of bromine increases with the depth of the water; it is at 984 feet from the surface that the largest proportion of this substance is found. The fragments of bitumen, which float on the surface of the water, and have gained for the basin the name of Lake Asphaltites, also proceed from the springs in its bed. As regards the saltness properly so called, it must have naturally increased by the gradual concentration of the water. When the latter extended over a larger surface of the country, the proportion of sea-salt dissolved in the liquid mass must have been much less. When the sea retires, it of course leaves a saline sediment; but this sediment is conveyed into it partly by streams and by the Jordan itself, which empties into the lake about 90 cubic yards of water a second (?), containing $6\frac{1}{2}$ bushels of sea-salt. At the present time the water of the Dead Sea, the specific gravity of which is in some places from 1·230 to 1·250, has almost reached the point of saturation; it deposits saline crystals at the bottom, and only dissolves to a very trifling extent the base of a cliff of rock-salt which overlooks the western coast.

All the great lakes of Asia Minor, situated at different altitudes between the two great depressions of the Dead Sea and the Caspian, are likewise rich in chemical substances. Lake Van, which covers an area of 1,544 square miles, especially contains sulphate of soda, which, during the dry season, when the waters are low, kills all the fish brought into it by the tributary streams. Lake Urimiyeh, still more extensive than Lake Van, is chiefly remarkable for the enormous quantity of sea-salt which it holds in a state of solution; in this respect, it is only equalled by the lagunes of the deserts and steppes, where the salt is so concentrated that it is deposited upon the bottom in thick beds. Of this kind is Lake Elton, to the north-west of the Caspian. The bed of this sheet of water consists of immense layers of salt, to which each day adds a fresh sediment. In winter, the rivulets which empty themselves into this small closed basin bring a certain quantity of brine, which afterwards evaporates during the heat, leaving upon the soil a bed of crystals several inches in thickness. In summer, when the shores are not covered with water, they appear to extend as far as one can see, like an immense field of snow. Every year more than 220,550,000 lbs. of salt are extracted from Lake Elton, and yet the saltness of its waters has not perceptibly diminished.

The Great Salt Lake of America is another Dead Sea, into which falls another Jordan, and, by a strange historical coincidence, the Mormons have established themselves upon the shores of this very lake; for this sect call themselves the successors of the Jews, and the chosen people of the New World. This inland sea, the real shape of which has only been known since 1850, by means of the explorations of Stansbury, is one of the most remarkable lacustrine sheets in the world; it is not less than 248 miles in circumference, but its depth is inconsiderable, and does not exceed 32 feet; the average is only about 6 feet.

The degree of saltness of the Great Lake varies according to the seasons and the duration of rains and drought; but it is always much more intense than that of the ocean. In fine weather, one might go to sleep on the waves of the lake without fear of being drowned; nevertheless, it is very difficult to swim in it on account of the effort which must necessarily be made in order to keep the legs below the surface. A single droplet falling into the eye causes the most cruel suffering, and the water when swallowed causes paroxysms of spasmodic coughing. Stansbury doubts if the most experienced swimmer could escape death if he were exposed far from the shore to the violence of the waves and wind. Although the Great Lake only contains a very small proportion of those salts so destructive to animal life which are found in the Dead Sea, yet neither fish nor molluscs exist in it; life is only represented by sea-weed of the *Nostoc* tribe, and by a small worm which here and there burrows in the sand of the shores. The trout which are carried into its waters by the Jordan perish immediately. Nevertheless, the surface of the lake affords hospitality to innumerable flocks of gulls, wild geese, swans, and ducks. Whole armies of young pelicans, tended by their old lame guardian-birds, contemplate the waves from the top of all the ledges of rock, whilst the parents go to fish in the Bear, Weber, and Jordan rivers, all of which abound in fish. Not a tree grows upon the shores of the lake nor in the adjacent plains; the only vegetation to be seen far and wide is tufts of *Artemisia*, and other plants which delight in a soil impregnated with saline substances.

The line of separation between the water and dry ground is generally undecided; it is impossible to tell where the shore begins or where the lake ends, as so much of the shore presents muddy banks upon which the water spreads in thin sheets and drifts about its flaky foam. Higher up the shore the mud dries in the sun and peels off in scales, which have the appearance of leather; sulphureous exhalations escape from cracks in the soil and diffuse an intolerable odour in the air. On the western side vast plains, nearly as level as the surface of the water, extend between the lake and a range of distant mountains. During some of the summer months these plains, which are crossed by rivulets loaded with chemical substances, are covered by an immense sheet of crystalline salt split up into innumerable furrows produced by contraction of the soil. Whenever rain falls, or even when the air is simply charged with moisture, the salt becomes deliquescent, and nothing is to be seen but an expanse of blackish clay, into which beasts of burden sink at every step they make.

Formerly the Great Salt Lake, like all other inland seas saturated with salt, spread over a much more considerable area. The parallel basins of the plateau of Utah, and the lateral valleys which run into them, were the gulfs, bays, and straits of the inland sea. At a great height above the present level of the lake, the former alluvial shores and cliffs surround the valleys with their concentric rings traced upon the sides of the mountains. Even in the plains some distance off, the surface of which exhibits a thin bed of vegetable earth, the sub-stratum is lake-clay saturated with sea-salt and the sulphates of lime and magnesia. Agriculture, therefore, is nearly impossible upon these ancient lacustrine beds. In the earliest years of colonisation the damp and virgin earth still produced crops to some extent, but subsequently the vegetable soil has lost its nutritive elements, and the clayey substratum coming in contact with the roots of the plants withers them up by means of its acrid properties.

Similar causes to those which led to the contraction of the Caspian and the Dead Sea have constantly tended to diminish the waters of Lake Utah, and also to saturate them with an enormous quantity of salt. The Great Basin is separated from the Pacific by high mountains of comparatively recent formation, which arrest the progress of the clouds, and prevent them from pouring upon the plateau the moisture derived from the sea. On the other hand, the evaporation is very considerable upon these high, rocky, and bare plains, and the winds which traverse them are but little impeded from carrying the vapours outside the basin of Utah. In consequence of this constant loss, the level of the Great Lake is become lower, the streams are dried up, the springs are exhausted, and the salt has concentrated more and more in the water. It is probable that, at the present time, an equilibrium is at length established between the annual fall of snow and rain and the mists which rise from the surface of the diminished lake. Since the establishment of the Mormons in the territory of Utah, the level of the lake has alternately risen and sunk.*

The various phenomena which take place in the waters of the

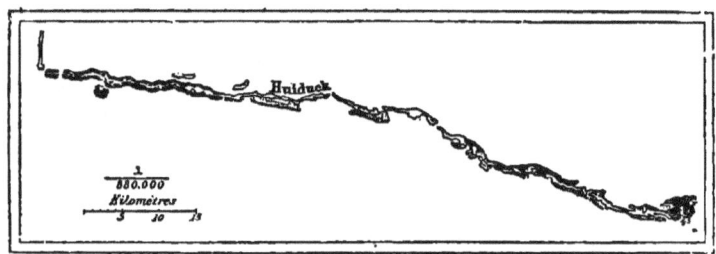

Fig. 167.—Lakes of Huidu.k.

Great Salt Lake, as well as in those of the Caspian, Lake Urimiyeh, and the Dead Sea, are also produced in a multitude of other lacustrine basins of less importance, with all the variations caused by the difference of climate, the nature of the soil, and the composition of the water. But as a great number of these lakes are situated in regions destitute of rain, and owe their saltness to the copious evaporation which has abstracted so large a part of their waters, they are, in consequence of this diminution, of very small area and have become converted into lagoons and marshes. Sometimes, indeed, they are reduced to surfaces which are alternately muddy and white with salt, when they have been either wetted by some casual rains or dried up by the solar rays. As a type of these salt-tracts may be mentioned the steppes of Huiduck in the Ponto-Caspian isthmus, and the Chott Melr'ir, a range of marshes which stretch from east to

* Fremont; Stansbury; Jules Remy; Engelmann.

west, over a length of more than 186 miles to the south of Djebel Aouress, formerly communicating with the gulf of Great Syrtes, by the Strait of Gabes, at present choked up by sand.* These marshes are separated one from another by isthmuses and islets of dry ground, and extend at unequal levels to 95, 118, 128, 213, 249, and even 279 feet below the sea.† During the rainy season they are sheets of shallow water, which spread far and wide into the plains; during the dry season they are fields of salt, over which the mirage throws its illusions.

* *Vide* below, the chapter on "Upheavals and Depressions."
† Dubocq, *Mémoire sur le Ziban et l'Oued-R'ir*.

CHAPTER LX.

MARSHES.—SWAMPS OF NORTH AMERICA.—PEAT-BOGS.—UNHEALTHINESS OF MARSHES.

MARSHES proper are shallow lakes, the waters of which are either stagnant or actuated by a very feeble current; they are, at least in the temperate zone, filled with rushes, reeds, and sedge, and are often bordered by trees, which love to plunge their roots into the muddy soil. In the tropical zone a large number of marshes are completely hidden by multitudes of plants or forests of trees, between the crowded trunks of which the black and stagnant water can only here and there be seen. Marshes of this kind are inaccessible to travellers, except where some deep channel, winding in the midst of the chaos of verdure, allows boats to attempt a passage between the water-lilies, or under some avenue of great trees with their long garlands of creepers waving in the shade. Whatever may be the climate, it would, however, be impossible to draw any distinction, even the most vague, between lakes and marshes, as the level of these sheets of water oscillates according to the seasons and years, and as the greater number of lakes, principally those of the plains, terminate in shallow bays which are perfect marshes. Some very important lacustral basins, among others Lake Tchad, one of the most considerable in all Africa, are entirely surrounded by swamps and inundated ground, which prohibit access to the lake itself and prevent its true dimensions from being known.

In like manner, a portion of the course of many rivers traverses low regions in which marshes are formed, either temporary or permanent, the uncertain limits of which change incessantly with the level of the current. The borders of great water-courses, when left in their natural state, are the localities in which these marshy reservoirs principally exist, which, in the absence of basins and artificial weirs, are of very great importance to the regulation of the fluviatile discharge. The most remarkable marshes of this kind are perhaps those crossed by the Paraguay and several of its tributaries; they consist of wet prairies and interminable sheets of water which

a from one horizon to the other. They have
Lakes Xarayes, Pantanal, &c. Farther south,
the Parana, the Maloya, the Batel, and the
he State of Corrientes from north-east to south-
t wide marshes, the water of which overflows
ss on the imperceptible slope of the territory.
of these marshes, the Laguna Bera, which
ŗ into the two great rivers of Parana and
:manent inundations, however, cannot fail to
ater, before the encroachments of cultivation.
s the low river-shores are frequently converted

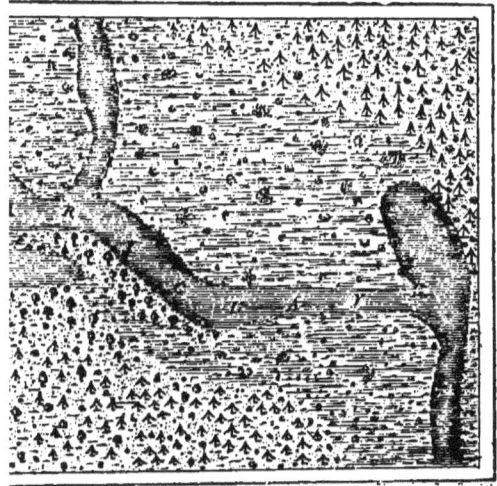

Fig. 168.—Salt Marshes of Paraguay.

ents of the sea-coasts when but slightly inclined
by marshes, which are generally separated from
ies of sand gradually thrown up by the waves.
it of which once formed a part of the sea and
t outline, the water presents the most varied
admixture. In some places, when evaporation
uid is much more salt than the sea itself; but
rsh, fed by fresh water which comes from the
rackish. The saltness of the water, however,
all parts of the marsh, according to the alter-

nations of flow and ebb and of rainy and dry weather. These half dried-up bays are rarely deep enough to allow of large vessels sailing in them, and their banks are generally overrun by the most luxuriant vegetation. The shore constantly keeps gaining upon them, and, thus tends to the increase of the mainland.

The coasts which surround the Caribbean Sea and the Gulf of

Fig. 169.—Marshes of Corrientes.

Mexico, and also the Atlantic shores of North America from the point of Florida to the mouth of the Chesapeake, are bordered by a very large number of marine marshes, forming a continued series over hundreds and thousands of miles in length. In this immense series of coast-marshes all kinds of vegetation seem to flourish, and

etter of the mud and water, and to convert them
) the south, upon the shores of Columbia and
) mangroves and other trees of like species plunge
of their aerial roots deep into the mud, crossing
n arch-like form, and retaining all the *débris* of
under the inextricable network of their natural
shores of the Gulf of Mexico, in Louisiana,
a, are bordered by cypress swamps, or forests of
'isticha) ; these strange trees, the roots of which,
w out above the layer of water which covers the
ttle cones, the business of which is to absorb the
f acres nearly all the marshy belt along the sea-
t an immense cypress swamp, with trees bare of
g in the wind their long hair-like fibres of moss.
trees and muddy soil give place to bays, lakes, or
ormed by a carpet of grass lying upon a soil of
)on the hidden water. In Brazil these buoyant
re frequently met with, and the significant name
en given to them : in Ireland these are called
The least movement of the traveller who ventures
le soil tremble to some yards' distance.
lorida, in the Carolinas and Virginia, the belt of
ntinue ; but in consequence of the change of
tion, the quaking-meadows are gradually con-
osses. Evaporation being much less active in
a in those situated farther to the south, and
g much less prolonged, the water arising from
n remains—as if in the pores of an immense
) interstices of the entangled mass of mosses,
e, and other aquatic plants. The whole marsh
:entre, because the droplets, divided by innumer-
spread out laterally, and are drawn by capillary
resh beds of plants which are formed above the
urface of the marsh is incessantly renewed by a
etation, while below, the dead plants, deprived of
y in the moisture which surrounds them : these
t which form upon the ground just as the layers
in previous geological epochs.
i side, the first great peat-bog of a well-defined
)ismal Swamp," which extends along the frontiers
nd Virginia. This spongy mass of vegetation rises

10 feet above the surrounding land. In the centre, and, so to speak, upon the summit of the marsh, lies Lake Drummond, the clear water of which is coloured reddish-brown by the tannin of the plants. A canal, which crosses the Dismal Swamp to connect it with the adjacent streams, is obliged to make its way along the marsh by means of locks. To the north of Virginia peat-bogs proper become more and more numerous; and in Canada, Labrador, &c., they cover vast expanses of country. All the interior of the island of Newfoundland, inside the enclosure formed by the forests on the shore, is nothing but a labyrinth—a great part of which is still unknown—of lakes and peat-bogs; even on the sides of the hills there are marshes on so steep an incline that the water from them would disappear and run off in a stream if it was not stopped by the thick carpet of plants which it saturates. Many a large peat-bog which may be crossed dry-shod contains more water than many lakes filling a hollow of the valley with deep water.

Opposite Newfoundland, on the other side of the Atlantic, Ireland is hardly less remarkable for the enormous development of its peat-mosses or bogs. These tracts of saturated vegetation, in which *Sphagnum palustre* predominates, comprehend nearly two and a half millions of acres—the seventh part of the whole island. The inhabitants continue to extract from them, every year, immense quantities of fuel. The spaces left by the spade in the vegetable mass are gradually filled up again by new layers. After a certain number of years, which vary according to the abundance of rain, the depth of the bed of water, the force of vegetation, and the slope of the soil, the turf " quarry " is formed anew. In Ireland it generally takes about ten years to entirely fill up again the trenches, measuring from nine to thirteen feet in depth, which are made in the bogs on the plains, when a fresh digging of turf may be commenced. In Holland, crops of this fuel may be gathered, on an average, every thirty years. In other peat-moss districts the period of regeneration lasts forty, fifty, and even a hundred years. In France, on the borders of the Seugne (Charente-Inférieure), it has been ascertained that ditches 5 feet deep and nearly 7 feet wide are completely obstructed by vegetation after the lapse of twenty years. As for the beds of peat which carpet the sides of mountains, they take centuries to form afresh.

As everything in nature is continually changing and modifying, peaty marshes, like lakes, are all either in a period of increase or a period of decay—some form while others disappear. Independently of the action exercised by the labour of man, the vegetation of peat-bogs may cease to be produced in any basin, either because the water flows away naturally through some wide outlet after the heavy rains, or because some river, by changing its course, has exhausted or immoderately swollen the mass of water necessary to the nourishment of the peat; or, again, because the rain, becoming either more rare or more frequent, has dried up the basin or converted it into an inundated marsh; lastly, the sinking or upheaving of the soil may also, according to the various conditions of the relief of the country, be the cause of the disappearance of the Flora of the peat-bogs. The same causes, acting in contrary directions, give rise to and increase these enormous masses of plants swollen with water. In Ireland, the Low Countries, the north of Germany and Russia, heaps of trunks of former forest-trees—oaks, beech, alder, and other trees —are frequently discovered, which by their decay have made way for the peat-mosses. The *Sphagnum*, too, often takes possession of ground of which man had previously made himself master, and in many places roads, remains of buildings, and other vestiges of human labour are found below the modern bed of vegetation by which they are now covered. Certain peat-bogs in Denmark and Sweden may be considered, on account of the curiosities which have been found in them, as perfect natural museums, in which the relics of the civilisation of ancient nations have been preserved for the *savants* of our own day.

The air above the peat-mosses of Ireland and other countries in the world is not often unhealthy, either because the heat is not sufficient to develope miasma, or else because the vegetation, by absorbing the water into its spongy mass, impedes the corruption of the liquid, and produces a considerable quantity of oxygen. Farther south, the peat-mosses, which are intermixed with pools of stagnant water, and especially marshes properly so-called, generate an impure air, which spreads fever and death over the surrounding country. Unless marshes are surrounded with dense forests, which arrest the dispersion of the gases, the latter exercise a most injurious influence on the general salubrity of the district; for during dry weather, a vast area of the bed of the marshes becomes exposed, and the heaps of organic *débris* lying on the bottom decompose in the heat and infect the whole atmosphere. The average of life is much shorter in all marshy countries than in the adjacent regions which are invigorated by running water. In Brescia, Poland, in the marshes of Tuscany, and in the Roman plains, the wan and livid complexion of the inhabitants, their hollow eyes, and their feverish skin announce at first

sight the vicinity of some centre of infection. There are som
marshes in the torrid zone where the decomposition of organi
remains goes on with a much greater rapidity than in temperat
climates; no one can venture on the edges of these districts withou
peril to his life. As Frœbel ascertained in his journey across Centra
America, the miasma is occasionally produced in such abundanc
that not only can it be smelt, but a distinct impression of it is lef
upon the palate. One of the most important works of civilisation i
to deal with these unwholesome regions, which are still, as it were
undecided between land and water, and to render them fit for culti
vation and to be the abode of man.

PART IV.
SUBTERRANEAN FORCES.

CHAPTER LXI.

ERUPTION OF ETNA IN THE YEAR 1865.—MUTUAL DEPENDENCE OF ALL TERRESTRIAL PHENOMENA.

THE Greek mythology, harmonising in this respect with the ideas of most nations which were acquainted with volcanoes, attributed to these mountains an origin altogether independent of the forces which are in action on the surface of the ground. According to the views of the Hellenes, water and fire were two distinct elements, and each had its separate domain, its genii, and its gods. Neptune reigned over the sea; it was he that unchained the storms and caused the waves to swell. The tritons followed in his train; the nymphs, sirens, and marine monsters obeyed his orders, and in the mountain valleys, the solitary naïads poured out to his honour the murmuring water from their urns. In the dark depth of unknown abysses was enthroned the gloomy Pluto; at his side Vulcan, surrounded by Cyclops, forged thunderbolts at his resounding anvil, and from their furnaces escaped all the flames and molten matter the appearance of which so appalled mankind. Between the gods of water and of fire there was nothing in common, except that both were the sons of Chronos, that is, of Time, which modifies everything, which destroys and renews, and, by its incessant work of destruction, makes ready a place for the innumerable germs of vitality which crowd on the threshold of life.

Even in our days, the common opinion is not much at variance with these mythological ideas, and volcanic phenomena are looked upon as events of a character altogether different from other facts of terrestrial vitality. The latter, the sudden changes of which are visible and easily to be observed, are justly considered to be owing principally to the position of the earth in respect to the sun and the alternations of light and darkness, heat and cold, dryness and moisture which necessarily result. As regards volcanoes, on the contrary, an order of entirely distinct facts is imagined, caused by the gradual cooling of the planet or the unequal tides of an ocean of lava and fire. Certainly, the eruptions of ashes and incandescent matter have not revealed the mystery of their formation, and in this respect numerous problems still remain unsolved by scientific men. Nevertheless, the facts already known warrant us in asserting that volcanic crises are connected, like all other planetary phenomena, with the general causes which determine the continual changes of continents and seas, the erosion of mountains, the courses of rivers, winds, and storms, the movements of the ocean, and all the innumerable modifications which are taking place on the globe. If, some day, we are to succeed in pointing out exactly and plainly how volcanoes likewise obey, either partially or completely, the system of laws which govern the exterior of the globe, the first and most important requisite is to observe with the greatest care all the incidents of volcanic origin. When all the premonitory signs and all the products of eruptions shall have been perfectly ascertained and duly classified, then the glance of science will be on the point of penetrating into, and duly reading, the secrets of the subterranean abysses where these marvellous convulsions are being prepared.

The last great eruption of Etna, that central pyramid of the Mediterranean, which the ancients named the "Umbilicus of the world," is one of the most magnificent examples which can be brought forward of volcanic phenomena; and as it has, moreover, been studied most precisely and completely, it well deserves to be described in some detail.

The explosion had been heralded for some long time by precursory signs. In the month of July, 1863, after a series of convulsive movements of the soil, the loftiest cone of the volcano opened on the side which faces the south. The incandescent matter descended slowly over the plateau on which stands the "Maison des Anglais;" and this building itself was demolished by the lumps of lava which were hurled from the mouth of the crater. In some places heaps of ashes several yards thick covered the slopes of the volcano. After this first explosion, the mountain never became completely calm; numerous fissures, which opened on the outer slopes of the crater, continued to smoke, and the hot vapour never ceased to jet out from the summit in thick eddies. Often, indeed, during the night, the reflection of the lava boiling up in the central cavity lighted up the atmosphere with a fiery red. The liquid, being unable to rise to the mouth of the crater, pressed against the external walls of the volcano, and sought to find an issue through the weakest point of the crust by melting gradually the rocks that opposed its passage. Finally, in the night of the 30—31 of January, 1865, the wall of the crater yielded to the pressure of the lava; some subterranean roaring was heard; slight agitations affected the whole of the eastern part of Sicily, and the ground was rent open for the length of a mile and a half to the north of Monte-Frumento, one of the secondary cones which rise on the slope of Etna. Through this fissure, which opened on a gently-inclined plateau, the pent-up lava violently broke through to the surface.

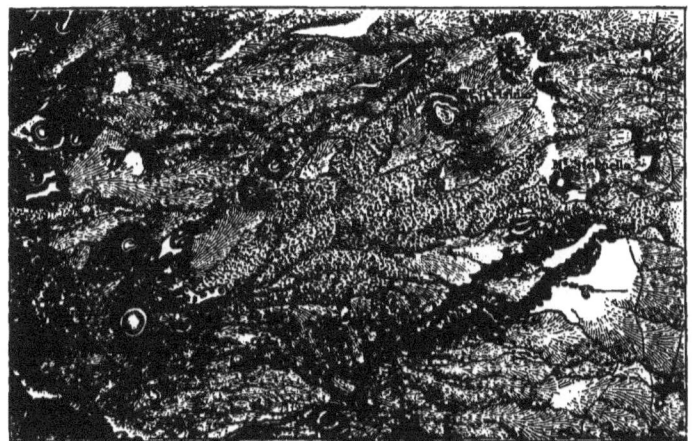

Fig. 170.—*Coulée* of Monte-Frumento.

The fissure which opened on the side of the mountain, and could be easily followed by the eye to a point about two-thirds of the height of Monte-Frumento, in the direction of the terminal crater of Etna, seems to have vomited out lava but for a very few hours. Being soon obstructed by the snow and the *débris* of the adjacent slopes, it ceased to retain its communication with the interior of the mountain, and now resembled a kind of furrow, as if hollowed out by the rain-water on the side of the cone. On the 31st of January all the volcanic activity of the crevice was concentrated on the gently inclined plateau which extends at the base of Monte-Frumento, in the midst of which several new hillocks made their appearance. On the lower prolongation of the line of fracture, all the phenomena of the eruption properly so-called were distributed in a perfectly regular way. Six principal cones of ejection were raised above the crevice, and gradually increased in size, owing to the *débris* which they threw out of their craters; these, gradually mingling their intervening slopes, and blending them one with another, absorbed in succession other smaller cones which had been formed by their sides, thus reaching a height of nearly 300 feet. Soon after the commencement of the eruption the two upper craters, standing close together on an isolated cone, vomited nothing but lumps of stone and ashes, while jets of still liquid lava were emitted by the lower craters, which were arranged in a semicircle round a sort of funnel-shaped cavity. In consequence of the specific gravities of the substances evacuated, a regular division of labour took place between the various points of the crevice. The projectiles which had solidified the triturated *débris*, and the more or less porous fragments which floated on the top of the lava, made their escape by the higher orifices; but the liquid mass, being heavier and more compact, could only burst forth from the ground by the mouths opening at a less elevation.

Two months after the commencement of the eruption, the cone which was the nearest to Frumento ceased to send out either scoriæ or ashes. The pipe of the crater was filled up with *débris*, and the internal activity was revealed by vapours either of a sulphurous character or charged with hydrochloric acid. These rose like smoke from the slope of the hillock. The second cone, situated on a lower part of the fissure, remained in direct communication with the central flow of lava; but it was not in a constant state of eruption, and rested after each effort as if to take breath. A crash like that of thunder was the forerunner of the explosion; clouds of vapour, rolling in thick folds, grey with ashes, and furrowed with stones, darted out from the mouth of the volcano, darkening the atmosphere, and throwing their projectiles over a radius of several hundreds of yards round the hillock. Then, after having discharged their burdens of *débris*, the dark clouds, giving way before the pressure of the winds, mingled far and wide with the mists on the horizon. The lower cones, which rose immediately over the lava-source, continued to rumble and to discharge molten matter outside their cavities. The vapour which escaped from the seething well of lava crowded in dark contortions round the orifice of the craters. Some of it was red or yellow, owing to the reflection of the red-hot matter, and some was variously shaded by the trains of *débris* ejected with it; but it was impossible to follow them with the eye, so rapid was their flight. An unintelligible tumult of harsh sounds simultaneously burst forth; they were like the noises of saws, whistles, and of hammers falling on an anvil. Sometimes one might have fancied it like the roaring of the waves breaking upon the rocks during a storm, if the sudden explosions had not added their thunder to all this uproar of the elements. One felt dismayed, as if before some living being, at the sight of these groups of hillocks, roaring and smoking, and increasing in size every hour, by the *débris* which they vomited forth from the interior of the earth. The volcano, however, then commenced to rest; the erupted matter did not rise much beyond 100 yards above the craters, whilst, according to the statement of M. Fouqué, at the commencement of the eruption it had been thrown to a height of 1,850 to 1,950 yards.

During the six first days the quantity of lava which issued from the fissure of Monte-Frumento was estimated at 117 cubic yards a second, equivalent to a volume twice the bulk of the Seine at low-water time. In the vicinity of the outlets the speed of the current was not less than 20 feet a minute; but lower down, the stream, spreading over a wider surface, and throwing out several branches into the side valleys, gradually lost its initial speed, and the fringes of scoriæ, which were pushed on before the incandescent matter, advanced on the average, according to the slope of the ground, not more than 1½ to 6 feet a minute.* On the 2nd of February the principal current, the breadth of which varied from 300 to 550 yards, with an average thickness of 49 feet, reached the upper ledge of the escarpment of Colla-Vecchia, or Colla-Grande, three miles from the fissure of eruption, and plunged like a cataract into the gorge below. It was a magnificent spectacle, especially during the night, to see this sheet of molten matter, dazzling red like liquid iron, making its way, in a thin layer, from the heaps of brown scoriæ which had gradually accumulated up above; then, carrying with it the more solid lumps, which dashed one against the other with a metallic noise, it fell over into the ravine, only to rebound in stars of fire. But this splendid spectacle lasted only for a few days; the fiery fall, by losing in height, diminished gradually in beauty. In front of the cataract, and under the jet itself, there was formed an incessantly increasing slope of lava, which ultimately filled up the ravine, and, indeed, prolonged the slope of the valley above. From the reservoir, which was more than 160 feet deep, the stream continued to flow to the east towards Mascali, filling up to the brink the winding gorge of a dried-up rivulet.

* These figures, borrowed from the account of the Professor Orazio Silvestri, are the results of measurements made by Viotti, the engineer.

By the middle of the month of February, the fiery stream, already more than six miles long, made but very slow progress, and the still liquid lava found it difficult to clear an outlet through the crust of stones cooled by their contact with the atmosphere; when, all of a sudden, a breaking out took place at the side of the stream, at a point some distance up, not far from the source. Then a fresh branch of the burning river, flowing towards the plains of Linguagrossa, swallowed up thousands of trees which had been felled by the woodman. This second inundation of lava did not, however, last long. The villages and towns situated at the base of the mountain were no longer directly menaced; but the disasters caused by the eruption were, notwithstanding, very considerable. A number of farm-houses were swept away; vast tracts of pasturage and cultivated ground were covered by slowly hardening rock, and—a misfortune which was all the worse on account of the almost general deforesting of Sicily—a wide band of forest, comprising, according to the various estimates that were made, from 100,000 to 130,000 trees—oaks, pines, chestnuts, or birches—was completely destroyed. When seen from the lower part of the mountain, all these burning trunks borne along upon the lava, as if upon a river of fire, singularly contributed to the beauty of the spectacle. As is always the case in the events of this world, the misfortune of some proved to be a source of gratification to others. During the earliest period of the eruption, whilst the villagers of Etna looked at it with stupor, and were bitterly lamenting over the destruction of their forests, hundreds of curious spectators, brought daily by the steamboats, from Catania and Messina, came to enjoy at their ease the contemplation of all the splendid horrors of the conflagration.

The aspect of the current of lava, as it appeared covered with its envelope of scoriæ, was scarcely less remarkable than the sight of the matter in motion. The black or reddish aspect of the *cheire* was all roughened with sharp-edged projections, which resembled steps, pyramids, or twisted columns, on which it was a difficult matter to venture, except at the risk of tearing the feet and hands. Some months after the commencement of the eruption, the onward motion of the interior of the molten stone, which, by breaking the outer crust in every direction, had ultimately given it this rugged outline, was still visibly taking place. Here and there cracks in the rock allowed a view, as if through an air-hole, of the red and liquid lava swelling up as it flowed gently along like some viscous matter. A metallic clinking sound was incessantly heard, proceeding from the fall of the scoriæ, which were breaking under the pressure of the liquid matter. Sometimes, on the hardening current of lava, a kind of blister gradually rose, which either opened gently, or bursting with a crash gave vent to the molten mass which formed it. *Fumerolles*, composed of various gases, according to the degree of heat of the lava which gave rise to them, jetted out from all the issues. Even on the banks of the river of stone the soil was in many places all burning and pierced with crevices, through which escaped a hot air, thoroughly charged with the smell of burnt roots.

On the slopes of the Frumento, quite close to the upper part of the fissure, at a spot where the liquid mass had flowed like a torrent, M. Fouqué noticed a remarkable phenomenon ;—sheaths of solidified lava were surrounding the trunks of pines, and thus showing the height to which the current of molten stone had reached. In like manner, the streams of obsidian which flow rapidly from the basin of Kilauea, in the isle of Hawaii, leave behind them on the branches of the trees numerous stalactites, like the icicles which are formed by melting snow which has again frozen. Below the escarpments of the Frumento, the torrent, which was there retarded in its progress, had not contented itself with bathing for a moment the trunks of the forest trees, but had laid them low. Great trunks of trees, broken down by the lava, lay stretched in disorder on the uneven bed of the stream, and, although they were only separated from the molten matter by a crust a few inches thick, numbers of them were still clothed with their bark ; several had even preserved their branches. At the edge of the *cheire*, some pine trees, which had perhaps been preserved from the fire by their moisture being converted by the heat into a kind of coating of steam, were surrounded by a wall of heaped-up lava, and their foliage still continued green ; it could not yet be ascertained if the sources of the sap had perished in their roots.

In some places, rows of firs very close together were sufficient to change the direction of the flow, and to cause a lateral deviation. Not far from the crater of eruption, on the western bank of the great *cheire*, a trunk of a tree was noticed which by itself had been able to keep back a branch of the stream, and to prevent it from filling up a glen which opened immediately below. This tree, being thrown down by the weight of the scoriæ, had fallen so as to bar up a slight depression in the ground which presented a natural bed to the molten matter. The latter had bent and cracked the trunk, but had failed in breaking it, and the stony torrent had remained suspended, so to speak, above the beautiful wooded slopes which it threatened to destroy completely.

Round the very mouths of the volcano, a vast glade was formed in the forest; the ground was covered everywhere with ashes which the wind had blown up into hillocks, like the dunes on the sea-coast; all the trees had been broken down by the volcanic projectiles, and burned by the scoriæ and small stones. The nearest trees that were met with, at unequal distances from the mouths of eruption, had had their branches torn off by the falling lumps of stone, or were buried in ashes up to their terminal crown. A spectator might have walked among a number of yellow branches which were once the tops of lofty pines. Thus, on the plateau of Frumento and the lower slopes, everything was changed both in form and aspect; we might justly say, that by the effects of the erupted matter, the outline of the sides of Etna itself had been perceptibly modified.

And yet this last eruption, one of the most important in our epoch, is but an insignificant episode in the history of the mountain; it was but a mere pulsation of Etna. During the last twenty centuries only, more than seventy-five eruptions have taken place, and in some of them the flows of lava have been more than twelve miles in length, and have covered areas of more than forty square miles, which were once in a perfect state of cultivation, and dotted over with towns and villages. In former ages, thousands of other lava-flows and cones of ashes have gradually raised and lengthened the slopes of the mountain. The mass of Mount Etna, the total bulk of which is three or four thousand times greater than the most considerable of the rivers of stone vomited from its bosom, is, in fact, from its summit to its base, down even to the lowest sub-marine depths, nothing but the product of successive eruptions throwing out the molten matter of the interior. The volcano itself has slowly raised the walls of its crater, and then extended its long slopes down to the waters of the Ionian Sea. By its fresh beds of lava and scoriæ incessantly renewed one upon the other, it has ultimately reared its summit into the regions of snow, and has become, as Pindar called it, the great "pillar of heaven."

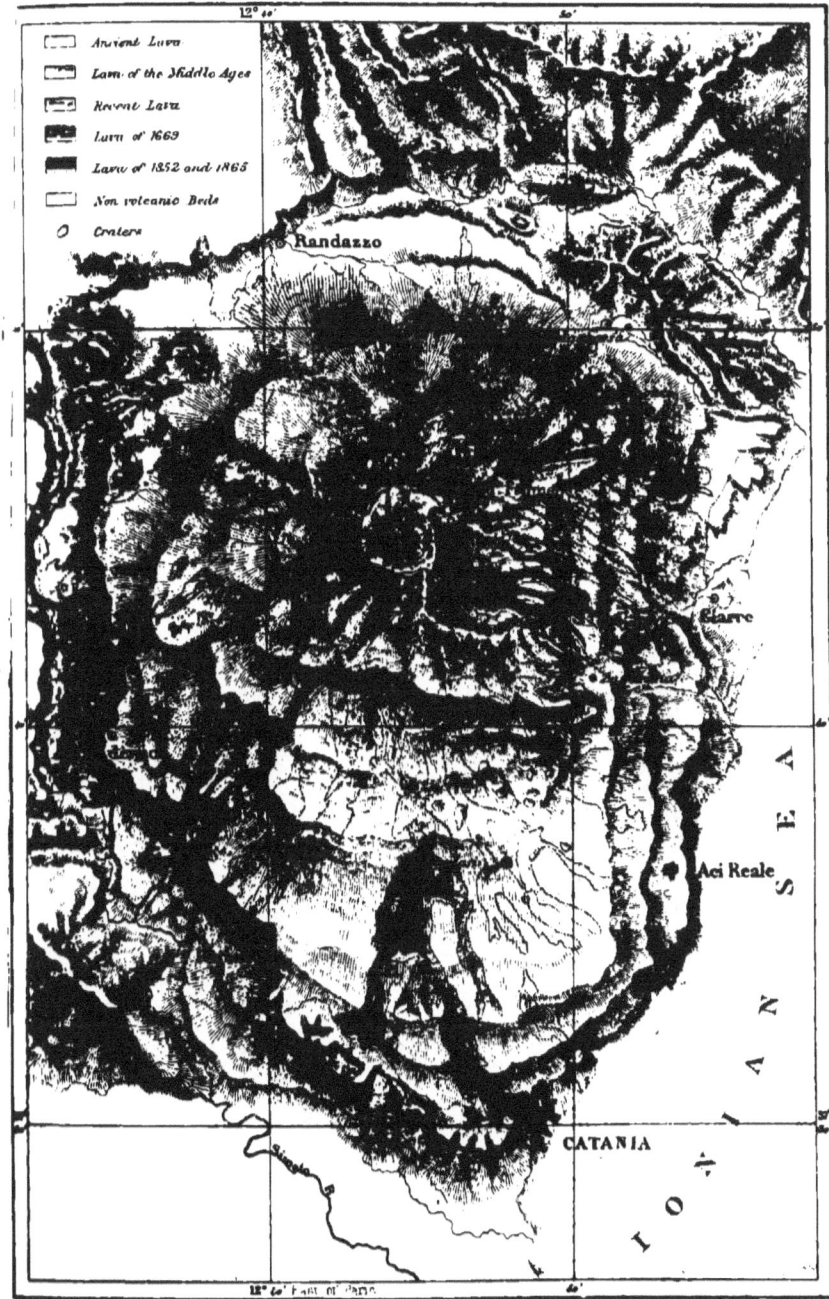

ERUPTIONS OF ETNA

Drawn by A. Vuillemin after Sartorius de Waltershausen. Engd by Erhard.

CHAPTER LXII.

SEA-COAST LINE OF VOLCANOES.—THE PACIFIC "CIRCLE OF FIRE."—VOLCANOES OF THE INDIAN OCEAN; OF THE ATLANTIC; OF THE MEDITERRANEAN; OF THE CASPIAN; OF CENTRAL ASIA.

THE earth being generally looked upon as immobility itself, it is a very strange thing to see it open to shoot out into the air torrents of gas, and shedding forth like a river the molten rocks of its interior. From what invisible source do all these fluid matters proceed which spread out in sheets over vast regions? Whence come those enormous bodies of steam, extensive enough to gather immediately in clouds round the loftiest summits, and sometimes indeed to fall in actual rain-showers? Science, as we have already said, has not completely answered these questions, the positive solution of which would be so highly important for our knowledge of the globe on which we live.

According to an ancient popular belief, Etna merely vomits forth, in the shape of vapour, the water which the sea has poured into the gulf of Charybdis. This legend, although clothed in a poetic garb, has in fact become the hypothesis which is thought beyond dispute by those *savants* who look upon volcanic eruptions as being a series of phenomena caused chiefly by water converted into steam.

The remarkable fact that all volcanoes are arranged in a kind of line along the coasts of the sea, or of inland lacustrine basins, is one of the great points which testify in favour of this opinion as to the infiltration of water, and give to it a high degree of probability. The Pacific, which is the principal reservoir of the water of our earth, is circled round by a series of volcanic mountains, some ranged in chains, and others very distant from one another, but still maintaining an evident mutual connection, constituting a "circle of fire," the total development of which is about 22,000 miles in length. This ring of volcanoes does not exactly coincide with the semicircle formed by the coasts of Australia, the Sunda Islands, the Asiatic continent, and the western coasts of the New World. Like a crater described within some ancient and more extensive outlet of eruption,

K K

the great circle of igneous mountains extends its immense curve in a westward direction across the waves of the Pacific, from New Zealand to the peninsula of Alaska; on the east, it is based on the coast of America, rising in the south so as to form some of the loftiest summits of the Andes.

The still smoking volcanoes of New Zealand, Tongariro and the cone of Whakari, on White Island, are, in the midst of the southern waters of the Pacific properly so called, the first evidence of volcanic activity. On the north, a considerable space extends in which no volcanoes have yet been observed. The group of the Feejee Islands, at which the volcanic ring recommences, presents a large number of former craters which still manifest the internal action of the lava by the abundance of thermal springs. At this point, a branch crossing the South Sea in an oblique direction from the basaltic islands of Juan Fernandez as far as the active volcanoes of the Friendly group, unites itself with the principal chain which passes round, in a northeast direction, the coasts of Australia and New Guinea. The volcanoes of Abrim and Tanna in the New Hebrides, Tinahoro in the archipelago of Santa Cruz, and Semoya in the Salomon Isles, succeeding one after the other, connect the knot of the Feejees to the region of the Sunda Islands, where the earth is so often agitated by violent shocks. This region may be considered as the great focus of the lava streams of our planet. On the kind of broken isthmus which connects Australia with the Indo-Chinese peninsula, and separates the Pacific Ocean from the great Indian seas, one hundred and nine volcanoes are vomiting out lava, ashes, or mud in full activity, destroying from time to time the towns and the villages which lie upon their slopes; sometimes, in their more terrible explosions, they ultimately explode bodily, covering with the dust of their fragments areas of several thousands of miles in extent. From Papua to Sumatra, every large island, including probably the almost unknown tracts of Borneo, is pierced with one or more volcanic outlets. There are Timor, Flores, Sumbawa, Lombok, Bali, and Java, which last has no less than forty-five volcanoes, twenty-eight of which are in a state of activity, and lastly, the beautiful island of Sumatra. Then, to the east of Borneo,—Ceram, Amboyna, Gilolo, the volcano of Ternata, sung by Camoëns, Celebes, Mindanao, Mindoro, and Luzon; these form across the sea, as it were, two great tracks of fire.

Northward of Luzon, the volcanic ring curves gradually so as to follow a direction parallel to the coast of Asia. Formosa, the Liou-Kieou archipelago, and other groups of islands stand in a line over

DISTRIBUTION OF VOLCANOES.

the submarine volcanic fissure; farther on, there are the numerous volcanoes of Japan, one of which, Fusiyama, with a cone of admirable regularity, is looked upon by the inhabitants of Niphon as a sacred mountain, from which the gods come down. The elongated archipelago of the Kuriles, comprising about a dozen volcanic orifices, unites Japan to the peninsula of Kamtchatka, in which no less than fourteen volcanoes are reckoned as being in full activity. To the east of this peninsula, the range of craters suddenly changes its direction, and describes a graceful semicircle across the Pacific, from Behring Island to the point of Alaska. Thirty-four smoking cones stand on this great transversal dike, extending from continent to continent. Ounimak, which rises on the extremity of the peninsula of Alaska, the peak of which is 7,939 feet in height, serves as the

Fig. 171.—Curve of Volcanic Islands.

western limit of the New World, and is also pierced by a crater in a state of full activity.

Eastward of the peninsula, the volcanic chain extends along the sea-coast of the continent. Mount St. Elias, one of the highest summits in America, often vomits lava from its crater, which opens at an elevation of 17,716 feet. Farther to the south, another active volcano, Mount Fair Weather, rises to a height of 14,370 feet. Next comes Mount Edgecumbe, in Lazarus Island, and the volcanic region of British Columbia. The whole chain of the Cascades, in Oregon, is well as the parallel ranges of the Sierra Nevada and the Rocky Mountains, are overlooked by a great number of volcanoes; but only a few of them continue to throw out smoke and ashes: these are Mount Baker, Renier, and St. Helens, enormous peaks 10,000 to

K K 2

16,000 feet high. In California and Northern Mexico, it is probable that the basaltic and trachytic mountains on the coast no longer present any outlets of eruption. Subterranean activity is not manifested with any degree of violence until we reach the high plateaus of Central Mexico. There a series of volcanoes, rising over a fissure crossing the continent, extends over the whole plateau of Anahuac, from the Southern Ocean to the Gulf of Mexico. The Colima, then the celebrated Jorullo, which made its appearance in 1759, the Nevado de Tolima, Istacihuetl, Popocatepetl, Orizaba, and Tuxtla are the vents for the furnace of lava which is boiling beneath the Mexican plateau. To the south, in Guatemala and the South American republics, thirty burning mountains, much more active and terrible than those of Anahuac, rise in two chains, one of which is parallel to the sea-coast, and the other crosses obliquely the isthmus of Nicaragua. Among these numerous volcanoes there are some, the names of which have become famous on account of the frightful disasters which have been caused by their eruptions. Such are the mountains del Fuego and del Agua, above the Ciudad-Antigua of Guatemala; the Pharo d'Isalco, which during the night lights up far and wide the plains of Salvador with its jets of molten stone and its column of red smoke; Coseguina, the last great eruption of which was probably the most formidable of modern times; the Viejo, Nuevo, Momotombo, and other mountains, which are almost worshipped from being so much dreaded.*

The depressions of the Isthmuses of Panama and Darien interrupt the series of volcanoes which borders the coast of the Pacific. The peak of Tolima, which rises to the great height of 17,716 feet, is the most northern of the active volcanoes of South America, and is also one of the most distant from the sea among all the fire-vomiting mountains, for the distance from its base to the Pacific coast is not less than 124 miles. South of Tolima, and the great plateau of Pasto, where there likewise exists a crater, stands the magnificent group of sixteen volcanoes, some already extinct and some still smoking, over which towers the proud dome of Chimborazo. Occupying an elliptical space, the greater axis of which is only about 112 miles long, this group, comprising the Tunguragua, Carahuirazo, Cotopaxi, Antisana, Pichincha, Imbabura, and Sangay, is often looked upon as but one volcano with several cones of eruption; it is the cluster which, on the southern coasts of the Isthmus of Panama, corresponds symmetrically to the volcanic group of Anahuac. South

* Moritz Wagner.

VOLCANOES OF AMERICA. 501

of Sangay, which is perhaps the most destructive volcano on the earth, the chain of the Cordilleras offers no volcanoes for a length of about 930 miles; but in Southern Peru the volcanic series recommences, and outlets of eruption still in action open at intervals amongst extinct volcanoes and domes of trachyte. The three smoking peaks of the inhabited part of Chili, the mountains of

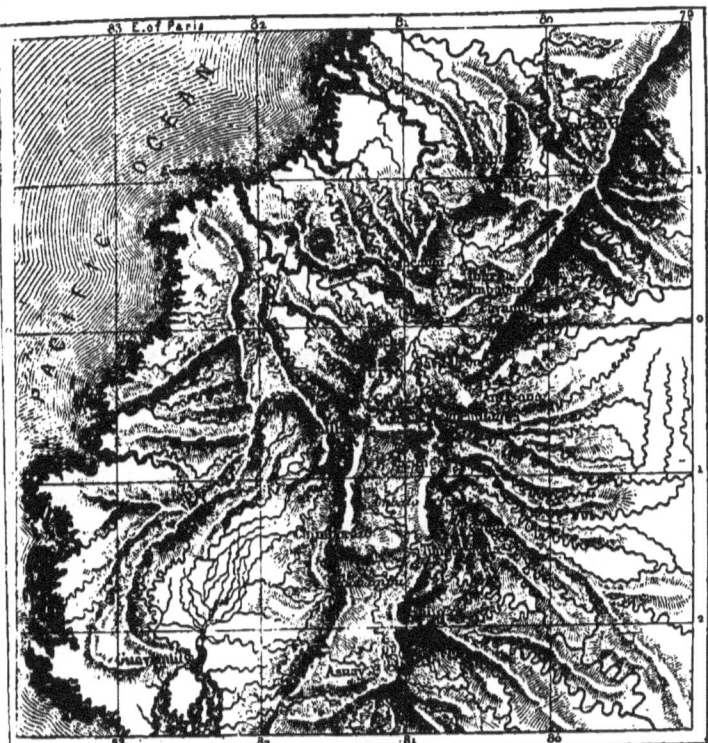

Fig. 172.—Equatorial Volcanoes.

Antuco, Villarica, and Osorno terminate the series of the great American volcanoes;* the activity of subterranean action is, however, disclosed by some other less elevated craters down to the extremity of the continent as far as the point of Terra-del-Fuego. This is not all; the South Shetland Islands, situated in the Southern Ocean, in a line with the New World, are likewise volcanic in their

* Philippi, *Mittheilungen von Petermann*, vol. iv., 1861.

character; and if the same direction be followed towards the pola·
regions, the line will ultimately touch upon the coasts of the land o:
Victoria, on which rise the two lofty volcanoes of Erebus and Moun
Terror, discovered by Sir John Ross. Stretching round the sphere o
the earth, the great volcanic circle is extended towards the nortl
by various islets of the antarctic, and ultimately rejoins the archi
pelago of New Zealand. Thus is completed the great ring of fir
which circles round the whole surface of the Pacific Ocean.

Within this immense amphitheatre of volcanoes a multitude o
those charming isles, which are scattered in pleiads over the ocean
are also of volcanic origin, and many of them can be distinguisher
from afar by their smoking or flaming craters. Of this kind are
some of the Marianne and Gallapagos Islands, which contain severa
orifices in full activity, and more than two thousand cones in a state
of repose. Among these we must especially mention the Sandwich
Islands, the lofty volcanoes of which rise in the middle of the central
basin of the North Pacific like so many cones of eruption in the
midst of a former crater changed into a lake. The Mauna-Loa and
Mauna-Kea, the two great volcanic summits of the island of Hawaii,
are each more than 13,000 feet in height; and the eruptions of the
first cone, which are still in full activity, must be reckoned among
the most magnificent spectacles of this kind. On the sides of the
Mauna-Loa opens the boiling crater of Kilauea, which is, without
doubt, the most remarkable lava-source which exists on our planet.

Round the circumference of the Indian Ocean the border of
volcanoes is much less distinct than round the Pacific; still it is
possible to recognise some of its elements. To the north of Java
and Sumatra, the volcanoes of which overlook the eastern portion of
the basins of the Indian seas, stretches the volcanic archipelago of
the Andaman and Nicobar Islands, in which there are several cones
of eruption in full activity. On the west of Hindostan, the peninsula
of Kutch, and the delta of the Indus, are often agitated by subterranean forces. Many mountains on the Arabian coast are nothing
but masses of lava; and, if various travellers are to be believed, the
volcanic furnace of these countries is not yet extinct. The Kenia,
the great mountain of Eastern Africa, has on its own summit a crater
still in action—perhaps the only one which exists on this continent.
Lastly, a large number of islands which surround the Indian Ocean
on the west and on the south — Socotora, Mauritius, Reunion,
St. Paul, and Amsterdam Island—are nothing but cones of eruption,
which have gradually emerged from the bed of the ocean.

The volcanic districts which are scattered on the edge of the Atlantic are likewise distributed with a kind of symmetry round three sides of this great basin. On the north, Jan Mayen, so often wrapt in mist, and the more considerable island of Iceland, pierced by numerous craters, Hecla, the Skapta-Jokul, the Kotlugaja, and seventeen other mountains of eruption separate the Atlantic from the the Polar Ocean. At about 1,500 miles nearer the equator, the peaks of the Azores, some extinct and some still burning, rise out of the sea. The archipelago of the Canaries, over which towers the lofty mass of the peak of Teyda, continues towards the south the volcanic line of the Azores, and is itself prolonged by the smoking summits of the Cape de Verde Islands. All the other mountains of lava which spring up from the bed of the Atlantic more to the south appear to have completely lost their activity, and on the coast itself there is, according to Burton, only one volcano still in action—that of the Cameroons. With regard to the "line of fire" along the western Atlantic, it is developed at the entrance of the Caribbean Sea with perfect regularity, like the range of the Aleutian Isles. Trinidad, Grenada, St. Vincent, St. Lucia, Dominica, Guadeloupe, Montserrat, Nevis, St. Kitts, and St. Eustatius are so many outlets of volcanic force, either through their smoking craters or their mud-volcanoes, their *solfataras* or their thermal springs. North and south of the Antilles, the eastern coast of America does not present a single vent of eruption. It is a remarkable fact that the two volcanic groups of the Antilles and the Sunda Islands are situated exactly at the antipodes one of the other, and also in the vicinity of the two poles of flattening, the existence of which on the surface of the globe has been proved by the recent calculations of astronomers.* More than this, these two great volcanic centres, which are undoubtedly the most active on the whole earth, flank, one on the west and the other on the east, the immense curve of volcanoes which spreads round the Pacific.

The Mediterranean is not surrounded by a circle of volcanoes; but there, as elsewhere, it is from the midst of the sea, or immediately on the sea-coast that the burning mountains rise—Etna, Vesuvius, Stromboli, Volcano, Epomeo, and Santorin. In like manner, the volcanoes of mud and gas of the peninsula of Apcheron, and the summit of Demavend, 14,436 feet high, rise at no great distance from the Caspian Sea. With regard to the volcanoes of Mongolia,—the Turfan, which is said to be still in action, and the Pe-chan, which,

* See above, p. 2.

according to Chinese authors, vomited forth up to the seventh century "fire, smoke, and molten stone which hardened as it cooled,"*—their existence is not yet absolutely proved; but even if these mountains, situated in the centre of the continent, should be in full activity, their phenomena might depend on the vicinity of extensive sheets of water, for this very region of Asia still possesses a large number of lakes, the remnants of a former inland sea, almost as vast as the Mediterranean.

What is the number of volcanoes which are still vomiting forth lava during the present period of the earth's vitality? It is difficult to ascertain, for often mountains have seemed for a long time to be extinct; forests have grown up in their disused craters, and their beds of lava have been covered up under a rich carpet of vegetation, when suddenly the sleeping force beneath is aroused and some fresh volcanic outlet is opened through the ground. When Vesuvius woke up from its protracted slumber to swallow up Pompeii and the other towns lying round its base, it had rested for some centuries, and the Romans looked upon it as nothing but a lifeless mountain like the peaks of the Apennines. On the other hand, it is very possible that some craters, from which steam and jets of gas are still escaping, or which have thrown out lava during the historic era, have entered decisively into a period of repose, ceasing somehow to maintain their communication with the subterranean centre of molten matter. The number of vents which serve for the eruption of lava can, therefore, be ascertained in a merely approximate way. Humboldt enumerates 223 active volcanoes; Keith Johnston arrives at the larger number of 270, 190 of which are comprehended in the islands and the Pacific "circle of fire;" but this latter estimate is probably too small. To the number of these burning mountains, standing nearly all of them on the sea-shore, or in the vicinity of some great fresh-water basin, must be added the *salses*, or mud-volcanoes, which are also found near large sheets of salt water. With regard to the thousands of extinct volcanoes which rise in various parts of the interior of continents, geology shows that the sea used formerly to extend round their bases.

* *Chroniques Chinoises.* Humboldt, *Tableaux de la Nature.*

The Earth. Vol. 1.

Eng^d by Erhard

CHAPTER LXIII.

TORRENTS OF STEAM ESCAPING FROM CRATERS.—GASES PRODUCED BY THE DECOMPOSITION OF SEA-WATER.—HYPOTHESES AS TO THE ORIGIN OF ERUPTIONS.—INDEPENDENCE OF THE SEVERAL VOLCANIC OUTLETS.

ONE of the most decisive arguments which can be used in favour of a free communication existing between marine basins and volcanic centres is drawn from the large quantities of steam which escapes from craters during an eruption, and composes, according to M. Ch. Sainte-Claire Deville, at least 999 thousandths of the supposed volcanic smoke. During the eruption of Etna, in 1865, M. Fouqué attempted to gauge approximately the volume of water which made its escape in a gaseous form from the craters of eruption. By taking as his scale of comparison the cone which appeared to him to emit an average quantity of steam, he found this mass, reduced to a liquid state, would be equivalent to about 79 cubic yards of water for each general explosion. Now, as these explosions took place on the average every four minutes during a hundred days, he arrived at the result, that the discharge of water during the continuance of the phenomenon might be estimated at 2,829,600 cubic yards of water —a flow equal to that of a permanent stream discharging 55 gallons a second. Added to this, account ought to have been taken of the enormous convolutions of vapour which were constantly issuing from the great terminal crater of Etna, and, bending over under the pressure of the wind, spread out in an immense arch around the vault of the sky. In great volcanic eruptions it often happens that these clouds of steam, becoming suddenly condensed in the higher layers of the atmosphere, fall in heavy showers of rain, and form temporary torrents on the mountain-sides. According to the statements of Sir James Ross, the mountain Erebus, of the antarctic land, is covered with snow, which it has just vomited forth in the form of vapour. It has besides been remarked that the vapour which issues from volcanoes is not always warm; often, according to Pœppig, it is of the same temperature as the surrounding air.

As was said long since by Krug von Nidda, a German *savant*, volcanoes must be looked upon as enormous intermittent springs. The basaltic flows may be compared to streams on account of the water which they contain. It is probable that most of the lava which flows from volcanic fissures owes its mobility to the innumerable particles of vapour which fill up all the interstices of the moving mass. Being composed in great measure of crystals already formed, as may be proved by an examination of the *cheires*, in the body of which may be noticed nodules and crystals rounded by friction, the lava would be unable to descend over the slopes if it were not rendered fluid by its mixture with steam; and the gradual slackening in speed and ultimate stoppage of the flow are chiefly caused by the setting free of the gases which served as a vehicle to the solid matter. Owing to this rapid loss of their humidity, basalts contain in their pores but a very slight quantity of water in comparison with other rocks.* Yet even old lavas themselves contain as much as 10 to 19 thousandths of water at the edges of the bed, and 5 to 18 thousandths at the centre.†

The various substances which are produced from craters also tend to show that sea-water has been decomposed in the great laboratory of lava. Ordinary salt or chloride of sodium, which is the mineral that is most abundant in sea-water, is also that which is deposited the first and most plentifully round the orifices of eruption. Sometimes, the scoriæ and ashes are covered for a vast space with a white efflorescence, which is nothing but common salt; one might fancy it a shingly beach which had just been left by the ebbing tide. After each eruption of Hecla, the Icelanders are in the habit, it is said, of collecting salt on the slopes. The lava from the eruption of Frumento, analysed by M. Fouqué, contained about 13 ten-thousandths of marine salt.

Almost all the other component parts of sea-water are likewise found in the gases and deposits of *fumerolles;* only the salts of magnesia have disappeared, but still are found under another form among the volcanic products. Being decomposed by the high temperature, just as they would be in the laboratory of a chemist, they go to constitute other bodies. Thus the chloride of magnesium is changed into hydrochloric acid and magnesia; the gas escapes in abundance from the *fumerolles*, whilst the magnesium remains fixed in the lava.‡

---

* Poulett Scrope, *Volcanoes*.
† Delesse, *Bulletin de la Société Géologique de France*, 1859.
‡ Fouqué, *Phénomènes Chimiques de l'Eruption de l'Etna en* 1865.

As M. Ch. Sainte-Claire Deville was the first to ascertain with certainty, four successive periods may be observed in every eruption, each of which periods assumes a distinct character owing to the exhalation of certain substances. After the first period, remarkable especially for marine salt and the various compounds of soda and potash, comes a second in which the temperature is lower, and during which brilliantly coloured deposits of chloride of iron are formed and hydrochloric and sulphurous acids are expelled. When the temperature is below 392° (Fahr.), there are ammoniacal salts and needles of sulphur, which are found in yellowish masses on the scoriæ of lava. Lastly, when the heat of the erupted bodies is below 212° (Fahr.), the *fumerolles* eject nothing but steam, azote, carbonic acid, and combustible gases. Thus the activity of the exhalations and deposits is in proportion to the incandescence of the lava. At the commencement of the eruption, the orifices throw out a large quantity of substances, from marine salt to carbonic acid; but by degrees the power of elaboration weakens simultaneously with the heat, and the gases ejected gradually diminish in number, and testify, by their increasing rarity, to the approaching cessation of volcanic phenomena. In consequence of the difference which is presented by the exhalations during the various phases of eruptions of lava, observers have, at first sight, thought that each volcano was distinguished by emanations peculiar to itself. Hydrochloric acid was looked upon as one of the normal products of Vesuvius, and sulphurous vapours as more special to Etna. It was stated (with Boussingault) that carbonic acid was exhaled especially by the volcanoes of the Andes; and, with Bunsen, it was believed that combustible gases prevailed in the eruptions of Hecla.*

In his beautiful investigations into the various chemical phenomena presented by Etna and the neighbouring volcanic outlets, such as Vesuvius and Stromboli, M. Fouqué appears to have established as a fact which must be henceforth beyond dispute, that the gradual series of these emanations is just that which would be produced by the decomposition of sea-water. Added to this, we also find in lava, iodine and fluorine, both of which we should expect to detect in it on account of their presence in sea-water. The salts of bromine, of which, however, only a slight trace is found in sea-water, have not yet been detected in volcanic products, which, no doubt, proceeds from the difficulty which chemists have experienced in separating such very small quantities.

* Fouqué, *Revue des Deux Mondes*, August, 1866.

The other matters ejected by eruptions are of terrestrial origin, and evidently proceed from rocks reduced by heat to a liquid or pasty state; they consist principally of silica and alumina, and contain besides lime, magnesia, potash, and soda. Oxides of iron also enter into the composition of lava, to the extent of more than one-tenth, which is a very considerable proportion, and warrants us in looking upon the volcanic flows as actual torrents of iron-ore; sometimes, indeed, this metal appears in a pure state. It is to this presence of iron that lava especially owes its reddish colour, and the sides of the crater their diversely coloured sides. Compounds of copper, manganese, cobalt, and lead are also met with in lava; but, in comparison with the iron, they are but of slight importance. Lastly, phosphates, ammonia, and gases composed of hydrogen and carbon, are discharged during eruptions. The presence of these bodies is explained by the enormous proportion of animal and vegetable matter which is decomposed in sea-water. Ehrenberg found the remains of marine animalcula in the substances thrown out by volcanoes.

Is the composition of the lava, and especially that of the vapour and gases, the same in those eruptions which take place at a great distance from the ocean? It is probable that, as regards this point, considerable differences might be established between the products of volcanoes placed on the sea-coast, such as Vesuvius and Etna, and those which rise far in the interior of the land, as Tolima, Jorullo, and Puracé. This comparative study, however, which would be calculated to throw light on the chemical phenomena of deep-lying beds, has as yet been made at only a few points. Eruptions are rare in volcanoes situated far from the coast, and when they do take place, scientific men do not happen to be on the spot to study the course of the occurrence. Popocatapetl, one of the most remarkable continental volcanoes, produces a large quantity of hydrochloric acid; the snow from it, which has a very decided muriatic taste, is carried by the rain into the Lake of Tezcuco, where, in conjunction with soda, it forms salt.*

When the water, either of sea or rivers, penetrates into the crevices of the terrestrial envelope, it gradually increases in temperature the same as the rocks it passes through. It is well known that this increase of heat may be estimated on the average, at least as regards the external part of the planet, at 1° (Fahr.) for every 54 feet in depth. Following this law, water descending to a point 7,500 feet below

* Virlet, *Bulletin de la Société Géologique de France*, May 1st, 1865.

the surface would show, in the southern latitudes of Europe, a temperature of about 212° (Fahr.) But it would not on this account be converted into steam, but would remain in a liquid state, owing to the enormous pressure which it has to undergo from the upper layers. According to calculations, which are based it is true on various hypothetical data, it would be at a point more than nine miles below the surface of the ground that the expansive force of the water would attain sufficient energy to balance the weight of the superincumbent liquid masses, and to be suddenly converted into steam at a temperature of 800° to 900° (Fahr.) These gaseous masses would then have force to lift a column of water of the weight 1,500 atmospheres; if, however, from any cause, they cannot escape as quickly as they are formed, they exercise their pressure in every direction, and ultimately find their way from fissure to fissure until they reach the fused rocks which exist in the depths. To this incessantly increasing pressure we must, therefore, attribute the ascent of the lava into vent-holes of volcanoes, the occurrence of earthquakes, the fusion and the rupture of the terrestrial crust, and finally, the violent eruptions of the imprisoned fluids.* But why should the vapour thus pervade the subterranean strata and upheave them into volcanic cones, when, by the natural effect of its overcoming the columns of water which press it down, it ought simply to rise towards the bed of the sea from which it descended? In the present state of science, this is a question to which it seems absolutely impossible to give a satisfactory answer,† and geologists must at least have the merit of candidly acknowledging their ignorance on this point. The discoveries of natural philosophy and chemistry, which have been the means of making known to us the enormous activity of steam in volcanic eruptions, will, doubtless, sooner or later explain to us in what way this activity is exercised in the subterranean cavities. But, at the present time, the phenomena which are taking place in the interior of our globe are not better known to us than the history of the lunar volcanoes.

Be this as it may, the direct observations which have been made on volcanic eruptions have now rendered it a very doubtful point whether the lavas of various volcanoes proceed from one and the same reservoir of molten matter, or from the supposed great central furnace which is said to fill the whole of the interior of the planet. Volcanoes which are very close to one another show no coincidence

---

* Buff, *Briefe über die Physik der Erde.*
† Otto Volger, *Erdbeben der Schweiz.*

in the times of their eruptions, and vomit forth, at different epochs, lavas which are most dissimilar both in appearance and mineralogical composition. These facts would be eminently impossible, if the craters were fed from the same source. Etna, the group of the Lipari Isles, and Vesuvius, have often been quoted as being volcanic outlets placed upon the same fracture of the terrestrial crust; and it is added, in corroboration of this assertion, that a line traced from the Sicilian volcano to that of Naples passes through the ever active

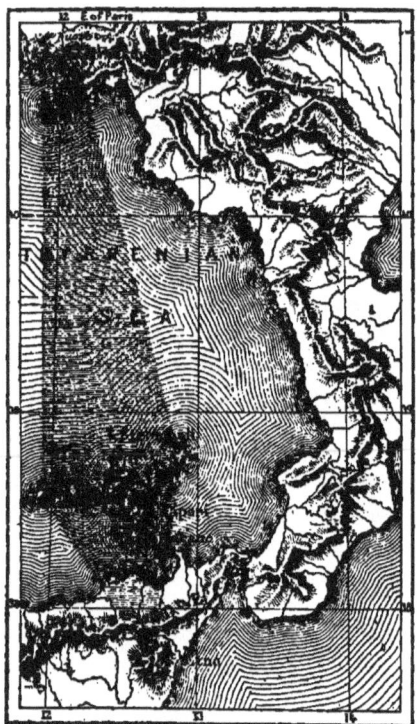

Fig. 173.—Line of Fracture between Etna and Vesuvius.

furnace of the Lipari Isles. Although the mountain of Stromboli, so regular in its eruptions, is situated on a line slightly divergent from the principal line, and, on the other side, the volcanic isles of Salini, Alicudi, and Felicudi tend from east to west, it is possible and even probable that Vesuvius and Etna are in fact situated on fissures of the earth which were once in mutual communication. But during the thousands of years in which these great craters have been at work, no connection between their eruptions has ever been positively certified.

Sometimes, as in 1865, Vesuvius vomits forth lava at the same time as Etna; sometimes it is in a state of repose when its mighty neighbour is in full eruption, and rouses up when the lava of Etna has cooled. There is nothing which affords the slightest indication of any law of rhythm or periodicity in the eruptive phenomena of the two volcanoes. The inhabitants of Stromboli state that, during the winter of 1865, at the moment when the sides of Etna were rent, the volcanic impulse manifested itself very strongly in their island by stirring up the always agitated waves of the lava-crater which commands their vineyards and houses. A comparative calm, however, soon succeeded this temporary effervescence, and in the adjacent island of Volcano, no increase of activity was noticed. If the shafts of Etna, Vesuvius, and the intervening volcanoes, take their rise in

Fig. 174.—Section of the Island of Hawaii.

one and the same ocean of liquid lava, all the lower craters must necessarily overflow simultaneously with the most elevated. Now, as has often been noticed, the lava may ascend to the summit of Etna, at a height of 10,827 feet, without a simultaneous flow of rivers of molten stone from Vesuvius, Stromboli, and Volcano, which are respectively but one-third, one-fourth, and one-tenth the height of the former. In like manner, Kilauea, situated on the sides of Mauna-Loa, in the Isle of Hawaii, in no way participates in the eruptions of the central crater opening at a point 9,800 higher up, and not more than 12 miles away. If there is any present geological connection between the volcanoes of one and the same region, it probably must be attributed to the fact of their phenomena depending on the same climatic causes, and not because their bases penetrate to one and the same ocean of fire. Volcanic orifices are not, therefore, "safety valves," for two centres of activity may exist on one mountain without their eruptions exhibiting the least appearance of connection.*

Isolated as they are, amidst all the other formations on the surface

* Dana, *Proceedings of the American Association*, 1849.

of the earth, lavas appear as if almost independent of the rest. Basalts, trachytes, and volcanic ashes, are the comparatively modern products which are scarcely met with in the periods anterior to the Tertiary age. Only a very small quantity of these lavas of eruption has been found in the Secondary and Palæozoic rocks. Formerly, most geologists thought that the granites and rocks similar to them had issued from the earth in a pasty or liquid state; they looked upon them as the "lavas of the past," and believed that these first eruptive rocks were succeeded in age after age by the diorites, the porphyries, the trap-rocks, then by the trachytes and the basalts of our own day, all drawn from a constantly increasing depth. They thought also that, in the future, when the whole series of the present lavas shall have been thrown up to the surface, volcanoes would produce other substances as distinct from the lavas as the latter are from the granite. Granites, however, differ so much from the trachytes and basalts as to render it impossible for us to imagine that they have the same origin; added to which, the labours of modern *savants* have proved that, under the action of fire, granite, and the other rocky masses of the same kind, would have been unable to assume the crystalline texture which distinguishes them. We are, then, still ignorant how volcanic eruptions commenced upon the earth, and how they are connected with the other great phenomena which have co-operated in the formation of the external strata of the globe.

CHAPTER LXIV.

GROWTH OF VOLCANOES.—THEORIES OF HUMBOLDT AND LEOPOLD VON BUCH AS TO THE UPHEAVAL OF CRATERS.—DISAGREEMENT OF THESE THEORIES WITH THE FACTS OBSERVED.

CONSIDERED singly, each volcano is nothing but a mere orifice, temporary or permanent, through which a furnace of lava is brought into communication with the surface of the globe. The matter thrown out accumulates outside the opening, and gradually forms a cone of *débris* more or less regular in its shape, which ultimately attains to considerable dimensions. One flow of molten matter follows another, and thus is gradually formed the skeleton of the mountain; the ashes and stones thrown out by the crater accumulate in long slopes; the volcano simultaneously grows wider and higher. After a long succession of eruptions, it at last mounts up into the clouds, and then into the region of permanent snow. At the first outbreak of the volcano the orifice is on the surface of the ground; it is then prolonged like an immense chimney through the centre of the cone, and each new river of lava which flows from the summit increases the height of this conduit. Thus the highest outlet of Etna opens at an elevation of 10,892 feet above the level of the sea; Teneriffe rises to 12,139 feet; Mauna-Loa, in Hawaii, to 13,943 feet; and, more gigantic still, Sangay and Sahama, in the Cordilleras, attain to 18,372 and 23,950 feet in elevation.

This theory of the formation of volcanic mountains by the accumulation of lava and other matters cast out of the bosom of the earth presents itself quite naturally to one's mind. Most *savants*, from Saussure and Spallanzani down to Virlet, Constant Prévost, Poulett Scrope, and Lyell, have been led, by their investigations, to adopt it entirely; indeed, in the present day it is scarcely disputed. It is true that Humboldt, Leopold von Buch, and, following them, M. Elie de Beaumont, have put forth quite a different hypothesis as to the origin of several volcanoes, such as Etna, Vesuvius, and the Peak of Teneriffe. According to their theory, volcanic mountains do not owe their present conformation to the long-continued accumulation of

L L

514   THE EARTH.

lava and ashes, but rather to the sudden upheaval of the terrestrial strata. During some revolution of the globe, the pent-up matter in the interior suddenly upheaves a portion of the crust of the planet into the form of a cone, and opens a funnel-shaped gulf between the

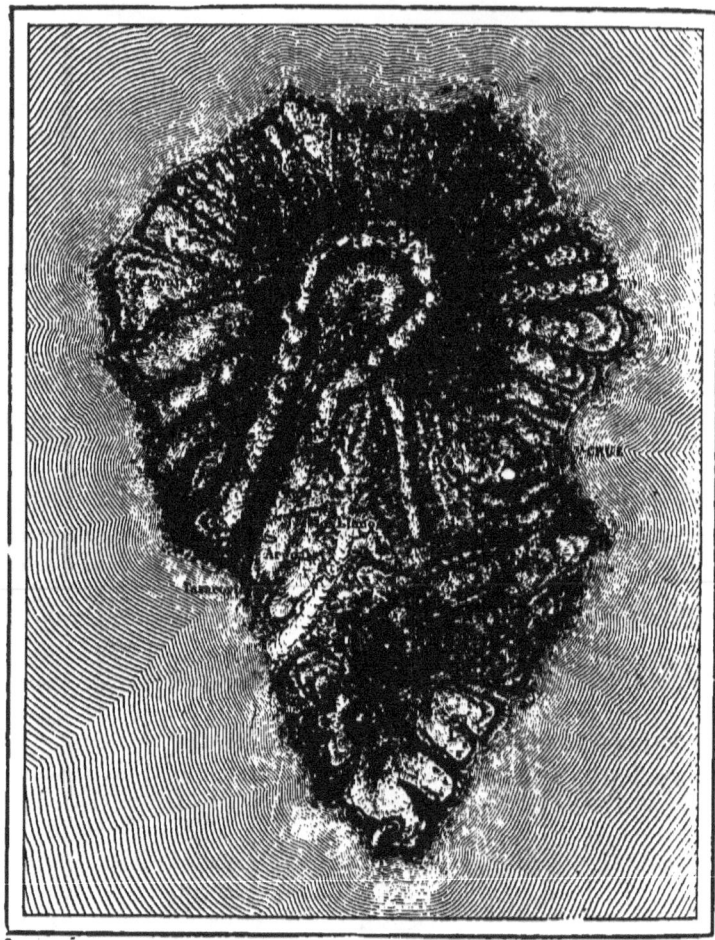

Fig. 175.—Isle of Palma

dislocated strata, thus by one single paroxysm producing lofty mountains, as we now see them. As an important instance of a crater thus formed by the upheaval and rupture of the terrestrial strata,

Leopold von Buch mentions the enormous abyss of the Isle of Palma, known by the natives under the name of "Caldron," or *Caldera*. This funnel-shaped cavity is of enormous dimensions, and is not less than four or five miles in width on the average; the bottom of it is situated about 2,000 feet above the level of the sea. Lofty slopes, from 1,000 to 2,000 feet in height, rise round the vast amphitheatre, and abut upon inaccessible cliffs, the upper ledges of which reach a total altitude of 5,900 to 6,900 feet in height. The highest point, the Pico-de-los-Muchachos, is covered by snow during the winter months; and although it penetrates into regions of the atmosphere which are of a very different character from those of the rest of the island, the slope that is turned towards the crater is so steep that blocks of stone falling from the summit roll down into the enclosed hollow.

The prodigious cavity in the Isle of Palma was perhaps the most striking instance that Leopold von Buch could bring forward in favour of his hypothesis; nevertheless, the exploration of this island, since carried out by Hartung, Lyell, and other travellers, is very far from confirming the ideas of the illustrious German geologist. The lofty side-walls of the hollow appear to be formed principally, not of solid lava, which constitutes scarcely a quarter of the whole mass, but of layers of ashes and scoriæ, regularly arranged like beds of sand on the incline of a talus. Basalts and strata of ashes lie one upon another in the greatest order round the enclosed hollow, which would be a fact impossible to comprehend if any sudden upheaval, acting in an upward direction with sufficient violence to break the terrestrial crust, had shattered and ruptured all the strata, and, by a mighty explosion, opened out the immense Caldron of Palma. Finally, if a phenomenon of this kind had taken place, star-formed cracks, like those produced in broken glass, would be visible across the thickness of the upheaved strata, and their greatest width would be turned towards the crater. Now, there are no fissures of this kind, and the ravines in the circumference of the volcano, which one might perhaps be tempted to confound with actual ruptures of the ground, become wider in proportion as they approach the sea. The enormous cavity in Palma is, therefore, a crater similar to those of volcanoes of less dimensions. It is, however, certain that the Caldera was once both shallower and less in extent, for the ashes and volcanic scoriæ are easily carried away by the rain, which is swallowed up in the bottom of the basin, and has hollowed out for itself a wide drainage-channel in a south-west direction.

M. Elie de Beaumont, as his chief support of Leopold von Buch's hypothesis, brought forward the fact that most of the strata of lava —a section of which may be seen on the sides of Etna, in the immense amphitheatre of the Val del Bove—are very sharply inclined. The celebrated geologist affirmed that thick sheets of molten matter could not run down steep slopes without being very soon reduced, in consequence of the acceleration of their speed, into thin layers of irregular scoriæ. If this were really the case, the position of the thick flows of lava in the Val del Bove must have changed since the date of the eruption; it would then be necessary to admit that they have been violently tilted up after having been originally deposited on the soil in sheets, which were either horizontal or very gently sloped. Nevertheless, the recent observations made by Sir C. Lyell, those of Darwin on the cones of the Gallapagos Isles, and of Dana on the lava flows of Kilauea; lastly, the remarks of the Italian *savants* who studied on the spot the volcanic phenomena of

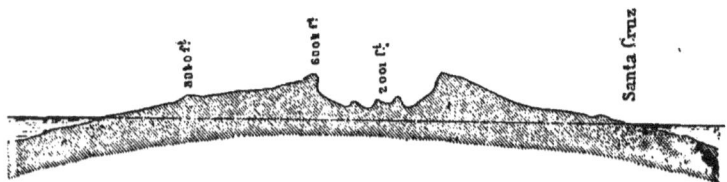

Fig. 176.—Section of the Island of Palma from South-east to North-west.

Vesuvius and Etna, have satisfactorily proved that, in modern times, a great number of rivers of lava, and especially that of the Val del Bove, in 1852 and 1853, have flowed over steep slopes varying in inclinations from 15 to 40 degrees. It must, besides, be understood that the lava which poured over the steepest slopes was exactly that portion which, not having experienced any cause of delay, or met with any obstacle, in its course, presented layers of the most uniform consistence and the most regular action.

One of the strongest arguments of scientific men in favour of the theory of upheaval is, that certain volcanic mountains, especially that of Monte-Nuovo, of Pouzzoles, and Jorullo, in Mexico, had been suddenly raised up by the swellings of the soil. Now, the unanimous testimony of those who, more than three centuries ago, witnessed the eruption of Monte-Nuovo is, that the earth was cleft open, affording an outlet to vapour, ashes, scoriæ, and lava, and that the hill, very much lower than some of the subordinate cones of Etna, gradually rose during four days by the heaping up of the matter thrown out. The total volume of this eruption was no doubt considerable, but compared with the amount of matter which flowed down upon Catania in 1669, or with the rivers of lava from Skaptar-Jokul, it is a mass of no great importance. Added to this, if the soil was really upheaved, how was it that the neighbouring houses were not thrown down, and that the colonnade of the Temple of Neptune, which stands at the foot of the mountain, kept its upright position? With regard

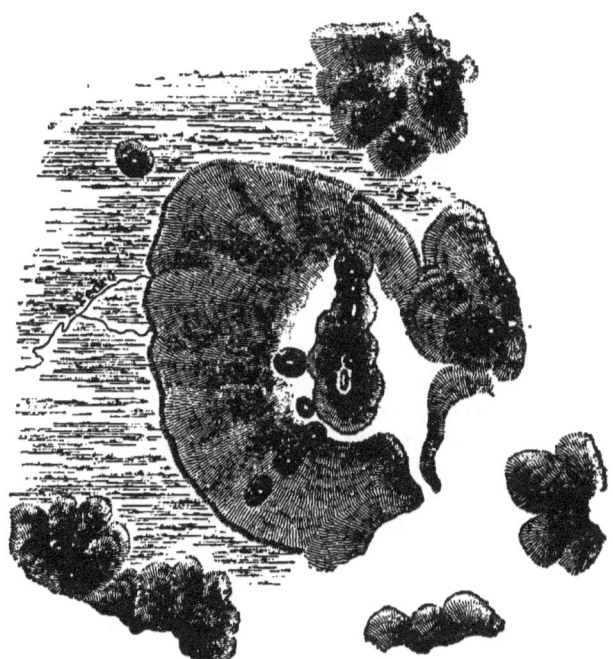

Fig. 177.—Volcano of Jorullo, Mexico.

to Jorullo, which rises to a height of more than 1,650 feet, the only witnesses of this volcano making its first appearance were the Indians, who fled away to the neighbouring heights, distracted with terror. We have, therefore, no authentic testimony on which we can base an hypothesis as to any swelling up of the ground in the form of a blister. Quite the contrary, the travellers who have visited this Mexican volcano since Humboldt have discovered beds of lava lying one over the other, as in all other cones of eruption; and more than this, they have also ascertained that none of the strata in the ground overlooked by the mountain have been at all tilted up.*

It is true enough that local swellings have often been observed in the burning matter issuing from the interior of the earth: in many places the lava is pierced by deep caverns, and entire mountains—especially that of Volcano—have so many hollows in the rocks on their sides that every step of the climber resounds on them as if on a vault. Besides, the lava itself, being a kind of impure glass, is so pervaded by bubbles filled with volatile matter that, when acted upon by fire, so as to expel the water and the gas, it loses on an average, according to Fouqué, two-thirds of its weight. But these caverns, these hollows and bubbles, proceed from the mixture of the lava with vapour which is liberated with difficulty from the viscous mass, or are caused by the longitudinal rupture of the strata during an eruption, and can in no way be compared to the immense blister-like elevation which would be formed by the strata of a whole district being tilted up to a height of hundreds, or even thousands, of yards, leaving at the summit, between two lines of fracture, room for an immense cavity.

None of these prodigious upheavals have been directly observed by geologists, and none of the legends invented by the fears of our ancestors, referring to the sudden appearance of volcanic mountains, have been since confirmed. Lastly, the very structure of the peaks which are said to have risen abruptly from the midst of the plains testifies to the gradual accumulation of material that has issued from the bowels of the earth. It is, therefore, prudent to dismiss definitively an hypothesis which marks an important period in the history of geology, but which, for the future, can only serve to retard the progress of science.

* Arnold Boscowitz, *Les Volcans et les Tremblements de Terre*.

CHAPTER LXV.

NUMBER AND ARRANGEMENT OF VOLCANIC OUTLETS.—FORM OF VOLCANIC CONES AND CRATERS.

As, when the burning matter seeks an outlet, the earth is generally cleft open in a straight line, the volcanic orifices are frequently distributed somewhat regularly along a fissure, and the heaps of erupted matter follow one another like the peaks in a mountain chain. In

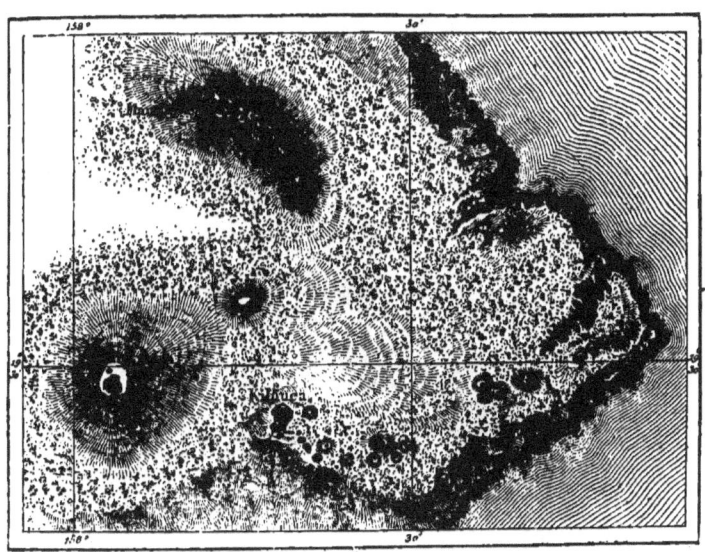

Fig. 178.—Series of Craters, Hawaii.

other places, however, the volcanic cones rise without any apparent order on ground that is variously cleft; just as if a wide surface had been softened in every direction, and had thus allowed the molten matter to make its escape, sometimes at one point, sometimes at another. From the town of Naples—which is itself built on half a crater in great part obliterated—to the Isle of Nisida, which is an old volcano of regular form, the Phlegræan Fields present a remarkable example of this confusion of craters. Some are perfectly rounded, others are broken into, and their circle is invaded by the waters of the sea: grouped for the most part in irregular clumps, even encroaching upon one another and blending their walls, they give to the whole landscape a chaotic appearance. As Mr. Poulett

Fig. 179.—Auckland, and its Volcanoes.

Scrope very justly remarks, the aspect of the terrestrial surface at this spot reminds one exactly of the volcanic districts of the moon, dotted over, as it is, with craters.

As the type of a region pierced all over with volcanic orifices, we may also mention the Isthmus of Auckland in New Zealand, where

Dr. Hochstetter has reckoned, in an area of 230 square miles, sixty-one independent volcanoes, 520 to 650 feet in height on the average. Some are mere cones of tufa; others are heaps of scoriæ, or even eruptive hillocks, which have shed out round them long flows of lava. At one time the Maori chiefs used to intrench themselves in these craters as if in citadels; they escarped the outer slopes in terraces, and furnished them with palisades. At the present day, the English colonists, having become lords of the soil, have constructed their farms and country houses on these ancient volcanoes, and are constantly bringing the soil under cultivation.*

The Safa, in the Djebel-Hauran, is also a complete chaos of hillocks and abysses. On this plateau of 460 square miles, which the Arabs call a "portion of hell," almost all the craters open on the surface of the ground, and not on the summits of volcanoes scattered here and

Fig. 180.—Cone of Tuff.   Fig. 181.—Cone of Tuff, and Crater of Scoriæ.

there on the black surface. In every direction there may be seen rounded cavities like the vacuities formed in scoriæ by bubbles of gas, only these cavities are 600 to 900 feet wide, and 65 to 160 feet deep. Some are isolated; some either touch or are separated by nothing but narrow walls like masses of red or darkish-coloured glass. One hardly cares to venture on these narrow isthmuses, bordered by precipices, and intersected here and there by fissures.†

The normal form of the volcanoes in which the work of eruption takes place is that of a slope of *débris* arranged in a circular form round the outlet. Whether the volcano be a mere cone of ashes or mud only a few yards high, or rise into the regions of the clouds, vomiting streams of lava over an extent of 10 or 20 miles, it none the less adheres to the regular form so long as the eruptive action is maintained in the same channel, and the *débris* thrown out falls equally on the external slopes.

The beauty of the cone is increased by that of the crater. The

* Ferd. von Hochstetter, *Neu-Seeland.*
† Wetzstein, *Zeitschrift für Erdkunde*, 1859.

terminal orifice from which the lava boils out, well deserves, from the purity of its outline, its Greek name of "cup," and the harmony of its curve contrasts most gracefully with the declivity of the slope. In some volcanoes the symmetry of the architectural lines is so complete, that the crater itself contains a cone placed exactly in the centre of the cavity, and pierced by a second crater in miniature, from which vapour makes its escape.

Volcanoes in which the eruptive action frequently changes its position—and these are the more numerous class—do not possess this elegance of outline. Very often the upheaved lava finds some weak

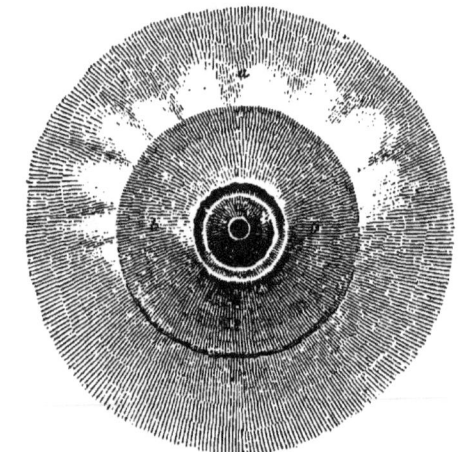

a. Declivities of Tuff.   *b.* Cone of Lava.   *c.* Pyramid of Scoriæ.

Fig. 184.—Plan and Section of the Volcano of Rangitoto.

place in the walls of the crater; it hollows them out at first, and then, bringing all its weight to bear on the rocks which oppose its passage, it ultimately completely breaks down the edge of the crater, leaving perhaps only one side standing. Among the European volcanoes, Vesuvius is the best example of these ruptured craters: before A.D. 79, the escarpments of La Somma, which now surround with their semicircular rampart the terminal cone of Vesuvius, were the real crater. The portion of it which no longer

FORM OF VOLCANOES.

exists disappeared and buried under its *débris* the towns of Herculaneum and Pompeii.

Active volcanoes, however, never cease to increase in all their dimensions, and sooner or later the breach is ultimately repaired; the remains of the former craters are gradually hidden under the growing slopes of the central cone. Thus, a former crater on Etna, which was situated at a point three miles in a straight line from the present outlet, at the commencement of the Val del Bove, has been

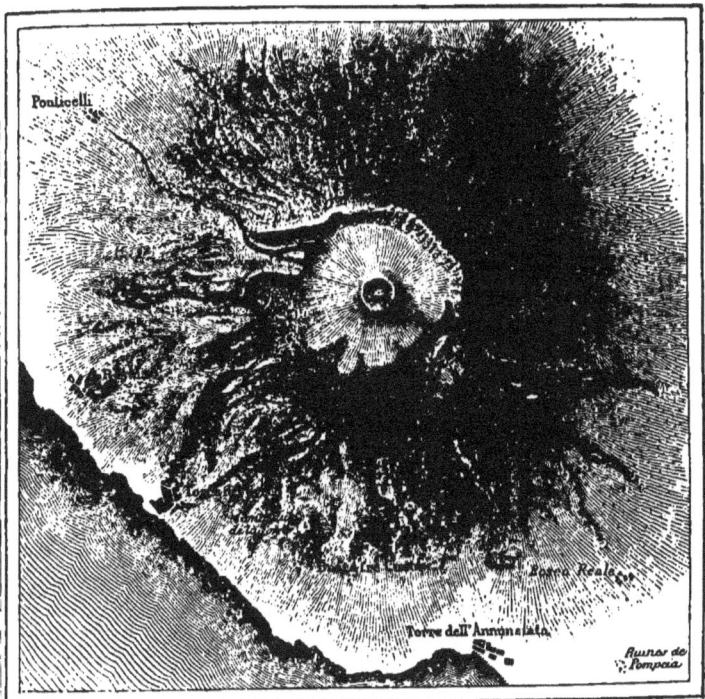

Fig. 183.—Mount Vesuvius.

gradually obliterated by the lava of successive eruptions: prolonged explorations on the part of MM. Seyell and Waltershausen have been necessary in order to find it out. The normal form of Etna is that of a cone of *débris*, placed upon a large dome with long slopes, becoming more and more gentle, and descending gracefully towards the sea. In fact, in most of the eruptions, the lava does not rise as far as the great crater, and breaks through the sides of the volcan so as to flow laterally over the flanks of Etna. These eruptions succeeding one another in the course of centuries, bring about th necessary result of gradually enlarging the dome which constitute

Fig. 134.—Section of Vesuvius from South to North.

the mass of the mountain, thus breaking the uniformity of the lateral talus. The same thing occurs with regard to Vesuvius on the side which faces the sea-coast. There, too, the terminal cone stands or a kind of dome, which has been gradually formed by the coats of lava

Fig. 135.—Section of Etna from West to East.

running one over the other. If Vesuvius continues to be the great volcanic outlet of Italy, and rises gradually into the sky by the superposition of lava and ashes, it cannot fail, some time or other, to assume a form similar to that of the Sicilian giant.

Fig. 136.—Mount Orizaba.

The volcanoes which present cones of almost perfect regularity are those which have their terminal outlet alone in a state of activity, and vomit out a large quantity of ashes or other matter which glides

FORM OF VOLCANOES. 525

readily over the slopes. Among this class of mountains, those which attain any considerable elevation are distinguished by their majesty from all other peaks. Stromboli, although it is not more than 2,600 feet in height, is one of the wonders of the Mediterranean. From its proud form, it will readily be understood that its roots plunge down into the sea to an enormous depth; the slope of *débris* may be seen,

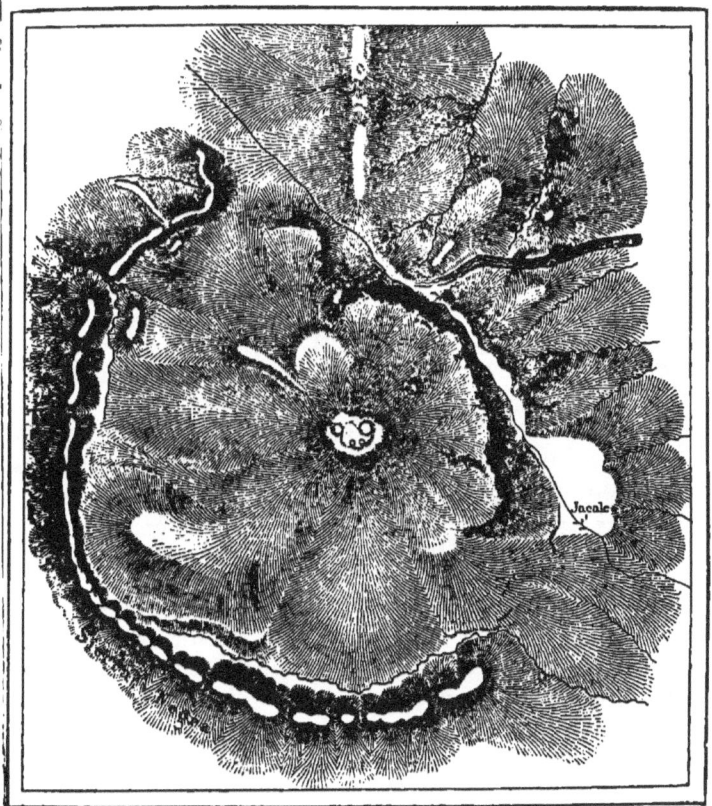

Fig. 187.—Profile of Orizaba.

to speak, prolonged under the water down to the abysses of 3,000 to 4,000 feet, which the sounding-line has reached at the bottom of the Æolian Sea. At sight of it, one feels as if suspended in the midst of the void, as if the ship was sailing in the air midway up the mountain. This feeling of admiration mingled with dread increases when this great pharos of the Mediterranean is approached during the night, over the dark-waved sea. Then the sky above the summit seems all lighted up by the reflection of the lava, and a misty band of clouds and vapour may be dimly seen girdling round the body of the volcano. In the daytime the impression made is of different character; but it is none the less deep, for the real grandeur of Stromboli consists, not so much in the immensity of the mass as in the harmony of its proportions.

Volcanic mountains of an ideal form are those which infant nations have most adored. Among these sacred mountains are the sublime Cotopaxi of the Andes, Orizaba of Mexico, Mauna-Loa of Hawaii, and Fusi-Yama of Japan. The volcanoes of Java, and

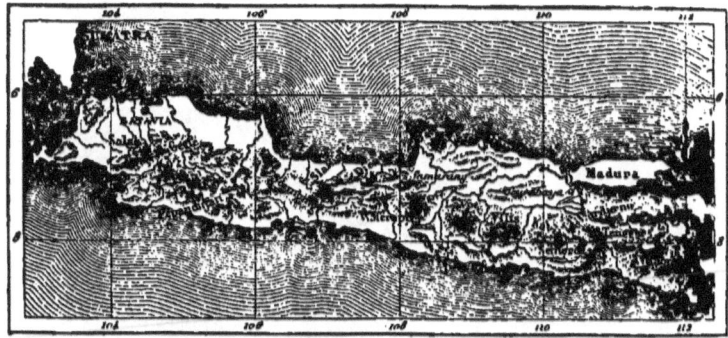

Fig. 188.—Volcanoes of Java.

chiefly those in the eastern portion of the island, also present a very majestic appearance on account of their isolation. Those on the western side are based upon an undulating plateau, which causes them to lose their appearance of height; but on the east all the volcanic mountains rise up from verdant plains like islands above the waves of the sea, and command the horizon far and wide with their enormous cones. Between the Merapi and Lavoe mountains lies a depression, the highest ledge of which exceeds the level of the sea by only 312 feet. Between Lavoe and Villis the plain is 230 feet in height. Lastly, the plains which separate the Villis and Kelœet mountains nowhere attain an elevation of more than 200 feet above the ocean.*

In the external details of their conformation many of the

* Junghuhn, *Java, seine Gestalt und innere Bauart.*

volcanoes of Java present a regularity of outline which is all the more striking, since they owe it in great part to the monsoon rains, the most destructive agents of the tropical regions. In beating against the mountains, the clouds let fall their burden of moisture on the slopes composed of ashes and loose scoriæ. The latter offer but a slight resistance to the action of the temporary torrents which carry them away, and crumbling down into the plains which surround the base of the volcano, are deposited in long slopes, like those caused by avalanches. In consequence of the fall of all this *débris*, the sides of the mountain are cut out at intervals by ravines or furrows, which gradually widen from the summit to the base of the mountains, and attain a depth of 200, 600, and 660 feet. There are some volcanoes, such as the Sumbing, in which these ravines assume so perfect a regularity, that the whole mountain, with its equidistant furrows and its intermediate walls, resembles a gigantic edifice based upon enormous buttresses, like the nave of a Gothic cathedral.*

Formerly the beauty of the island and the fury of its volcanoes were the cause of its being altogether dedicated to Siva, the god of destruction; and in the very craters of the burning mountains the worshippers of Terror and Death were in the habit of building their temples. In many spots the ruins of these sanctuaries are discovered in the midst of trees and thickets, which the Arab conquerors have left to grow in the formidable cavities of the volcanoes. Semeræ, the loftiest peak in the island, was the sacred mountain *par excellence*; the Sumbing, which rises in the centre of the island, was the "nail which fastens Java to the earth." Even in our own time some faithful followers of Siva inhabit a sandy plain, more than four miles wide, which was once the crater of the Tengger volcano; every year they proceed solemnly to pour rice on the summit of an eruptive cone, into the roaring mouth of the monster. In like manner, in New Zealand, the ever-smoking orifice of Tongariro was considered as the only place worthy of receiving the dead bodies of their great chiefs: when cast into the crater, the heroes went to sleep among the gods.

But the volcanic divinities, like most of the other rulers invoked by nations, did not content themselves with the fruits of the earth or the companionship of a few warriors; they also demanded blood, both by their subterranean roarings, by their thundering eruptions, and their devastating rivers of lava. Innumerable sacrifices have been offered to volcanoes to appease their anger: impelled by a

* Arnold Buscowitz, *Volcans et Tremblements de Terre*.

mingled feeling of fear and ferocity, the priests of not a few religions have cast victims with great pomp into the gaping hollows of these immense furnaces. Scarcely three centuries ago, when the disciples of Christianity were exterminated over the whole length and breadth of Japan, the followers of the new religion were thrown by hundreds into one of the craters of the Unsen, one of the most beautiful volcanoes of the archipelago; but this offering to the offended gods did not appease their anger, for, towards the end of the eighteenth century, this very same mountain and the neighbouring summits caused by their eruptions one of the most frightful disasters of any that are mentioned in the history of volcanoes. Actuated by a feeling of dread very similar to that exhibited by the Japanese priests, the Christian missionaries in America recognised in the burning mountains of the New World, not the work of a god, but that of the devil, and went in procession to the edge of the craters to exorcise them. A legend tells how the monks of Nicaragua climbed the terrible volcano of Momotombo, in order to quiet it by their conjurations; but they never returned: the monster swallowed them up.

CHAPTER LXVI.

COMPOSITION OF LAVAS; TRACHYTES; PUMICE-STONE; OBSIDIAN; BASALTS; BASALTIC COLONNADES.

LAVA is the most important product of the volcanic fires. The various kinds of lava differ very much in their external appearance, in the colour of their substance, and in the variety of their crystals, but they are all composed of silicates of alumina or magnesia, combined with protoxide of iron, potash or soda, and lime. When the feldspathic minerals predominate, the rock is generally of a whitish, greyish, or yellowish hue, and receives the name of trachyte. When the lava contains an abundance of crystals of augite, hornblende, or titaniferous iron, it is heavier, of a darker colour, and often more compact; it then takes the generic denomination of basalt. Numerous varieties, diversely designated by geologists, belong to this group.

Of all the lavas, trachyte is the least fluid in its form. In many places rocks of this nature have issued from the earth in a pasty state, and have accumulated above the orifice in the shape of a dome, "just like a mass of melted wax."* In this way were formed the great domes of Auvergne, the Puys de Dôme and de Sarcouy. In this district the flows of trachytic lava are very inferior in length to the basaltic *cheires;* the most important do not exceed four or five miles in length. At the present day, eruptions of trachyte are much more rare than those of other lavas; so much so, that certain authors class all the trachytic rocks among the formations of anterior ages. It is, however, ascertained that most of the American volcanoes and those of the Sunda Archipelago vomit out lava of this nature; the last eruptions of the Æolian Isles, Lipari, and Volcano, likewise produced only trachyte and pumice-stone.

This latter substance resembles certain white, yellow, or greenish scoriæ, which issue like a frothy dross from the furnaces of our ironworks, and is, like the compact trachyte, of a feldspathic nature. Some mountains are almost entirely composed of it; among others, the Monte Bianco of Lipari, which, viewed from a distance, appears

* Poulett Scrope, *Volcanoes: the Character of their Phenomena, &c.*

as if covered with snow. Long white flows, like avalanches, fill up all its ravines, from the summit of the mountain to the shore of the Mediterranean; the slightest movement caused by the tread of an animal or a gust of wind detaches from the surface of the slope hundreds of stones, which bound down to the foot of the incline, and are borne away by the waves which bathe the base of the mountain. In the southern part of the Tyrrhenean Sea, and especially in the vicinity of the Lipari (Æolian) Islands, the water is sometimes covered with these floating stones, almost like flakes of foam. In the Cordilleras the currents of fresh water convey the morsels of pumice to considerable distances. The river Amazon drifts down large quantities of pumice as far as its mouth, more than 3,000 miles from the place where it fell into the river. Bates says that the Indians, who live too far away from the volcanoes even to know of their existence, assert that these stones, floating down the river by the side of their canoes, are assuredly solidified foam.

The external appearance of various lavas differs even more than their chemical composition. The more or less perfect state of fluidity, and the presence in them of a greater or less quantity of bubbles of vapour, give a very different texture to rocks which are composed of the same elements. Pumice-stone has the appearance of sponge; obsidian looks like black glass, and sometimes even it is semi-transparent. It is entirely liquid, and issues from the interior of the earth like a stream flowing rapidly over the steeper slopes, and coagulating slowly in large sheets in the low ground and on the gentle inclines whither its own weight has drawn it. The surface of obsidian—for instance, that of Teneriffe—shines with a vitreous glitter; the cleavage of the rock is clean and sharp.

Some less degree of fluidity in the current of lava gives it sometimes the appearance of resin; this is the stone which is called *pechstein* (pitch-stone). When the rock, issuing in a state of fusion from the bosom of the mountain, becomes still cooler, it contains innumerable perfectly-formed crystals, and only owes its fluidity to the particles of vapour in its pores. The external layer of the lava is also immediately covered with scoriæ which float in flakes on the fiery stream. These scoriæ, too, assume a great variety of shapes; some are mammillated, others are exceedingly rough and irregular. In the Djebel-Hauran, near the crater of Abu-Ganim, there is an infinity of needles of red lava, about a yard high on the average, and bent in various directions towards the surface of the plateau; one might often fancy them flames half beaten down under the pressure of the wind. According to M. Wetzstein, these strange stone needles proceed from an eruption of flaky lava. In the Sandwich Islands, and in the Island of Réunion, certain crystals of a ferruginous appearance are grouped at the outlet of the crater in herbaceous forms of the most curious and sometimes elegant character. Some of the products of the volcano of Mauna-Loa and Kilauea resemble the tow of hemp! These are the whitish filaments which are sometimes carried away by the wind; the Kanakes used to consider them as the hair of Pele, the goddess of fire.

Among the old basaltic lavas there are some to which the name of *basalt* is more specially applied, which present a columnar disposition with wonderful regularity. These form the enormous monu-

Fig. 180.—Flow of Vitreous Lava at Mauna-Loa.

ments, much more imposing than those of man, which seem as if they had been constructed by giant builders, turning their mighty hands to the noble art of architecture, which is still practised, though on a smaller scale, by us their feeble descendants. These magnificent colonnades of basalt are everywhere attributed to giants. In Ireland, on the coast of Antrim, the summits of 40,000 prisms, levelled pretty regularly by the waves of the sea, and resembling a vast paved quay, have received the name of the Giant's Causeway. In Scotland, the beautiful cave of the Isle of Staffa, hollowed out by the action of the waves between two ranges of basaltic shafts, is celebrated as the work of Fingal, the demigod. In the Sicilian Sea, the Faraglioni Isles, or Isles of the Cyclopes, situated not far from Catania, at the base of Etna, are looked upon by tradition as the rocks cast by Polyphemus on the ships of Ulysses and his companions. Many of these prisms are from 100 to 160 feet high, and are not less than 6 to 10 feet in thickness. Near Fair Head and the Giant's Causeway some of the shafts connected with the perpendicular cliff of the headland are nearly 400 feet in height. In the Isle of Skye, some of the columns, according to M'Culloch's statement, are still higher. On the other hand, there are also colonnades in miniature, each shaft of which is not more than three-quarters of an inch to an inch from the summit to the base; instances of these are found in the basalts of the hill of Morven in Scotland.

Some geologists have thought that basaltic columns could not be formed except under the pressure of enormous masses of water; but a comparative study of these rocks in different parts of the world has proved that several beds of lava are arranged in columns at heights considerably above the level of the sea. In this colonnade-like formation of lava there is, however, no phenomenon which is entirely peculiar to basalt. Trachyte, also, sometimes assumes this form, and M. Fouqué has discovered a magnificent instance of it in the Island of Milo, in which there is a cliff composed of prismatic shafts 320 feet in height. Masses of mud when dried in the sun, the alluvium of rivers, beds of clay or tufa, and, in general, all matter which, in consequence of the loss of its moisture, passes from a pasty to a solid state, either in a state of nature or in our manufactories and dwellings, likewise assume a columnar structure similar to that of the basaltic lava. In fact, the entire mass, when gradually losing the moisture which swelled out its substance, cannot contract so as to shift the position of all its particles towards the centre; certain points remain fixed, and round each of these the contraction of a portion of the mass takes place. In basalt, in particular, it is the lower layer which assumes the columnar structure; for these alone cool gently enough to allow the phenomena of contraction to follow the normal course. The highest portion of the mass, being deprived, immediately after its issue from the earth, of the caloric and the steam which filled its pores, is almost immediately transformed into a more or less rough and cracked mass. But this very crust protects the rest of the lava against any radiation, and serves as a covering to the semi-crystalline columns which, by the continual contraction of their particles, are slowly separated from the rest of the mass. When a section of a bed of basaltic lava has been laid bare by the water of a river, the waves of the ocean, or earthquakes, the rough stone of the top layers may be seen lying, with or without any gradual transition, on a forest of prisms, sometimes rudimentary in their shape, but often no less regular than if they had been carved out by the hand of man. Most are of an hexagonal form; others, which were probably subject to less favourable conditions, have four, five, or seven faces; but all are definitely separated from one another by their particles gathering round the central axis. Mr. Poulett Scrope describes a fact which proves the enormous power of this contractile force. The colonnade of Burzet, in Vivarais, contains numerous nodules of olivine, many of which are as large as a man's fist; and, in spite of their extreme hardness, have been divided into two pieces, each fixed in one of two adjacent columns. Although the two corresponding surfaces have been polished by the infiltration of water, it is impossible to doubt that the two separate portions were not once joined in the same nodule.

As natural philosophers have verified by experiments on various viscous substances, basaltic shafts are always formed perpendicularly to the surface of refrigeration. Now, this surface being inclined, according to the locality, in a diversity of ways, the result is, that the columns may assume a great variety of directions in their position. Although most of them are vertical, on account of the cooling taking place in an upward direction, others, as at St. Helena, take a horizontal direction, and resemble trunks of trees heaped upon a woodpile. In other places, as at the Coupe d'Ayzac in Auvergne, the columns of a denuded cliff are arranged in the form of a fan, so as to lean regularly on the wall of the cliff as well as on the ground of the valley. At Samoskœ, in Hungary, a sheet of columnar basalt, very small at its origin, spreads out from the top of a rock like the water of a cascade, and hangs suspended over a precipice, resembling a cupola which has lost its base. Elsewhere, masses of basaltic pillars radiate in every direction, like the weapons in an immense trophy of arms.

An exact prismatic form is not, however, the only shape assumed by the cooling lava. The phenomenon of contraction takes place in different ways, according to the nature of the erupted matter, the declivity of the slopes, and all the other surrounding circumstances. Thus, in consequence of the sinking of the rock, most basaltic prisms exhibit at intervals a kind of joint, which gives the columns a resemblance to gigantic bamboos. In some lavas these joints are so numerous, and the edges of the stone are so eaten away by the weather, that the shafts are converted into piles of spheroids of a more or less regular form. At the volcano of Bertrich, in the Eifel, one might fancy them a heap of cheeses; whence comes the name "Cheese Cave," which is given to one of the caverns which opens in the flow of lava. Sometimes, too, crystals scattered about in the midst of the mass have served as nuclei to globular concretions formed of numerous concentric layers. Lastly, many currents of molten matter present a tabular or schistose structure, caused, like that of slate, by the pressure of the superincumbent masses.

CHAPTER LXVII.

SOURCES OF LAVA; STROMBOLI; MASAYA; ISALCO; KILAUEA.—LATERAL CRE-
VICES IN VOLCANOES.—ERUPTION AND MOTION OF LAVA.

ALTHOUGH lava, when cooled, is easy enough to study, it is more difficult to observe with any exactitude the molten matter immediately on its exit from the craters or fissures; besides this, the opportunities for study which are offered to *savants* are sometimes very dangerous. Long years often elapse before an inquirer can notice at his ease, and without fear of sudden explosions, the mouths of Etna or Vesuvius filling up to the brink with boiling lava.

Stromboli is the only volcano in Europe in which this phenomenon occurs regularly at closely-recurring intervals, sometimes of only five minutes, or even more frequently. When an observer stands on the highest edge of the crater, he sees, about 300 feet below him, the waves of a matter which shines like molten iron, and tosses and boils up incessantly; sometimes it swells up like an enormous blister, which suddenly bursts, darting forth eddies of vapour accompanied by solid fragments. For centuries past the lava has never ceased to boil in the cavity of Stromboli, and it is but very rarely that a period of even a few hours elapses without molten matter overflowing. Thus the crater, which, during the day, is white with steam, and during the night red with the glare of the lava, has served as a lighthouse for mariners ever since the first vessel ventured upon the Tyrrhenian Sea.

In Nicaragua, to the north of the Great Lake, the volcano of Masaya (or "Devil's Mouth") presents a spectacle similar to that of Stromboli, but grander, and perhaps still more regular. After having remained in a state of repose for nearly two centuries, from 1670 to 1853, the monster—which has received the name it bears from the frightful turbulence of its burning waves—resumed all its former activity. In this crater the enormous bubbles of lava, which ascend from the bottom of the abyss and throw out a shower of burning stones, break forth in a general way every quarter of an hour.

The volcano of Isalco, not far from Sonsonate, in the State of San

Salvador, is also one of the most curious on account of its regularity. Its first breaking out was noticed on the 29th of March, 1793, and since this date* it has almost always continued to increase in size by throwing outside its cavity ashes and stones. Some of its eruptions, remarkable for their comparative violence, have been accompanied by flows of lava; but, generally, the crater of Isalco confines itself to hurling burning matter to a height of 39 to 46 feet above its crater: explosions follow one another at intervals of every two minutes.† The total elevation of the cone of *débris*, above the village of Isalco, being 735 feet, and the slope of the side of the mass being, on the average, 35 degrees, M. von Seebach, one of the observers of the volcano, has been able to calculate approximately the bulk and

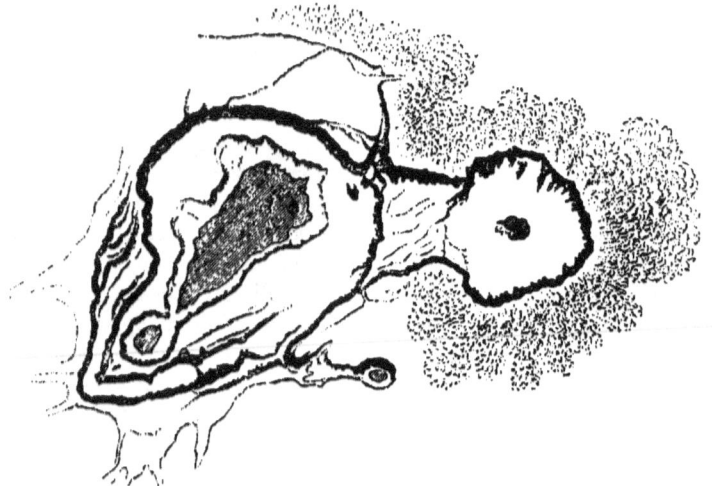

Fig. 190.—Craters of Kilauea.

regular increase of the mountain. In 1865 the mass of *débris* was about 35,000,000 of cubic yards, giving an increase of about 491,000 cubic yards every year, or 56 cubic yards every hour. The volcano, therefore, might be looked upon as a gigantic hour-glass.‡

Of all the craters in the world, the one which most astonishes those who contemplate it is the crater of Kilauea, in the Island of Hawaii. This volcanic outlet opens at more than 3,900 feet of elevation on the sides of the great mountain of Mauna-Loa, which is itself

* M. Squier gives another date, the 23rd of February, 1770, but it is probably an error.
† Moritz Wagner; Carl von Scherzer.
‡ *Zeitschrift für Erdkunde*, 1866.

Fig. 191.—Crater of Mauna-Loa.

crowned by a magnificent funnel-shaped crater, 2,735 yards across from one brink to the other. The elliptical crater of Kilauea is no less than 3 miles in length and 7 miles in circumference. The hollow of this abyss is filled by a lake of lava, the level of which varies from year to year, sometimes rising and sometimes falling like water in a well. In a general way, it lies about 600 to 900 feet below the outer edge, and in order to study its details, it is necessary to get on to a kind of ledge of black lava which extends round the whole circumference of the gulf; this is the solidified edge of a former sheet of molten matter, similar to those circular benches of ice which, in northern countries, border the banks of a lake, and even in spring still mark the level the water has sunk from. The surface of the sea of fire is generally covered by a thick crust over its whole extent; here and there the red lava-waves spring up like the water of a lake through the broken ice. Jets of vapour whistle and hiss as they escape, darting out showers of burning scoriæ, and forming cones of ashes on the crust, 60 to 100 feet in height, which are so many volcanoes in miniature. Intense heat radiates from the

Fig. 102.—Section across the Craters of Kilauea.

immense crater, and a kind of hot blast makes its way through all the chinks in the vertical walls of the sides. In the midst of the hot vapours, one feels as if lost in a vast furnace. During the night-time an observer might fancy himself surrounded with flames; the atmosphere itself, coloured by the red reflection of the vent-holes of the volcano, seems to be all on fire.

The level of the fire-lake of Kilauea is incessantly changing. In proportion as fresh lava issues forth from the subterranean furnace, the broken crust affords an outlet to other sheets of molten matter and fresh heaps of scoriæ, and gradually the boiling mass rises from ledge to ledge, and ultimately reaches the upper edge of the basin. Sooner or later, however, the level rapidly sinks. The fact is, that the burning mass contained in the depths of the abyss gradually melts the lower walls of solid lava; these walls ultimately give way at some weak points in their circumference, a crevice is produced in the outer face of the volcano, and the liquid matter "drawn off," like wine from a vat, rushes through the opening made for it. The flow increases the orifice by the action of its weight on the sill of the opening, and by melting the rocks which oppose its passage, and then running down over the slopes, flows into the sea, forming promontories on the shore. In 1840 the crater was full to the brink, when a crack suddenly opened in the side of the mountain. This fissure extended to a distance of 131 feet from its starting-point, and vomited forth a stream of lava 37 miles long and 16 miles wide, which entirely altered the outline of the sea-coast, and destroyed all the fish in the adjacent waters. Mr. Dana estimated the total mass of this enormous flow as equal to 7,200,000 cubic yards; that is, to a solid body fifty times as great as the quantity of earth dug out in cutting through the Isthmus of Suez. The enormous basin of Kilauea, 1,476 feet deep, remained entirely empty for some time, and the former lake of lava left no other trace of its existence than a solid ledge like those which had been formed at the time of previous eruptions. Since this date the great caldron of lava has been several times filled and several times emptied, either altogether or in part.

Almost all the volcanoes which rise to a great height get rid, like Kilauea, of their overflow of lava through fissures which open in their side walls. In fact, the column of molten matter which the pressure of the gas beneath raises in the pipe of the crater is of an enormous weight, and every inch it ascends towards the mouth of the crater represents an expense of force which seems prodigious. The more or less hypothetical calculations which have been made as to the degree of pressure necessary for the steam to be able to act on the lava-furnace lead to the belief that the outlet-conduits of volcanoes, and consequently the mass of liquid stone to be lifted, are not less than nine miles in depth.* Various geologists—amongst others, Sartorius von Waltershausen, the great explorer of Etna—believe that the volcano-shafts are of a still more considerable depth. The rocks of the terrestrial surface, limestone, granite, quartz, or mica, are of a specific gravity two and a half times superior to that of water, whilst the planet itself, taken as a whole, weighs nearly five and a half times as much as the same mass of distilled water; the density of the interior layers must therefore increase from the circumference to the centre. With regard to the proportion of this increase, it is established by a calculation, the whole responsibility of which must rest

* Buff, *Physik der Erde*.

upon its authors. Baron Waltershausen has ascertained, by means of a great number of weighings, that the lava of Etna and that of Iceland have a specific gravity of 2·911. The presumed consequence of this fact is that the rocks thrown out by the volcanoes of Sicily and Iceland proceed from a depth of 77 to 78 miles (?). Thus, the shaft which opens at the bottom of the crater of Etna would be no less than 77 miles deep, and the lava which boils in this abyss would be lifted by a force of 36,000 atmospheres, an idea altogether incomprehensible by our feeble imaginations. There would, then, be nothing astonishing in the fact that a mass of lava, which is sufficiently heavy to balance a pressure of this kind, should, in a great many eruptions, melt and break through the weaker parts of its walls instead of ascending some hundreds or thousands of feet higher so as to run out over the edge of the upper crater.

When the side of the mountain opens, and affords a passage to the lava, the fissure is always perceptibly vertical, and those which are continued to the summit pass through the very mouth of the volcano. In a general way these fissures of eruption are of considerable length, and are sufficiently wide to form an impassable precipice. Before these fissures become obliterated by the lava, or by other *débris*—such as the snow and earth of avalanches—they may be traced out by the eye as deep furrows hollowed out on the mountain side. In 1669 the lateral fissure of Etna extended over more than two-thirds of the southern side—from the plains of Nicolosi to the terminal gulf of the great crater. In like manner, in the Isle of Jan Mayen, the volcano of Beerenberg, 7,513 feet high, presents from top to bottom a long depression filled up with snow, which is nothing else than a fissure of eruption. On other mountains, especially in Montserrat, Guadaloupe, and Martinique, these fissures have assumed such dimensions that the peaks themselves have been completely split in two.

Through outlets of this kind the lava jets out, first making its appearance at the upper part, where the declivity is generally steeper, then springing out below on the more gentle slopes of the lower regions of the mountain.

At the source itself the lava is altogether fluid, and flows with considerable speed—sometimes, on steep slopes, faster than a horse can gallop; but the course of the molten stone soon slackens, and the liquid, hitherto dazzling with its light, is covered by brown or red scoriæ, like those of iron just come out of a furnace. These scoriæ come together, and, combining, soon leave no interstices between them beyond narrow vent-holes, through which the molten matter escapes. The scoriæ then form a crust, which is incessantly breaking with a metallic noise, but gradually consolidates into a perfect tunnel round the river of fire; this is the *cheire*,[*] thus named on account of the asperities which bristle on its surface. Any one may safely venture on the arch-shaped crust, although only a few inches above the mass in state of fusion, without any fear of being burnt—just as in winter we trust ourselves on the sheets of ice which cover a running stream. The pressure of the lava succeeds in breaking through its shell only at the lower parts of its flow, in spots where the waves of burning stone fall with all their weight. Then the envelope is suddenly ruptured, and the mass springs out like water from a sluice, pushing before it the resounding scoriæ, and swelling out gently in the form of an enormous blister; it then again becomes covered with a solid crust, which is again broken through by a fresh effort of the lava. Thus the river, surrounding itself with dikes which it constantly breaks through, gradually descends over the slopes, terrible and inexorable, so long as the original stream does not cease to flow. The only means of diverting the current is to modify the incline in front of it—either by opposing obstacles to it to throw it to either side, or by preparing a road for it by digging deep trenches, or by opening up above some lateral outlet for the pent-up lava. In 1669, at the time of the great eruption which threatened to swallow up Catania, all these various means were adopted in order to save the town. On one side the inhabitants worked at consolidating the rampart, and placed obstacles across the path of the current to turn it towards the south. Other workmen, furnished with shovels and mattocks, ascended along the edge of the flow, and, in spite of the resistance offered by the peasants, tried to pierce through the shell of scoriæ, and thus, by tapping the stream, to open fresh outlets for the molten matter. These means of defence partly succeeded, and the terrible current which, at its source near Nicolosi, had been able to melt and pierce through the volcanic cone of Monpilieri at its thickest point (this cone standing in its path), was turned from its course towards the centre of Catania, and destroyed nothing but the suburbs.

The radiation from the lava being arrested by the crust of scoriæ, which is a very bad conductor of heat, the temperature of the air surrounding a flow of lava rises but very slightly. The Neapolitan

[*] Or *serre*, in Italian *sciara*: these are synonyms of the word *scie* (saw) in the French of the present day.

guides have no fear in approaching the Vesuvian lava in order to stamp the rough medals made of it, which they sell to foreigners. At a distance of a few yards from the vent-holes in the *cheire*, the trees of Etna continue to grow and blossom, and some clumps, indeed, may be seen flourishing on an islet of vegetable earth lying between two branches of a flow of burning lava. And yet, by a contrast which at first sight seems incomprehensible, it sometimes happens that trees which are distant from any visible flow of molten matter suddenly wither and die. Thus, in 1852, at the time of the great eruption from the Val del Bove, on the eastern slopes of Mount Etna, vineyards and vines, covering a considerable area, and situated at a distance of more than half a mile below the front of the flow, were suddenly dried up, just as if the blast of a fire had burnt up their foliage. In order to explain this curious phenomenon it is necessary to admit that some rivulets of the great lava-river must have penetrated under the earth through the fissures of the soil, and have filled up a subterranean cavity in the mountain exactly below the vineyards that were destroyed: the roots being consumed, or deprived of the necessary moisture, the trees themselves could not do otherwise than perish.*

On lofty mountains in a state of eruption, the masses of snow and ice, which are covered by the fiery currents which issue from the volcanic fissures, do not always melt, and some have been preserved under the scoriæ for centuries, or even thousands of years. Lyell has discovered them under the lava of Etna, American geologists under the masses thrown out by the crater of Mount Hooker, Darwin under the ashes in Deception Island, in the Tierra del Fuego, M. Philippi under the flows of the volcano Nuevo de Chillan,† which, in 1861, erupted through a glacier. There every bed of snow which falls during the winter remains perfect under the coat of burning dust which is ejected from the outlet of eruption, and sections made through the mass of *débris* show for a great depth the alternate black and white strata of the volcanic ashes and the snow. In 1860 the crater of the mountain of Kutlagaya, in Iceland, hurled out simultaneously into the air lumps of lava and pieces of ice, all intermingled together.‡

In like manner, the immense flows of lava in Iceland have left in a perfect state of preservation the trunks of the *Sequoias*, and other

---

* Lyell, *Philosophical Transactions*, 1858.
† *Mittheilungen von Petermann*, vol. vii., 1863.
‡ Wallich, *North Atlantic Sea-bed*.

American trees, which adorned the surface of the island during the ages of the Tertiary epoch, at a time when the mean temperature of this country was 48° (Fahr.), that is, 42° to 44° above that which it is at present.* Although the radiation from the lava is so slight that it neither melts the ice nor burns the trunks of buried trees, yet, on the other hand, the heat and fluidity of the lava are maintained in the central part of the flow for a very considerable number of years. Travellers state that they have found deeply-buried lava which was still burning after it had remained for a century on the mountain side.

Although the lava covers up and often preserves the snow and the

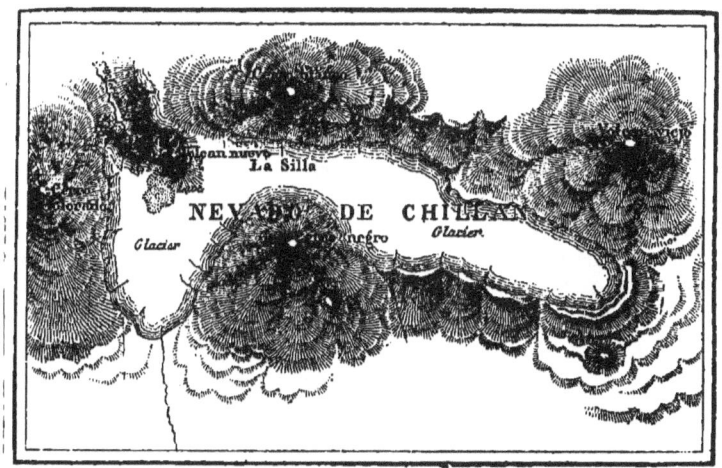

Fig. 193.—Nevado de Chillan.

ice, which are doubtless defended against the heat by a cushion of spheroidal particles of humidity, it immediately converts into steam the water with which it comes in contact. The liquid mass, being suddenly augmented to about 1,800 times its former volume, explodes like an enormous bomb-shell, and hurls away, like projectiles, all the objects which surround it. A serious occurrence of this kind is recorded, which took place in 1843, a few days after the formation of a fissure in Mount Etna, from which a current of molten matter issued, making its way towards the plain of Bronte. A crowd of spectators, who had come from the town, were examining from a distance the threatening mass, the peasants were cutting down the

* Carl Vogt, *Nordfahrt*.

trees in their fields, others were carrying off in haste the goods from their cottages, when suddenly the extremity of the flow was seen to swell up like an enormous blister, and then to burst, darting forth in every direction clouds of steam and volleys of burning stones. Everything was destroyed by this terrible explosion,—trees, houses, and cultivated ground; and it is said that sixty-nine persons, who were knocked down by the concussion, perished immediately, or in the space of a few hours. This disaster was occasioned by the negligence of an agriculturist, who had not emptied the reservoir on his farm; the water, being suddenly converted into steam, had caused the lava to explode with all the force of gunpowder.

The quantity of molten matter which is ejected by a fissure in one single eruption is enormous. It is known that the current of Kilauea, in 1840, exceeded 6,550 millions of cubic yards. That which proceeded from Mauna-Loa in 1835 produced a still larger quantity of lava, and extended as far as a point 76 miles from the crater. Flows of this kind are certainly rare; but there are some recorded in the earth's history which are still more considerable. Thus, the volcano of Skaptar-Jokul, in Iceland, was cleft asunder in 1783, and gave vent to two rivers of fire, each of which filled up a valley; one attained a length of 50 miles, with a breadth of 15 miles; the other was of less dimensions, but the depth of the mass was in some places as much as 492 feet. A subterranean fissure, 99 miles in length, which cleaves in two the ground of Iceland, was doubtless filled up with lava along its entire length, for hillocks of eruptions sprung up on various points of this straight line. It has been calculated that the whole of the lava evacuated by the Skaptar in this great eruption was not less in bulk than 655,000 millions of cubic yards, a mass equivalent to the whole volume of Mont Blanc; it would be a quantity sufficient to cover the whole earth with a film of lava 0·0393 inch in thickness. As to the celebrated flow from the Monti-Rossi, which threatened to destroy Catania in 1669, it seems very trifling in comparison; it contained a mass of molten stone, which was estimated at 1,310 millions of cubic yards. On how trifling a scale, therefore, are these ordinary eruptions compared with the surface of the globe! They are, however, phenomena perceptible enough to man, in all his infinite littleness.

CHAPTER LXVIII.

VOLCANIC PROJECTILES.—EXPLOSIONS OF ASHES.—SUBORDINATE VOLCANOES.—
MOUNTAINS REDUCED TO DUST.—FLASHES AND FLAMES PROCEEDING FROM
VOLCANOES.

THE lava swelling up in enormous blisters above the fissures from which it flows in a current over the slopes is far from being the only substance ejected from volcanic mountains. When the pent-up vapour escapes from the crater with a sudden explosion, it carries with it lumps of molten matter, which describe their curve in the air, and fall at a greater or less distance on the slope of the cone, according to the force with which they were ejected. These are the volcanic projectiles, the immense showers of which, traced in lines of fire on the dark sky, contribute so much during the night-time to the magnificent beauty of volcanic eruptions. These projectiles have already become partially cooled by their radiation in the air, and when they fall are already solidified on the outside, but the inside nucleus remains for a long time in a liquid or pasty state. The form of these projectiles is often of an almost perfect regularity. Each sphere is in this case composed of a series of concentric envelopes, which have evidently been arranged in the order of their specific gravity during the flight of the projectile through the air. The dimensions of these projectiles vary in each eruption; some of them are one or more yards in thickness; others are nothing but mere grains of sand, and are carried by the wind to great distances.

In most eruptions, these balls of lava, still in a fluid and burning state, constitute but a small part of the matter thrown out by the mountain. The largest proportion of the stones ejected proceed from the walls of the volcano itself, which break up under the pressure of the gas, and fly off in volleys, mingled with the products of the new eruption. This is the origin of the dust or ashes which some craters vomit out in such large quantities, which, too, are the cause of such terrible disasters.

When the impetus of the gas confines itself to forming a fissure in the side of the mountain, the fragments of rocks which are broken

N N

up and reduced to powder are comparatively small in quantity. They are projected in clouds out of the fissure, and falling like hail round the orifice, are gradually heaped up in the form of a cone on the side of the mountain from which they arose. In Europe the enormous circumference of Etna presents more than 700 of these subordinate volcanoes, some scarcely higher than an Esquimaux hut, and others, like the Monti-Rossi, Monte-Minardo, Monte-Ilici, several hundred yards high, and more than half a mile wide at the base. There are some which are entirely sterile or covered only by a scanty vegetation of broom, and are marked out by a red, yellow, or even black colour on the main body of Etna; those situated on the lower slopes are covered with trees or planted with vines, and sometimes contain admirable crops in the very cavity on their summit. These cones of ashes, springing up like a progeny on the vast sides of their mother-mountain, give to Etna a singular appearance of vital personality and of creative energy. The same phenomenon occurs on the volcanoes of Hawaii, which carry on their declivities thousands of subordinate cones.

In the formation of these hillocks a real division of labour takes place. The rocks and heavier stones fall either on the edge of the crater or in the gulf itself. The ashes and light dust are shot up to a much greater height, and, hurried along by the impulse of the wind, fall far and wide, like the chaff of corn winnowed in a threshing-floor. Thus the slope of the cone towards which the wind directs the ashes is always more elongated, and rises to a greater height on the edge of the crater. On Etna, where the wind generally blows in the direction of west to east, the eastern slope of the hillocks is more developed than on the opposite side. It must, perhaps, be attributed to the action of the wind blowing on the heights, and not, as Siemsen,* the geologist, supposes, to the obliquity of the shaft of the crater, that all the scoriæ and ashes fall to the north of the orifice of the volcano Nuevo de Chillan, in Chili.

The phenomena which take place when the ashes issue from the mouth of the crater itself do not differ from those which are observed at the outlets in fissures. In the former case, however, the mass of rocks reduced to powder is so considerable that the rain of ashes assumes all the proportions of a cataclysm. It has sometimes happened that, during a paroxysm of volcanic energy, the whole summit of a mountain, for a depth of several thousands of feet, has been hurled into the air, mingled with a cloud of vapour and the smoke of

* *Mittheilungen von Petermann*, vol. vii., 1863.

burning lava. Thus, Etna, if we are to believe Ælianus, was once much loftier than it is in our time, and on the north of the present terminal cone there may, in fact, be noticed a kind of platform which seems to have been the base of a summit twice as high as the present crest. The whole of the Val del Bove is probably an empty space left by the disappearance of a former cone.

With regard to Vesuvius, it is known that, in the year 79 of the present era, the whole of that part of the mountain which was turned towards the sea was reduced to powder, and that the *débris* of the cone, nothing of which now remains except the semicircular inclosure of La Somma, buried three towns and a vast extent of plain. The ashes and dust, mingled with white vapour rising in thick eddies, ascended in a column to a point far above the summit of the volcano, until, having reached those regions of the atmosphere where the rarefied air could no longer sustain them, they spread out into a wide umbrella-like shape, the falling dust of which obscured the sky. Pliny the Younger compared this vault of ashes and smoke to the

Fig. 194.—Regular Cone of Ashes. Fig. 195.—Cone of Ashes modified by the Wind.

foliage of an Italian pine curving at an immense height over the mountain. Since this memorable epoch the height of the column of vapour has been measured which has issued from Vesuvius at the time of several great eruptions, and it has been sometimes found that it reached 23,000 to 26,000 feet; that is, six times higher than the summit of the volcano itself.

One of these explosions of entire summits which caused most terror in modern times was that of the volcano of Coseguina, a hillock of about 500 feet high, situated on a promontory to the south of the Bay of Fonseca, in Central America. The *débris* hurled into the air spread over the sky in a horrible arch several hundreds of miles in width, and covered the plains, for a distance of 25 miles, with a layer of dust at least 16 feet thick. At the very foot of the hill the headland advanced 787 feet into the bay, and two new islands, formed of ashes and stones falling from the volcano, rose in the midst of the water, several miles away. Beyond the districts close round the crater, the bed of dust, which fell gradually, became thinner, but it was carried by the wind more than 40 degrees of longitude towards the west, and the ships sailing in those waters penetrated with difficulty the layer of pumice-stone spread out on the sea. To the north, the rain of ashes was remarked at Truxillo, Honduras, and at Chiapas, in Mexico; on the south, it reached Carthagena, Santa Martha, and other towns of the coast of Grenada; to the east, being carried by the counter-current of the trade winds, it fell on the plains of St. Ann's, in Jamaica, at a distance of 800 miles. The area of land and water on which the dust descended must be estimated at

Fig. 196.—Cone of Etna, and Val del Bove.

1,500,000 square miles, and the mass of matter vomited out could not be less than 65,500 million cubic yards.

The uproar of the breaking up of the mountain was heard as far as the high plateaux of Bogota, situated 1,025 miles away in a straight line. Whilst the formidable cloud was settling down round the volcano, thick darkness filled the air. For forty-three hours nothing could be seen except by the sinister light of the flashes darting from the columns of steam, and the red glare of the vent-holes opening in the mountain. To escape from this prolonged night, the rain of ashes, and the burning atmosphere, the inhabitants who

ERUPTION OF COSEGUINA. 549

dwelt at the foot of Coseguina fled in all haste along a road running by the black water of the Bay of Fonseca. Men, women, children, and domestic animals travelled painfully along a difficult path, through quagmires and marshes. So great, it is said, was the terror of all animated beings during this long night of horror, that the animals themselves, such as monkeys, serpents, and birds, joined the

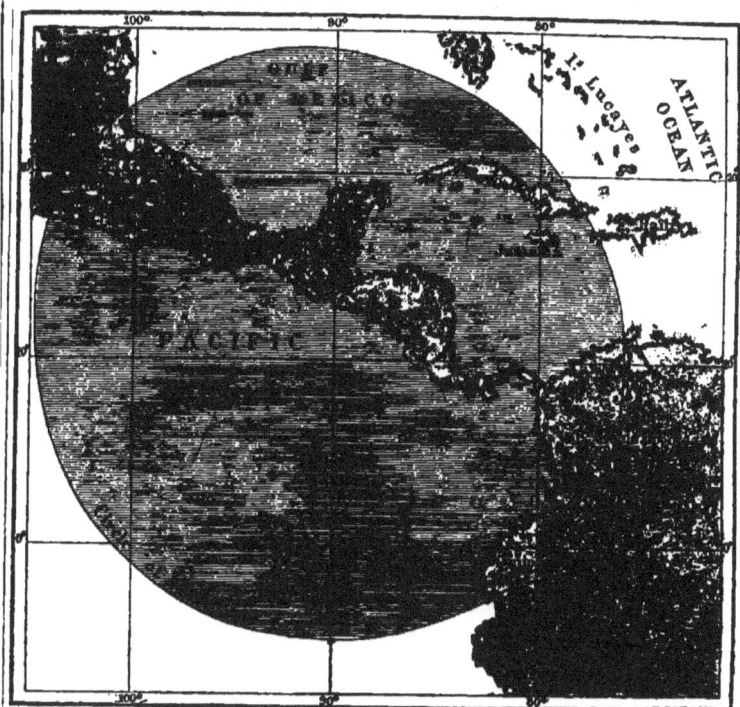

Fig. 197.—Eruption of Coseguina.

band of fugitives, as if they recognised in man a being endowed with intelligence superior to their own.*

A large number of volcanoes have diminished in height, or have, indeed, entirely disappeared, in consequence of explosions, which reduced their rocks to powder, and distributed them in thick sheets on the ground adjacent. Mount Baker, in California, and the Japanese volcano of Unsen, have thus raised the level of the surrounding plains at the expense of a diminution in their own volume.

* Landgrebe, *Naturgeschichte der Vulkane.*

In 1638, the summit of the peak of Timor, which might be seen like a lighthouse from a distance of 270 miles, exploded, and blew up into the air, and the water collecting, formed a lake in the enormous void caused by the explosion. In 1815, Timboro, a volcano in the Island of Sumbara, destroyed more men than the artillery of both the armies engaged on the battle-field of Waterloo. In the Island of Sumatra, 550 miles to the west, the terrible explosion was heard, and, for a radius of 300 miles round the mountain, a thick cloud of ashes, which obscured the sun, made it dark like night even at noonday. This immense quantity of *débris*, the whole mass of which was, it is said, equivalent to thrice the bulk of Mont Blanc—that is, 2,358,000 millions of cubic yards (?)—fell over an area larger than that of Germany. The pumice-stone which floated in the sea was more than a yard in thickness, and it was with some difficulty that ships could make their way through it. The popular imagination was so deeply impressed by this cataclysm, that at Bruni, in the Island of Borneo, whither heaps of the dust vomited out by Timboro, 870 miles away to the south, had been carried by the wind, they date their years from "the great fall of ashes." It is the commencement of an era for the inhabitants of Bruni, just as the flight of Mahomet was for the Mussulmen.

The friction of the steam against the innumerable particles of solid matter which are darted out into the air is the principal cause of the electricity which is developed so plentifully during most volcanic eruptions. In consequence of this friction, which operates simultaneously at all points in the atmosphere which are reached by the volcanic ashes and vapour, sparks flash out which are developed into lightning. The skies are lighted up not only by the reflection from the lava, but also by coruscations of light which dart from amidst the clouds. When the vast canopy of vapour spreads over the summit of the mountains, numerous spirals of fire whirl round on each side of the clouds, which, as they unroll, resemble the foliage of a gigantic tree. Doubtless, also, the encounter of two aërial currents may contribute to produce lightning in the columns of vapour; yet, when the latter are slightly mingled with ashes, they are rarely stormy.*

Although the evolution of electricity in the columns of vapour and ashes vomited out by volcanoes has never been called in question, the appearance of actual flames at the time of volcanic eruptions was for a long time disputed. M. Sartorius von Waltershausen, the patient observer of Etna, has maintained that neither this mountain, nor

* Arago, *Œuvres Complètes*, vol. i.

FLAMES FROM VOLCANOES.

Stromboli, nor any other volcano has ever presented among its phenomena any fire properly so called, and that the supposed flames were nothing more than the reflection of the red or white lava that was boiling in the crater. On the other hand, Élie de Beaumont, Abich, and Pilla positively assert that they have seen light flames on the summit of Vesuvius and Etna. It would, however, be very natural to believe that imflammable gases might be liberated and

Fig. 198.—Eruption of Timboro.

take fire at the outlet of those immense shafts, which place the great subterranean laboratory of lava in communication with the outer air.

This question was, however, resolved in the affirmative at the time of the recent eruption of Santorin, and popular opinion was right in opposition to most men of science. All those who were able to witness, at its commencement, the upheaval of the lava at Capo Georges and Aphroessa, have certified to the appearance of burning gas dancing above the lava, and even on the surface of the sea. All round the upheaved hillocks, bubbles of gas, breaking forth from the waves, became kindled as they came in contact with the burning mass, and were diffused over the water in long trains of white, red, or greenish flames, which the breeze alternately raised or beat down; sometimes a smart puff of wind put out the fire, but it soon recommenced to run over the breakers: by approaching it carefully, fragments of paper might be burnt in it, which lighted as they dropped. On the slopes of the volcano of Aphroessa, fire, rendered of a yellowish hue by salts of soda, sprung out from all the fissures, and rose to a height of several yards. On the rather older lava of Cape Georges the trains of flame were less numerous; there, however, bluish glimmers might be seen flitting about in some spots over the black ridges of lava.*

Added to this, are not the flames at Bakou, on the coast of the Caspian Sea, produced by the volcanic action of the ground? The "growing mountains" in the neighbourhood are mud-volcanoes, and we must, doubtless, attribute to the same subterranean activity the production of the hydrogen gas which burns in an "eternal flame" in the temples of the Parsi.† During some of the evenings in autumn, when the weather is fine and the sun has heated the surface of the ground, the flames occasionally make their appearance on the hills, and for several hours may be seen the marvellous spectacle of a train of fire stretching along the country, without burning the ground, and even without scorching a blade of grass.

* Fouqué, *Revue des Deux-Mondes*, August 15, 1866; Dekigallas; Schmidt.
† Arnold Boscowitz, *Volcans et Tremblements de Terre.*

CHAPTER LXIX.

STREAMS OF MUD EJECTED BY CRATERS.—MUD-VOLCANOES.

NEXT to lava and ashes, streams of water and mud are the most considerable products of volcanic activity, and the catastrophes which they have caused are perhaps amongst the most terrible which history has to relate. By means of these sudden deluges, towns have been swept away or swallowed up, whole districts dotted over with habitations have been flooded with mud or converted into marshes, and the entire face of nature has been changed in the space of a few hours.

The liquid masses which descend rapidly from the mountain height do not always proceed from the volcano itself. Thus the local deluge may be caused by a rapid condensation of large quantities of steam which escape from the crater and fall in torrents on the slopes. A phenomenon of this kind must evidently take place in a great many cases, and it was doubtless by a cataclysm of this kind that the town of Herculaneum, at the foot of Vesuvius, was buried. As regards the lofty snow-clad volcanoes of the tropical and temperate zones, and also those of the frozen regions, the torrents of water and *débris*—the " water-lava," as the Sicilians call them—may be explained by the rapid melting of immense masses of snow and ice, with which the burning lava, the hot ashes, or the gaseous emanations of the volcanic furnace have come in contact. Thus, in Iceland, after each eruption, formidable deluges, carrying with them ice, scoriæ, and rocks, suddenly rush down into the valleys, sweeping away everything in their course. These liquid avalanches are the most terrible phenomena which the inhabitants of the island have to dread. They show three headlands formed of *débris*, which the body of water descending from the sides of Kutlugaya in 1766 threw out far into the sea, in a depth of 246 feet of water.*

Other deluges no less formidable are caused by the rupture of the walls which pen back a lake in the cavity of a former crater, or by the formation of a fissure which affords an outlet to liquid masses

* Olafsen and Povelsen, *British Quarterly Review*, April, 1861.

contained in subterranean reservoirs. It would be difficult to explain otherwise the mud-eruptions of several trachytic volcanoes of the Andes—Imbambaru, Cotopaxi, and Carahuarizo. In fact, the mud (*lodozales*) which comes down from these mountains often contains a large quantity of organised beings, aquatic plants, infusoria, and even fish, which could only have lived in the calm waters of a lake. Of this kind is the *Pimelodes cyclopum*, a little fish of the tribe of the *Siluridæ*, which, according to Humboldt, has hitherto been found

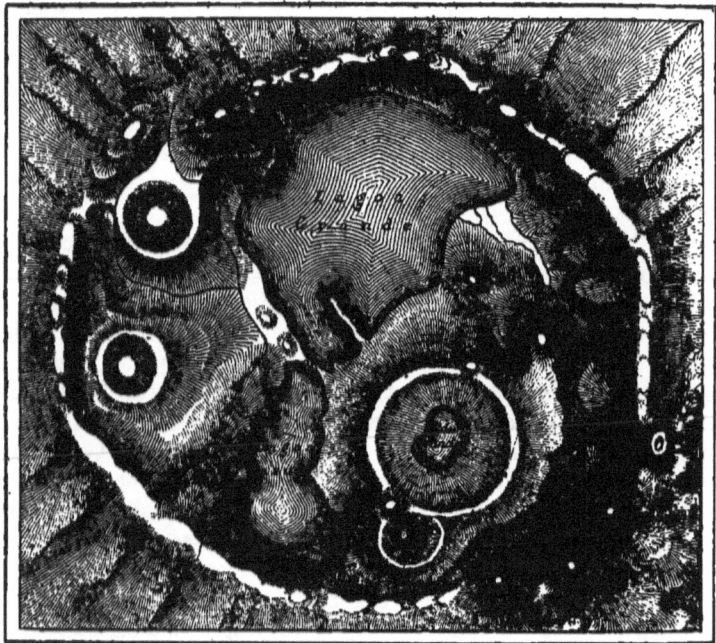

Fig. 109.—Crater of Sete Cidades.

nowhere except in the Andini caverns, and in the rivulets of the plateau of Quito. In 1691 the volcano of Imbambaru vomited out, in combination with mud and snow, so large a quantity of these remains of organisms that the air was contaminated by them, and miasmatic fevers prevailed in all the country round. The masses of water which thus rush down suddenly into the plains amount sometimes to millions, or even thousands of millions, of cubic yards.

Although, in some cases, these eruptions of mud and water may be looked upon as accidental phenomena, they must, on the contrary, as regards many volcanoes, be considered as the result of the normal action of the subterranean forces. They are, then, the waters of the sea or of lakes which, having been buried in the earth, again make their appearance on the surface, mingled with rocks which they have dissolved or reduced to a pasty state. A remarkable instance of these liquid eruptions is that presented by Papandayang, one of the most active volcanoes in Java. In 1792 this mountain burst, the summit was converted into dust and disappeared, and the *débris*, spreading far and wide, buried forty villages. Since this epoch a copious rivulet gushes out in the very mouth of the crater, at a height of 7,710 feet, and runs down into the plain, leaping over the blocks of trachyte. Round the spring, pools of water fill all the clefts in the rocks, and boil up incessantly under the action of the hot vapours which rise in bubbles; here and there are funnel-shaped cavities, in which black and muddy water constantly ascends and sinks with the same regularity as the waves of the sea; elsewhere, muddy masses slowly issuing from small craters flow in circular slopes over mounds of a few inches or a yard in height; lastly, jets of steam dart out of all the fissures with a shrill noise, making the ground tremble with the shock. All these various noises, the roaring of the cascades, the explosion of the gaseous springs, the hoarse murmur of the mud-volcanoes, the shrill hissing of the *fumerolles*, produce an indescribable uproar, which is audible far away in the plains, which, too, has given to the volcano its name of Papandayang, or "Forge," as if one could incessantly hear the mighty blast of the flames and the ever-recurring beating of the anvils.

In volcanoes of a great height it is rarely found that eruptions of water and mud are constant, as in the Papandayang; but temporary ejections of liquid masses are frequent, and there are, indeed, some volcanoes which vomit out nothing but muddy matter. The volcano of Aqua (or *water*), the cone of which is gently inclined like that of Etna, and rises to about 13,000 feet in height, into the regions of snow, has never vomited anything but water; and it is, indeed, stated that lava and other volcanic products are entirely wanting on its slopes.* Yet, in 1541, this prodigious intermittent spring hurled into the air its terminal point (*coronilla*), and poured over the plains at its base, and over the town of Guatemala, so large a quantity of water, mingled with stones and *débris*, that the inhabitants were compelled to fly with the greatest haste, and to reconstruct their

* Juarros, Landgrebe, *Naturgeschichte der Vulkane*, vol. i.

capital at the foot of the volcano of Fuego. This new neighbour, however, showed that he was as much or more to be dreaded than their former one, for the violent eruptions from the mountain compelled the inhabitants of the second town to again migrate, and to rebuild their capital at a point 20 miles to the north-west.

Several volcanoes in Java and the Philippines also give vent, during their eruptions, to large quantities of mud, sometimes mingled with organic matter in such considerable proportions that they have been utilised as fuel.* In 1793, a few months after the terrible eruption of Unsen, in the Island of Kiousiou,† an adjacent volcano, the Miyi-Yama, vomited, according to Kampfer, so prodigious a quantity of water and mud that all the neighbouring plains were inundated, and 53,000 people were drowned in the deluge; unfortunately, we have no historical details of this catastrophe. Of all the eruptions of mud, the best known is that of Tunguragua, a volcano in Ecuador, which rises to the south of Quito to 16,400 feet in height. In 1797, at the time of the earthquake of Riobamba, a whole side of the mountain sank in an immense downfall, with the forests which grew on it; at the same time, a flow of viscous mud issued from the fissures at its base, and rushed down into the valleys. One of these currents of mud filled up a winding defile, which separated two mountains, to a depth of 650 feet, over a width of more than 1,000 feet, and damming up the rivulets at their outlet from the side valleys, kept back the water in temporary lakes: one of these sheets of water remained for eighty-seven days.

The volcanic mud, therefore, has this point of resemblance with the lava—that it sometimes flows out through the crater, as on Papandayang; sometimes through side craters, as on Tunguragua. Doubtless, when the volcanic muds have been better studied, we shall be enabled to trace the transition which takes place by almost imperceptible degrees between the more or less impure water escaping from volcanoes, and the burning lava more or less charged with steam. This transition is, however, already noticed in the ancient matter which the water has carried down and deposited in strata at the foot of volcanic mountains. These rocks, known under the name of *tufa*, *trass*, or *peperino*, are nothing but heaps of pumice, scoriæ, ashes, and mud, cemented together by the water into a species of mortar or conglomerate, and gradually solidified by the evaporation of the humidity which they contained. Of this kind, for instance, is the

* Otto Volger, *Das Buch der Erde*, vol. i.
† *Vide* above, p. 527.

hardened stone which, for eighteen centuries, has covered the city of Herculaneum with a layer of 50 to 150 feet in thickness. Among rocks of various formations there are but few which exhibit a more astonishing diversity than the tufas. They differ entirely in appearance and physical qualities, according to the nature of the materials which have formed them, the quantity of water which has cemented them, the greater or less rapidity with which their fall and desiccation take place; lastly, the number and distribution of the chinks which are produced across the dried mass, and have been filled up with the most different substances. Many kinds of tufa resemble the most beautiful marble.

The small hillocks, which are specially called mud-volcanoes, or *salses*, on account of the salts which are frequently deposited by their waters, are cones which differ only in their dimensions from the mighty volcanoes of Java or the Andes. Like these great mountains, they shake the ground, and rend it, in order to discharge their pent-up matter; they emit gas and steam in abundance, add to their slopes by their own *débris*, shift their places, change their craters, throw off their summits in their explosions; lastly, some of these *salses* are incessantly at work, whilst others have periods of repose and activity. In nature, transitions merge into one another so perfectly, that it is difficult to discover any essential difference between a volcano and a *salse*, and between the latter and a thermal spring.*

Mud-volcanoes exist in considerable numbers on the surface of the earth, and, like the volcanoes of lava, the neighbourhood of the seacoast is the principal locality where we find their little cones. In Europe, the most remarkable are those which are situated at the two extremities of the Caucasus, on the coasts of the Caspian Sea, and on both sides of the Straits of Yenikale, which connect the Sea of Azof with the Black Sea. On the east, the mud-springs of Bakou are especially distinguished by their combination with inflammable gases; on the west, those of Taman and Kertch flow all the year round, but especially during times of drought, pouring out large quantities of blackish mud. One of these mud-volcanoes, the Gorela, or Kuku-Oba, which, in the time of Pallas, was called the "Hell," or Prekla, on account of its frequent eruptions, is no less than 246 feet in height, and from this crater, which is perfectly distinct, muddy streams have flowed, one of which was 2,624 feet long, and contained about 850,000 cubic yards.†

* Humboldt, *Cosmos*, vol. i.
† Ansted, *Intellectual Observer*, January, 1866.

The *volcancitos* of Turbaco, described by Humboldt, and the *maccalube* of Girgenti, which have been explored, since Dolomieu, by most European *savants* who have devoted themselves to the study of subterranean forces, are also well-known examples of mud-springs, and may serve as a type to all the hillocks of the same character. In winter, after a long course of rains, the plain of the *maccalube* is a surface of mud and water forming a kind of boiling paste, from which steam makes its escape with a whistling noise; but the warmth of spring and summer hardens this clay into a thick crust, which the steam breaks through at various points and covers with increasing hillocks. At the apex of each of these cones a bubble of gas swells up the mud like a blister, and then bursts it, the semi-liquid flowing in a thin coat over the mound; then a fresh bubble ejects more mud, which spreads over the first layer already become hard, and this action continues incessantly until the rains of winter again wash away all the cones. This is the ordinary course of action of the *salse*, sometimes interrupted by violent eruptions.* On the coasts of Mekran the mud-volcanoes are not only subject to the action of the seasons, but also depend on the action of the tides, although many of them are from 9 to 12 miles from the Indian Ocean. At the time of the flow the mud rises in great bubbles, accompanied by a hoarse murmur, like the distant roar of thunder. The highest cone is not more than 246 feet high, and stands 7 miles from the shore.†

In a general way, the expulsion of mud and gas is accompanied by a discharge of heat; but in some *salses*, like those of Mekran, the matter ejected is not higher in temperature than the surrounding air, as if the expulsion of the mud from the ground was an entirely superficial phenomenon. Occasionally, in peat-bogs, the ground cracks, and cold mud is ejected from the fissure; and then, after this kind of eruption, the spongy soil sinks and again levels down. Is this eruptive phenomenon similar to that presented by the mud-volcanoes, and caused by the fermentation of gases in the midst of substances in a state of putrefaction? This is M. Otto Volger's idea; and it would be difficult to give any other explanation of this phenomenon.

* Arnold Boscowitz, *Volcans et Tremblements de Terre*.
† Walton, *Nautical Magazine*, February, 1863.

CHAPTER LXX.

VOLCANIC THERMAL SPRINGS.—GEYSERS.—SPRINGS IN NEW ZEALAND.—FUMEROLLES.—SOLFATARAS.—CRATERS OF CARBONIC ACID.

VOLCANOES, both of lava and mud, all have, either on their sides or in the vicinity of their base, thermal springs, which afford an outlet to their surplus water, gas, and vapour. Most even of those mountains which are at present tranquil, but which were once centres of eruption, continue to manifest their activity by vapours and gas, like furnaces in which the flames are extinct, but the smoke is still rising. Although lava and ashes no longer make their escape from the crater or lateral fissures, yet numerous *fumerolles* and hot springs, formed by the condensation of the steam, generally serve as a vehicle for the gas pent up in the depths of the mountain. We may reckon by hundreds and thousands the "geysers," the "vinegar springs," and other thermal springs in countries once burning with volcanoes, the fires of which are extinct, or at least quieted down for a period more or less protracted. Thus the former volcanoes of Auvergne; the mountains of the Eifel, on the Rhine, the craters of which contain nothing but lakes or pools; the Demavend, with its mouth filled up with snow—all still exhale here and there, through springs and *fumerolles*, as it were, a feeble breath of their once mighty vitality.

The volcanic regions of the earth where thermal springs gush out are very numerous: in Europe we have Sicily, Iceland, Tuscany, and the peninsula of Kertch; in America—that land so rich in volcanoes—the springs warmed by subterranean vapour are still more numerous, and there are some on the sides of the volcano Nuevo de Chillan which gush out through a thick bed of perpetual snow.[*] A lateral gorge of the valley of Napa, in California, called the "Devil's Cañon," may be quoted as one of the most striking examples of the active production of thermal waters. The narrow ravine, filled with vapour rising in eddies, opens on the side of a red and bare mountain, that one might fancy was scorched by fire. The entry to the ravine follows the course of a rivulet, the boiling waters

[*] Philippi, *Mittheilungen von Petermann*, vol. vii., 1863.

of which are mingled with chemical substances horrible to the taste. Innumerable springs—some sulphureous, others charged with alum or salt—gush out at the base of the rocks. There are both warm and cold springs, and hot and boiling; some are blue and transparent, others white, yellow, or red with ochre. In a cavity which is called the "Sorcerers' Caldron" a mass of black and fetid mud boils up in great bubbles. Higher up, the "Devil's Steamboat" darts out jets of gaseous matter, which issue puffing from a wall of rock : *fumerolles* may be seen by hundreds on the sides of the mountain. All these various agents either murmur, whistle, rumble, or roar, and thus a tempest of deafening sounds incessantly fills the gorge. The burning ground, composed of a clayey mud—in one

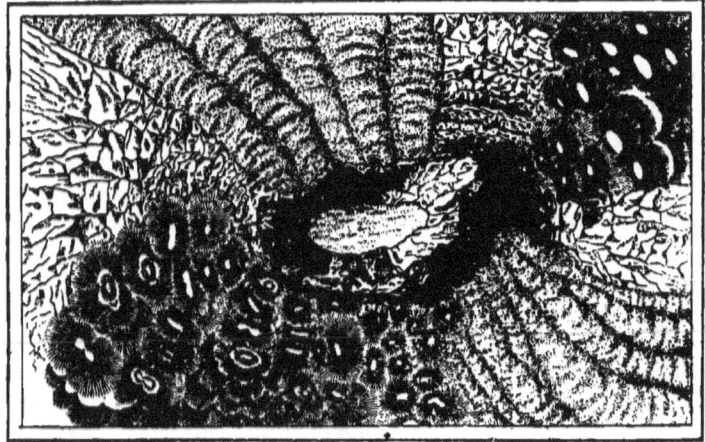

Fig. 200.—Crater of Demavend.

spot yellow with sulphur, and in another white with chalk—gives way under the feet of the traveller who ventures on it, and gives vent to puffs of vapour through its numberless cracks. The whole gorge appears to be the common outlet of numerous reservoirs of various mineral waters, all heated by some great volcanic furnace.*

The ravine of Infernillo (Little Hell), which is situated at the base of the volcano of San Vincente, in the centre of the Republic of San Salvador, presents phenomena similar to those of the "Devil's Cañon." There, too, a multitude of streams of boiling water gush from the soil, which is calcined like a brick, and eddies of vapour spring from

* Henry Auchincloss, *Continental Monthly*, September, 1864.

the fissures of the rock with a noise like the shrill whistle of a locomotive. The most considerable body of water issues from a fissure 32 feet in width, which opens under a bed of volcanic rocks at a slight elevation above the bottom of the valley. The liquid stream, partially hidden by the clouds of vapour which rise from it, is shut out to a distance of 130 feet as if by a force-pump, and the whistling of the water pent up between the rocks reminds one of the furnace of a manufactory at full work. One might fancy that it was the respiration of some prodigious being hidden under the mountain.

The hottest springs which gush out on the surface of the ground, such as those of Las Trincheras and Comangillas, do not reach the temperature of 212° (Fahr.);* but we have no right to conclude from this that the water in the interior of the earth does not rise to a much more considerable heat. It is, on the contrary, certain that water descending into the deepest fissures of the earth, although still maintaining a liquid state, may reach, independently of any volcanic action, a temperature of several hundred degrees; being compressed by the liquid masses above it, it is not converted into steam. At a depth which is not certainly known, but which various *savants* have approximately fixed at 49,000 feet, water of a temperature exceeding 750° (Fahr.) ultimately attains elasticity sufficient to overcome the formidable weight of 1,500 atmospheres which presses on it; it changes into steam, and in this new form mounts to the surface of the earth through the fissures of the rocks.† Even if this steam, passing through beds of a gradually decreasing temperature, is again condensed and runs back again in the form of water, still it heats the liquid which surrounds it, and increases its elasticity; it consequently assists the generation of fresh jets of steam, which likewise rise towards the upper regions. Thus, step by step, water is converted into steam up to the very surface of the earth, and springs out from fissures in the shape of *fumerolles*.

In Iceland, California, New Zealand, and several other volcanic regions of the world, jets of steam mingled with boiling water are so considerable as to rank among the most astonishing phenomena of the planet. The most celebrated, and certainly the most beautiful, of all these springs is the Great Geyser of Iceland. Seen from afar, light vapours, creeping over the low plain at the foot of the mountain of Blafell, point out the situation of the jet of water and of the neighbouring springs. The basin of siliceous stone which the Geyser itself has formed during the lapse of centuries is no less than 52 feet

* *Vide* above, p. 270. † *Vide* above, p. 509.

in width, and serves as the outer inclosure of a funnel-shaped cavity, 75 feet deep, from the bottom of which rise the water and steam. A thin liquid sheet flows over the edges of the basin, and descends in little cascades over the outer slope. The cold air lowers the temperature of the water on the surface, but the heat increases more and more in all the layers beneath; every here and there bubbles are formed at the bottom of the water, and burst when they emerge into the air. Soon bodies of steam rise in clouds in the green and transparent water, but, meeting the colder masses on the surface, they again condense. Ultimately they make their way into the basin, and cause the water to bubble up; steam rises in different places from the liquid sheet, and the temperature of the whole basin reaches the boiling-point; the surface swells up in foamy heaps, and the ground trembles and roars with a stifled sound. The caldron constantly gives vent to clouds of vapour, which sometimes gather round the basin, and sometimes are cleared away by the wind. At intervals, a few moments of silence succeed to the noise of the steam. Suddenly the resistance is overcome, the enormous jet leaps out with a crash, and, like a pillar of glittering marble, shoots up more than 100 feet in the air. A second and then a third jet rapidly follow; but the magnificent spectacle lasts but for a few minutes. The steam blows away; the water, now cooled, falls in and round the basin; and for hours, or even days, a fresh eruption may be waited for in vain. Leaning over the edge of the hole, whence such a storm of foam and water has just issued, and looking at the blue, transparent, and scarcely-rippled surface, one can hardly believe, says Bunsen, in the sudden change which has taken place.

The slight deposits of siliceous matter which are left by the evaporation of the boiling water have already formed a conical hillock round the spring, and, sooner or later, the increasing curb of stone will have so considerably augmented the pressure of the liquid mass in the spring, that the waters must ultimately open a fresh outlet beyond the present cone. From the experiments and observations made by Forbes as to the formation of the layer of incrustations round the jet, this spring must have commenced its eruptions ten centuries and a half ago, and they will probably cease in a much shorter space of time. Not far from the Geyser, the mound of deposits from which is not less than 39 feet in height, there are a number of pools which once acted as basins for springs which gushed up through them, but are now nothing but cisterns filled with blue and limpid water, at the bottom of which may be seen the mouth of a former channel of eruption. A shifting in the position of the centre

THERMAL SPRINGS.

of activity takes place in the Geyser, just as in mud-volcanoes and incrusting springs. Several springs lying on the same terrestrial fissure as the great *jet d'eau*, the Strokkr, the Small Geyser, and

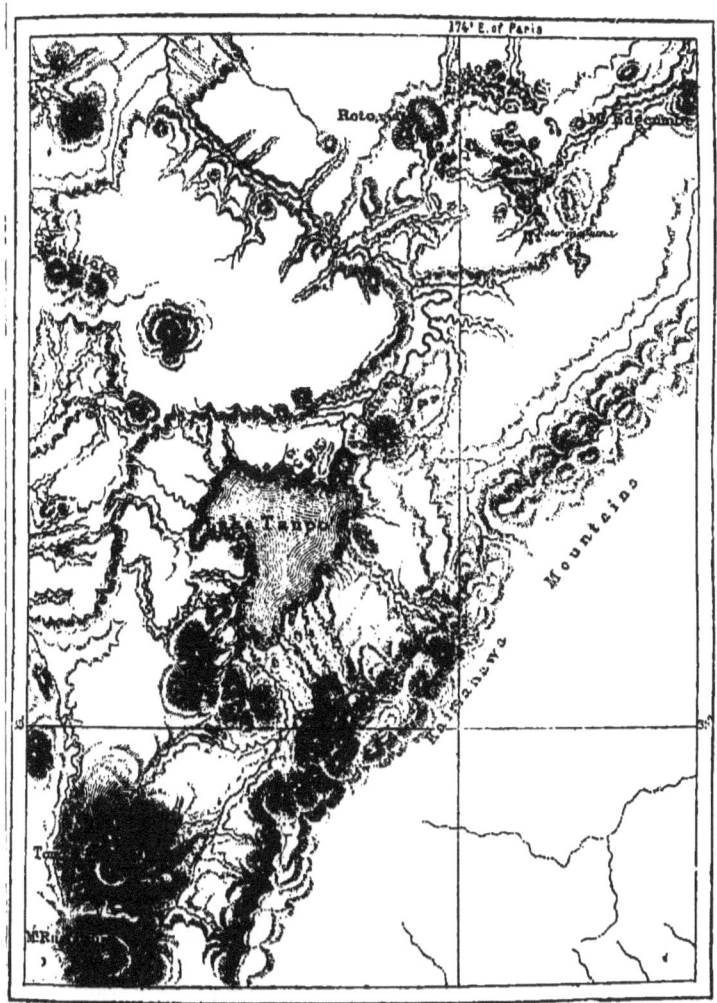

Fig. 201.—Volcanic Region of New Zealand.

some others, present phenomena which are nearly similar, and are evidently subject to the action of the same forces. The vicinity of the active volcanoes of Iceland warrants us, however, in supposing that the water produced by the melting of the snow on Blafell does not require to descend many thousands of yards into the earth in order to be converted into steam. There is no doubt that, at no very great depth below the surface, they come in contact with burning lava, which gives them their high temperature. By reproducing in miniature all the conditions which are thought to apply to the Icelandic springs—that is, by heating the bases of tubes of iron filled with water and surmounted by a basin—Tyndall succeeded in producing in his laboratory charming little geysers, which jetted out every five minutes.

About the centre of the northern island of New Zealand the activity of the volcanic springs is manifested still more remarkably even than in Iceland. On the slightly-winding line of fissure which extends from the south-west to the north-east, between the ever-active volcano of Tongariro and the smoking Island of Whakari, in Plenty Bay, thermal springs, mud-fountains, and geysers rise in more than a thousand places, and in some spots combine to form considerable lakes. In some localities the hot vapours make their escape from the sides of the mountains in such abundance that the soil is reduced to a soft state over vast surfaces, and flows down slowly to the plains in long beds of mud. For a distance of more than a mile a portion of the Lake of Taupo boils and smokes as if it was heated by a subterranean fire, and the temperature of its water reaches on the average to 100° (Fahr.). Further to the north, the two sides of the valley, through which flows the impetuous river of Waikato after its issue from the Lake Taupo, present, for more than a mile, so large a number of water-jets, that in one spot as many as seventy-six are counted. These geysers, which rise to various heights, play alternately, as if obeying a kind of rhythm in their successive appearances and disappearances. Whilst one springs out of the ground, falling back into its basin in a graceful curve bent by the wind, another ceases to jet out. In one spot a whole series of *jets d'eau* suddenly become quiet, and the basins of still water emit nothing but a thin mist of vapour. Further on, however, the mountain is all activity; liquid columns all at once shine in the sun, and white cascades fall from terrace to terrace towards the river. Every moment the features of the landscape are being modified, and fresh voices take a part in the marvellous concert of the gushing springs.*

* F. von Hochstetter, *Neu-Seeland*.

About the middle of the interval which separates the Lake of Taupo from the coast of Plenty Bay several other volcanic pools are dotted about, all most remarkable for their thermal and jetting springs. One of them, however, is among the great wonders of the world. This is the Lake of Rotomahana, a small basin of about 120 acres, the temperature of which, being raised by all the hot springs which feed it, is about 78° (Fahr.). Dr. von Hochstetter has not even attempted to count the basins, the funnels, and the fissures from which the water, steam-mud, and sulphurous gases make their escape. Here and there, indeed, he noticed, altogether, *salses, solfataras, fumerolles*, and springs. The most magnificent of all these jets is the Tetarata, about 82 feet above the eastern bank of the lake. The basin, from the centre of which the water and steam spout out, is a kind of crater, 286 feet in circumference, which is surrounded by ramparts of red clay 32 feet in height, resembling the sides of a crater. The basin is full of clear water, which has entirely covered the former with a coating of silex, white as marble. The water in this dazzling basin assumes a delicious blue shade, which is rendered

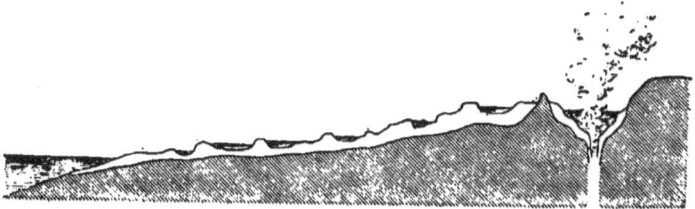

Fig. 202.—Section across the terraced Basins of Tetarata.

more beautiful by the reflection of the steam unrolling its spiral clouds. The liquid which flows from the basin runs into another pool, likewise covered by a coating of silex, and, falling from terrace to terrace, thus reaches the level of the lake. These glittering steps, over which the water spreads in thin sheets, and then falls in cascades, form a wonderful spectacle of splendour and grace. Sometimes—say the natives—the whole body of liquid in the upper basin is suddenly upheaved in an enormous column, and the pool empties 30 feet of its depth; in this case nothing can be wanting to complete the grandeur of the picture.

Have these wonderful springs existed for any great length of time, and are they destined to be preserved for centuries in all their beauty? These are points as yet unknown. When the New Zealand springs have been studied for a considerable number of years, it will perhaps be possible to describe the various modifications which are taking place in consequence of the increase or relaxation of the subterranean action in these regions. In several parts of Europe thermal springs have gradually lost both their heat and their mineral qualities owing to the cooling of the furnace which heated them, and are becoming more and more similar to ordinary springs: of this kind for example, are the springs of Bertrichbad, in Luxembourg.*

The gases which make their escape from the *fumerolles* situated above hidden beds of lava do not differ from those produced by great volcanic eruptions; they are the hydrochloric, sulphuric, and carbonic acids, either in a state of purity or in combination with alkaline, earthy, or metallic bases, and, as in the moving currents of lava, they indicate by their composition the degree of intensity of the subterranean heat. The most precise indications on these points have been furnished by the analyses of MM. Bunsen, Ch. Sainte-Claire Deville, and Fouqué.

When they issue from the earth, *fumerolles* deposit on the edges of the fissure various substances, such as sulphur, alum, and borax, which have been sublimated in the subterranean laboratory; the vapours then spread out into wide eddies, and are lost in the air. There is no spot in Europe where these *fumerolles* can be better studied than in the former crater of the Isle of Volcano. The cavity, at the bottom of which all these chemical operations take place, is more than a mile in circumference, and its southern sides rise to 980 feet in height; the bottom of the abyss may be about 320 feet wide. Through the mist of vapour which fills the immense caldron a spectator sees the lofty cliffs, red or golden yellow in colour, and streaked here and there with most varied hues. On the slopes leading to the bottom of the gulf the crumbling stones give way under the tread, and yet it is necessary to descend at a running pace, for in some places the hollow soil is burning like the arch over an oven. Vapour creeps along slowly over the slopes. The air is saturated with hydrochloric and sulphurous acid gases, and is difficult to breathe. An incessant noise of dull sobbings and whistlings fills the hollow, and on every side there are small orifices between the stones, whence jets of steam spring eddying out. This is the spot whither the workmen—who seem accustomed to live in the fire, like the legendary salamanders—resort to gather the stalactites of gilded sulphur which still crackle from the effect of heat, and the fine needles of boracic acid, white as swan's-down. At night the masses

* See chapter on Springs.

FUMEROLLES. 567

of vapour accumulated above the crater are coloured with a red glare, as if from the reflection of an immense fire.

Sometimes the rain which falls into the amphitheatre forms there

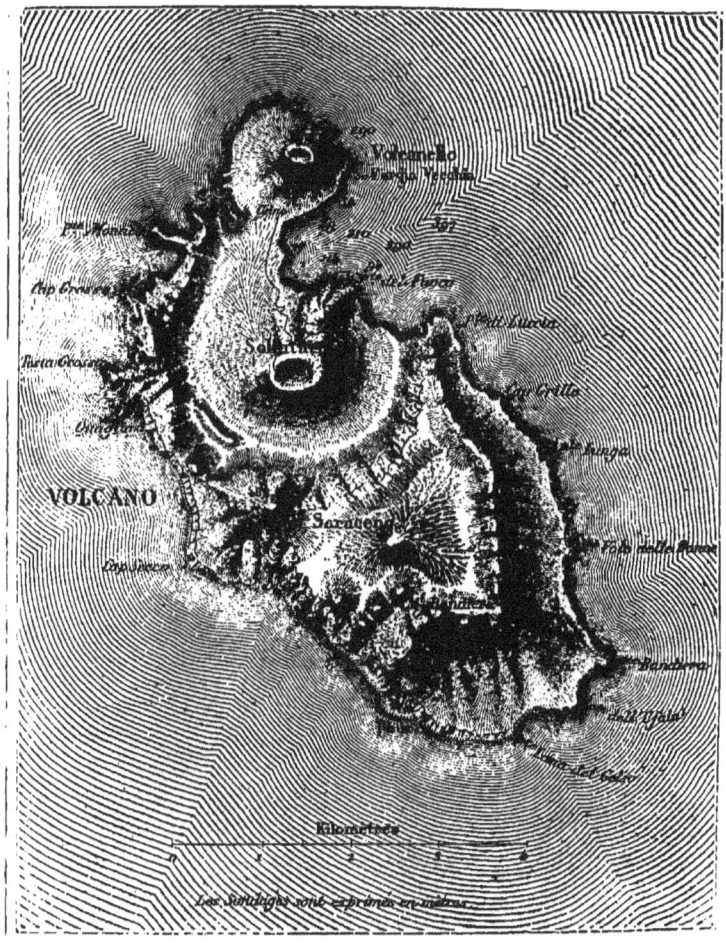

Fig. 203.—Island of Volcano.

a temporary lake; but a great part of the water makes its escape through the fissures in the ground, and flows in a stream over the external slopes, the remainder being rapidly evaporated by the heat of the mountain. Some of the *fumerolles*, the gases of which were analysed in 1865 by M. Fouqué, attain a temperature above 680° (Fahr.). Other jets of less heat make their appearance in various parts of the island, and even in the waters of the bay. From the edges of the great crater these vapours may be seen at the base of the slopes rising from the bed of the sea, and developing into wide, whitish-coloured spirals, similar in hue to the mud of potters' clay. In certain places the temperature of the sea-water heated by these gases is so high that voyagers can afford themselves the childish satisfaction of boiling their eggs in it.

The *solfatara* in the Isle of Volcano produces scarcely 10 tons of sulphur every year. This does not seem much; but if we were to calculate by centuries and geological periods the quantities deposited by the *fumerolles*, it would seem really enormous. Thus, the mines of Sicily, which have been worked for some centuries, furnish every year to commerce no less than 300,000 tons of sulphur; and yet the mining operations in these districts are altogether of a primitive character. A large number of beds are unexplored, and nearly a third of the sulphur brought to the surface evaporates in the furnaces or is lost in the *débris*. These almost inexhaustible veins appear to have been all produced, like the deposits of the Isle of Volcano, by jets of vapour saturated with sulphur.

Sometimes carbonic acid gas is the only gas which is discharged from the caverns and craters of volcanoes. This fluid, being much heavier than the atmosphere, does not ascend, like the other gases, and become lost in the air, but accumulates in weighty masses round the outlet. Plants which are bathed in this mephitic air rapidly wither, and all animated beings die from asphyxia, unless their heads rise up above the deadly atmosphere. In the Island of Java there is a small crater called Pakereman, or "Valley of Death," the amphitheatre of which, after the heavy tropical rains, is entirely filled with carbonic acid gas. No plant grows in this vast cavity. According to the statements of Loudon, the ground is strewn with the skeletons of animals. At one time there might have been seen in it the remains of human beings who had been doomed to perish from asphyxia in the poisoned air. In Europe there is no spring of carbonic acid which can be compared to that of Java; those which have been studied in Italy, Auvergne, and on the banks of the Rhine are most of them nothing but very inconsiderable emanations, filling, perhaps, a confined cave or the lower part of a small cavity well sheltered from the wind. Insects, and sometimes a few birds, are the only victims of this destructive gas. The famous cave in the vicinity of Naples is well known, into which, to satisfy the idle curiosity of travellers, the guides are cruel enough to bring some wretched dogs, and forcing them into the layers of carbonic acid gas which hang over the soil, cause them to pant for breath, and ultimately die. At one time the crater which is now filled up by the gloomy Lake of Avernus, which the ancients looked upon as the entrance to the infernal regions, gave vent to so large a quantity of carbonic acid gas, that birds flying over the lake fell as if struck by lightning; hence is derived the Greek name of the lake, "without birds."

CHAPTER LXXI.

SUBMARINE VOLCANOES.

The bed of the sea is, of course, inaccessible to our observation, but still there can be no doubt that the phenomena of submarine volcanoes resemble those of the burning mountains which tower so high above the ocean. Whenever the crest of an isle of scoriæ rises above the waves, all that we see of its eruptions suffices to prove that they take place exactly in the same way as those of the volcanoes on the mainland. There exist, however, in several places, former submarine craters which, since their period of activity, have been upheaved, with the plains surrounding them. The history of these craters, written legibly in their strata, is just the same as that of other volcanic outlets. Still, as would necessarily be looked for, on account of the pressure exercised by the immense body of sea-water, the cones of *débris* which have been formed under the pressure of the waves are, without any exception, more flattened than mountains of the same nature which have taken their rise on *terra firma*. With regard to the lava which issues from submarine fissures, it sooner becomes solidified, and does not flow to such great distances. Often, also, it is converted, under the superincumbent weight of the water, into basaltic colonnades. We can, too, readily understand that the vapour must rapidly condense while passing through the mass of cold water. It is only when the mouth of the crater is very close to the surface of the ocean that columns of vapour are freely discharged and ascend into the air.

Most submarine eruptions which have been known in modern times have taken place at no great distance from some of the great insular or continental volcanoes. The Sicilian Sea, the Greek Archipelago, the waters near St. Michael in the Azores, the portion of the Caspian which surrounds the Peninsula of Apcheron, the seas of Japan and of the Aleutian Isles, the Gulf of Darien, and the coast of Iceland, form, probably, a part of the same subterranean regions of activity as the volcanoes situated on the adjacent shores. There is, however, one volcanic district which is exclusively submarine; this is the narrowest part of the Atlantic, embraced between the two extreme points of the coasts of Guinea and Brazil. In this tract of the ocean the water is often agitated by violent shocks, and ships tremble as if they had run upon a sand-bank. Smoke, as if from a conflagration, rises above the waves, and pumice-stone and other light scoriæ are floated about by the current. Even islands of ashes have been noticed to emerge from the midst of the sea, which, being washed away by the waves, have diminished, and ultimately disappeared.*

The submarine cones proceeding from the bottom of the sea which resist the action of the water are those which have their layers composed of lava. The group of Aleutian Isles contain at least two mountains which have emerged from the water within a recent period. According to the accounts of the Chinese and Japanese chroniclers, several volcanoes have risen from the bed of the sea on the coasts of Japan and the Corea during the historical period. In the year 1007 a roar of thunder announced the appearance of the volcano of Toinmoura, or Tanlo, on the south of the Corea, and then, after seven days and seven nights of profound darkness, the mountain was seen; it was no less than four leagues in circumference, and towered up like a block of sulphur to a height of more than 1,000 feet.† More than this, the celebrated Fusi-Yama itself, the highest mountain in Japan, is said to have been upheaved in a single night (?) from the bosom of the sea twenty-one and a half centuries ago. With respect to marine volcanoes of a more recent epoch, we may mention all the isles of either extinct or still burning lava which rise in pyramids like Stromboli, or extend in a graceful semicircle like the crater of St. Paul in the Indian Ocean. Lakes also exhibit a number of volcanoes of the same kind. Such are Momotombo of Nicaragua, and the Taal of the Lake of Bongbong, in the Island of Luzon.

The sudden appearance of lava which has most impressed the popular imagination for many years past, and has also been best studied by scientific men, is the formation of hillocks of lava in the Santorin group. This circular archipelago is doubtless the remains of a great cone, 30 miles in circumference, which once rose in the midst of the sea. In consequence of the action of the waves, earthquakes, and subsidences, the sides of this cone were broken in upon and finally ruptured by the water. When the Hellenes established themselves in the islands of the Ægean Sea, the shattered mountain was divided into three fragments: one fragment assuming the form

* Daussy, Darwin, Poulett Scrope.
† Stanislas Julien, *Comptes Rendus*, vol. x., 1840.

of a crescent, and comprehending the larger portion of the former volcano, constitutes the Island of Thera, or Santorin; on the west, the little Isle of Therasia bends round in continuation of the ring of land; and, on the south-west, a great rock, the Islet of Aspro, rises as if to bear witness to the rampart of lava now disappeared. The outer slopes of Santorin and Therasia, partially covered with pumice-

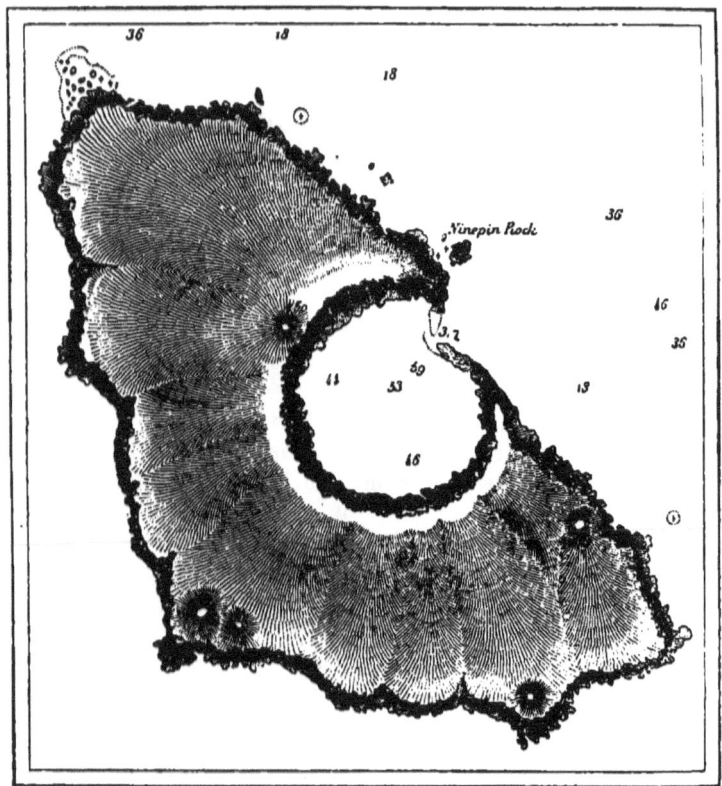

Fig. 204.—Isle of St. Paul: the measurements are in mètres.

stone somewhat resembling a coating of snow, are rather gently inclined towards the sea, but the escarpments, turned in the direction of the basin, which was once the crater, are in some places nearly perpendicular for a height of 650 and even 1,300 feet. The various layers, which correspond on the two sides of the gulf—on the sides of Therasia and of Santorin—are vividly marked out in bands of a red, yellow, blue, black, or white colour.

In the centre of this crater, after a series of upheavals and eruptions, an islet of lava emerged about the year 196 B.C. It was named Hiera (Holy), and on the summit a temple to Neptune was built. This islet, which was several times increased in the centuries which followed—in the years 46, 713, 726, and 1427 A.D.—is now known

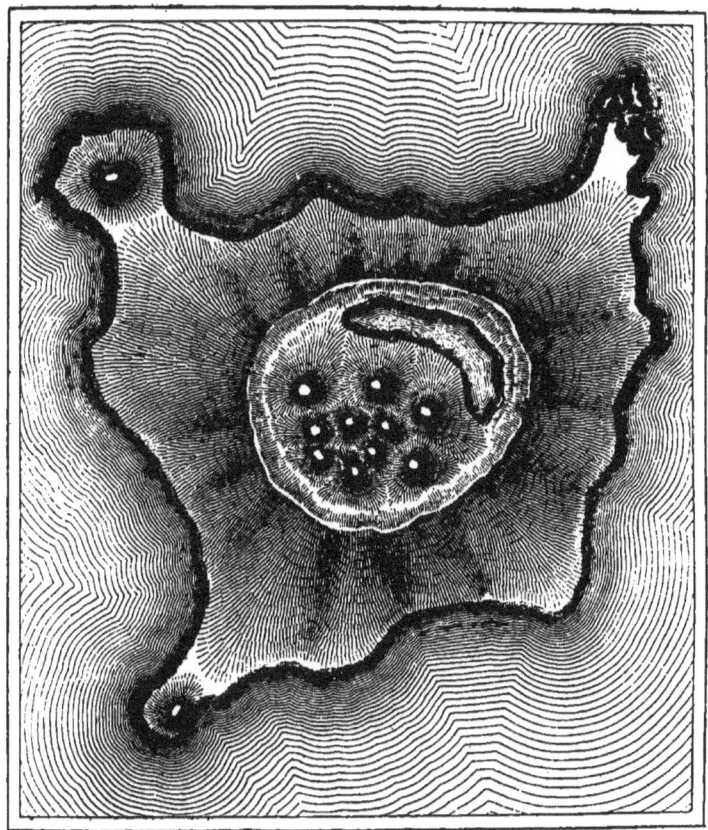

Fig. 205.—Volcano of Taal.

under the name of Palæo-Kaïmeni. Not far from this isle, another smaller one, Mikro-Kaïmeni, rose in 1570 or 1573. At the commencement of the eighteenth century, from 1707 to 1711, another mass of lava, Neo-Kaïmeni, an island no less than four miles in circumference, emerged within the annular crater formed by the crescent-shaped Santorin and the Isle of Therasia. In 1768 Neo-Kaïmeni was shaken by a violent eruption. From this date to 1866 no visible change took place above the water, but soundings proved that the bed of the sea had gradually risen. At a point near Mikro-Kaïmeni, a bank of rocks, situated in 1794 at a depth of 80 to 100

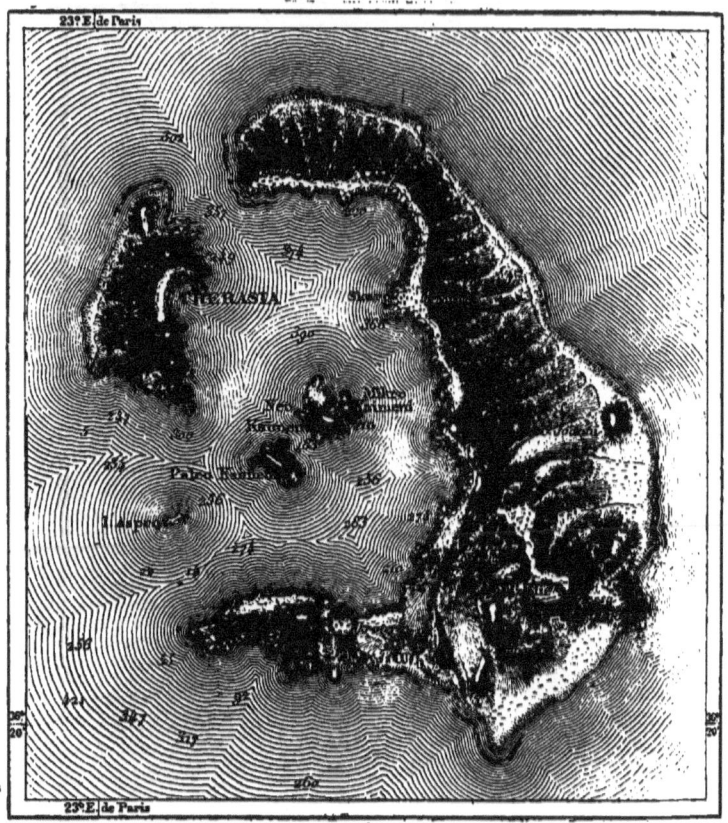

Fig. 206.—Santorin.

feet, was in 1835 only 13 feet from the surface. "Such is the singular fecundity of Santorin," says an historian of this volcano, "that isles seem to grow up round it like fungi in a wood."*

At the end of January, 1866, subterranean roarings, a gradual

* Pègues, *Histoire des Phénomènes du Volcan de Santorin*.

sinking of the ground, and the colouring of the water, announced an approaching eruption. The centre of the shock was felt just below the little village of Vulcano, situated on the edge of a creek where ships used to cast anchor in order that the water, mingled as it was with acid gases, might destroy the molluscs and sea-weed attached to their keels. On the 3rd of February a brilliantly black mass of lava was seen, which rose slowly from the bottom of the water, and increased in size every day with perfect regularity. This mound, designated by the name of George's Isle, attained in a few weeks a height of 164 feet, and ultimately became united to Neo-Kaïmeni, having filled up the creek of Vulcano. On the 7th of February, another mound, the Isle of Aphroessa, rose at a little distance to the south, and gradually filling up the channel which separated it from Neo-Kaïmeni, became converted into a promontory. Afterwards other islets made their appearance by the side of the two principal cones. All round them the sea, agitated by the gases which rose in great bubbles, exhibited in turn the most diversified hues. It was in succession reddish-coloured, milk-white, or shaded with green or a chemical blue; it was, besides, generally tepid or hot. The fish, either asphyxiated or killed by the heat, floated in multitudes on the surface, and mariners avoided steering their ships anywhere near the eruption, for fear the pitch between their planks should melt in the water heated to a temperature of 160° to 170° (Fahr.). The summit of the crater of Neo-Kaïmeni was shaken by frequent explosions. On the 20th of February lumps weighing 220 pounds were suddenly darted out to a distance of several hundred yards, showers of ashes fell upon the vines of Santorin, and clouds of *débris*, making their escape from the crater, shot up to a height of 6,500 to 9,800 feet. It was not before the end of the year 1866 that the cone of George's Island opened out for itself a crater by an explosion which destroyed the whole of the upper portion of the mound. But the intensity of the eruptive phenomena soon after gradually diminished. Currents of lava more than half a mile long were in some places more than 300 feet in depth. It is evident that the reason why the molten matter has thus formed in heaps close to the outlet, instead of flowing in long streams, as on the sides of Vesuvius and Etna, is that the lava, soon becoming cool by its contact with the salt water, was necessarily compelled to arrest its progress.*

The cones of shifting ashes which are thrown out by the eruptions of submarine volcanoes are not able to resist the action of the waves

* Fouqué, *Revue des Deux-Mondes*, August, 1866.

for any length of time; and however great may be their dimensions, they are certain ultimately to disappear, unless they are based on

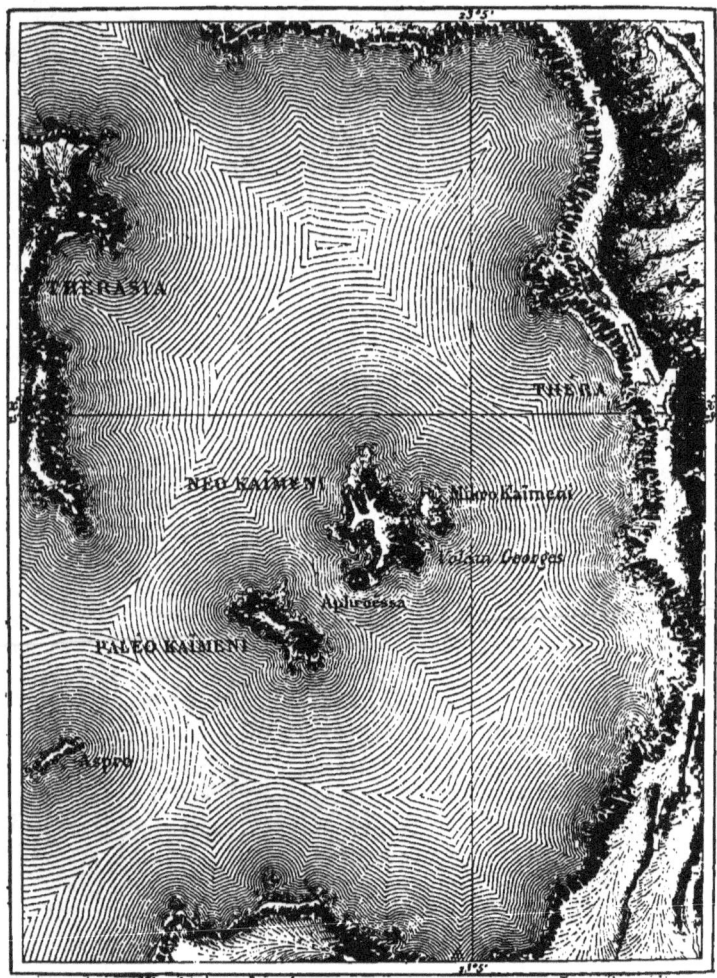

Fig. 207.—Kaïmeni Group.

solid foundations which have entirely emerged from the water. As an example of this we may mention the celebrated Julia or Graham's Island, which appeared in July, 1831, throwing up heaps of

ISLE OF JULIA.

dark-coloured scoriæ about 25 miles south of the shores of Sclimonte, in Sicily. An English captain, proud of being able to increase the British domains, hurried to hoist his flag over the smoking stones. The islet gradually increased round the crater, and before long it measured as much as four miles round. But, at the conclusion of the volcanic eruption, the work of demolition commenced, and, in conformity with M. Constant Prévost's prediction, the slope of *débris* was gradually undermined at the base by the waves and currents. In October nothing remained but a little mound. Six months after, the newly-discovered island, which was also claimed by the King of Naples through his diplomatists, was nothing more than an oval reef about half a mile long. A few years later the sounding-line

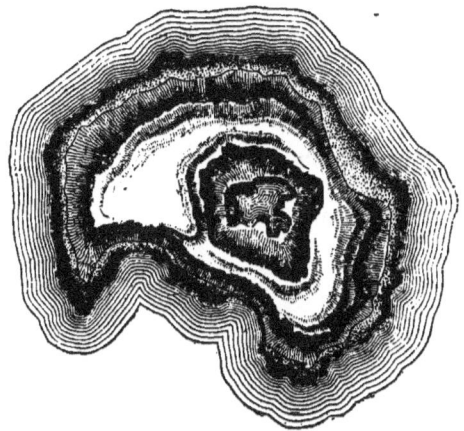

Fig. 209.—Graham's Island, or Isle of Julia.

showed a depth of 787 feet. In July, 1863, the isle again appeared, and, in a few weeks, rose to the height of 200 to 260 feet; but it existed only for a short space of time, for it soon again sank, being demolished stone by stone by the dash of the waves. It appears, however, that, previously to 1831, the volcano had made its appearance on one or two other occasions. It is said that a smoking island existed in this spot about the year 1801, and the shoal is pointed out in old charts. This island, first rising up above the waves and then sinking down again into the depths of the sea, seems to recall the recollection of that mysterious country mentioned in the "Arabian Nights," which plunged down into the ocean just at the moment when the voyagers were going to land on it.

Near the Island of San Miguel, one of the Azores, there is another submarine volcano, which likewise vomits out at each of its great eruptions a temporary cone of scoriæ, which rises above the level of the sea. In the years 1658, 1691, 1720, and 1812, a smoking islet made its appearance, and on each occasion it was destroyed after a few weeks of existence. Nevertheless, in 1812, an Englishman found time to take possession of it in the name of his Government, and to give it the name of "Sabrina." In 1867 a fresh eruption took place at about the same spot, but the volcanic cone did not reach the surface of the sea, and M. Fouqué, the indefatigable explorer of volcanoes, noticed nothing but floating scoriæ, gases thrown out, and flames dancing over the water. In the Icelandic seas, the land of Nynoë, which appeared in 1783, at a point 31 miles to the south-west of Cape Rekianess, enjoyed a much longer life, for it lasted a whole year. There, just as in the seas of Sicily and the Azores, the ashes washed away by the waves are deposited at the bottom of the water in vast layers of tufa mingled with shells and other *débris*. Some day or other, these beds of volcanic stone will appear on the surface, just as the tufas of the Euganean hills in the valley of the Po, and the Phlegræan fields at Naples, and will tell the tale of the submarine eruptions of the present period.

CHAPTER LXXII.

PERIODICITY OF ERUPTIONS.—INFLUENCE OF TEMPERATURE ON VOLCANIC PHE-
NOMENA.—EXTINCTION OF FURNACES OF LAVA.

ONE of the most important questions in respect to volcanoes is that of the order in which their phenomena take place. Everything in nature acts in sure conformity to regular laws of periodicity; but when this rule has to be applied both over vast spaces and long intervals, it may readily remain unknown to us for a very long time. The fact is, that the rhythm of the great volcanic revolutions has not yet been discovered, and, up to the present time, the whole of these events appear to us as if they were a perfect chaos. Even in the volcanic region which is the best known, and which *savants* have the best studied, that is, in the area which includes Southern Italy, Sicily, and the Lipari Islands, the succession of movements (except as regards Stromboli) has taken place with the greatest apparent want of order. Etna and Vesuvius have remained in a state of repose for centuries, and then suddenly, and after unequal intervals, they have experienced sometimes one convulsion only, sometimes a series of shocks more or less violent and numerous. Besides, as we have seen, there is no necessary coincidence between the phenomena of the two volcanoes. Sometimes, indeed, it happens that a period of extraordinary activity in one volcano is contemporary with protracted tranquillity in the other. True enough, during the dark night of the Middle Ages, phenomena like this were almost entirely lost sight of in history; but documents collected since the sixteenth century evidently prove that the geologist must limit himself to studying the symptoms of periodicity in each smoking mountain by itself.

In this last respect there is no want of evidence which renders highly probable the existence of some sympathetic tie between the eruptions of volcanoes and the more or less regular phenomena of the surrounding atmosphere. This, indeed, is a proverbial article of faith with mariners as regards a large number of active craters. The fishermen of Stromboli have for centuries all agreed in stating that the mountain serves as a barometer for them in pointing out the approach of winds and storms by a great increase of violence in its eruptions. At the time of the autumnal equinox, as well as in winter, the boiling lava in the crater runs out much more often through the outlet, and sometimes opens out a way of issue in the sides of the mountain, so as to flow down to the sea. The fact is, that the atmospheric pressure above the column of lava being diminished during stormy weather, does not act with the same energy on the compressed mass; the molten matter rises more rapidly in the crater, and flows out in greater abundance. Added to this, as the smoke of the volcano generally mounts up a thousand yards above the level of the sea, and sometimes rises in a column to the higher regions of the atmosphere, the contests which take place among the aërial currents may often be seen more or less clearly; thus, by experience, a knowledge may be obtained beforehand of the changes of weather which will gradually make their way down from the sky above. "By the smoke of this volcano," said Pliny, eighteen centuries ago, "the natives can predict the winds three days in advance, which leads them to believe that these vapours are in obedience to Æolus."

The inhabitants of Lipari say, also, that the vapours from the crater of Volcano form clouds of much greater volume when storms are brewing. This, then, would be another gigantic barometer, regularly indicating the diminished pressure of the atmosphere. According to both travellers and natives, the epoch of the equinoxes, when the monsoons blow, is the time when the eruptions of the Peak of Ternate, Taal, and other volcanoes of the Indian Archipelago are most terrible. In like manner, in the Isle of Chiloe, Fitzroy was told that the eruptions of Osorno always announced the approach of fine weather, and in consequence an increase in the weight of the column of air. With regard to phenomena other than the oscillations of the aërial masses, these may also be foretold by means of the volcanic movements, if we are to put faith in the statements of those who live at the very foot of the smoking mountains. Thus, in Japan, the volcanoes generally begin to vomit out lava about the time of the flow of the tide, and inundations caused by exceptionally high tides always take place after eruptions and earthquakes.* In 1827, Scuderi, a Sicilian, tried to prognosticate all the meteorological phenomena by the form and direction taken by the masses of vapour which were vomited out by the crater of Etna.†

* Siebold, quoted by Perrey.
† *Memorie dell' Academia Gioenia.*

M. Emile Kluge, by classifying all the known eruptions in seasons and months, has ascertained that these crises take place in summer especially, and that earthquakes are more frequent in winter. According to this geologist, there is no doubt that the eruptions of volcanoes depend on the changes of seasons. The melting of the snow and ice, the falls of rain, the oscillations in temperature, and the weight of the air are, in his idea, the real causes of the subterranean fires; the latter, too, being the mere result of chemical reaction,

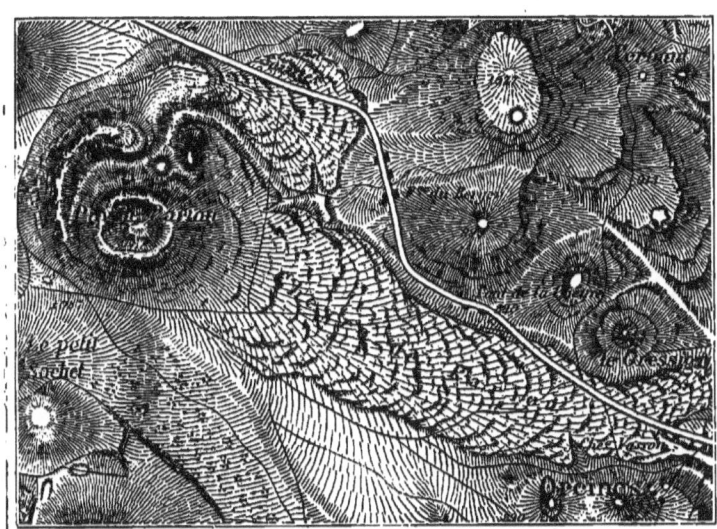

Fig. 209.—*Coulée* of Puy de Pariou.

and among the purely external phenomena of the planet. Not, therefore, in the abysses of a sea of fire, but at a depth of six to nine miles at most, must we seek for the furnaces in which the burning matter is elaborated.*

Nevertheless, although we may be warranted in looking upon the year itself, with its return of seasons and atmospheric phenomena, as a kind of *day* in the life of volcanoes, we are none the less in complete ignorance in respect to the great secular periods of plutonic activity in the various parts of the world. Like all terrestrial phenomena, that of the eruption of lava and ashes is limited in its duration. The volcanoes which rise up out of the ground are

* *Ueber die Periodicität vulcanischer Ausbrüche; Neues Jahrbuch für Mineralogie*, 1862

extinguished either after a short existence, or after thousands of eruptions. Over the whole surface of the earth there are a large number of former volcanoes the nature of which has been disclosed to us by modern geology alone. Some date from an epoch so remote that beds of modern formation have to a great extent covered their sides and filled up their craters. Others, like the hills of Auvergne are still visible, with their cones of scoriæ or their domes of trachyte their fissures and *cheires* of lava, such as they once were when the plains at their bases were arms of the sea. Among these well-preserved volcanoes there are, indeed, some—especially that of Denise, near the Puy-en-Velay—which were in a state of full activity at an epoch when man had already established himself in the adjacent districts, for in the tufa of their eruptions skeletons have been found of the ancient inhabitants of Gaul. Lastly, numerous volcanoes which have burnt during the historical period are now extinct, and perhaps for ever. The cycles of these lava-furnaces are doubtless connected with those of the continents themselves; the configuration of land and sea must change before these furnaces can again be lighted up. Continents and oceanic basins must also shift their position on the surface of the globe before they can be extinguished.

CHAPTER LXXIII.

EARTHQUAKES.—VIBRATIONS OF THE GROUND.—VARIOUS HYPOTHESES.

As volcanic eruptions sufficiently show, this planet is not the immovable mass which our imaginations depict it, when comparing it to the atmosphere which surrounds it, to the ever-mobile waves of the ocean, or even to the animated beings which wander over its surface. On the contrary, the ground which we tread under our feet vibrates very frequently. Without alluding to those great shocks which overthrow cities, bring down the sides of mountains, and open vast cracks across plains, there are other less violent vibrations, which have been recorded by the annalists of geological history, and may be reckoned by thousands as regards civilised countries and modern times alone. There can be, however, no doubt that the great majority of slight earthquakes pass unnoticed, being blended, especially in towns, with the confused rumbling of noises and murmurs. Various *seismological* instruments, recently invented or brought to perfection, reveal a large number of oscillations which it would be impossible to discover in any other way. An attentive observation of the levels of air-bubbles and of micrometrical threads reflected on the surface of a bath of mercury has, indeed, warranted M. d'Abbadie in asserting that the earth is in a state of constant vibration. The intervals of immobility which he has noted have never exceeded thirty hours.

What is the cause of these trepidations of the ground? A great number of *savants*, who have accepted the hypothesis of a *pyriphlegeton*, or central fire, do not hesitate to look upon these vibrations of the earth as the repercussion of the undulations of the great burning sea. Each of the shocks which are felt on the surface of the earth would then take its rise below the envelope of the planet, and would be at first produced in the form of a current or tide in the burning mass which is supposed to exist. As Humboldt says, an earthquake would be " the reaction of the liquid nucleus against the outer crust." Added to this, most geologists who base their arguments on the hypothesis of a central fire, admit that earthquakes must necessarily be in connection with volcanic phenomena, and that they invariably proceed from the same cause. Following out the comparison which must always present itself to the mind, the mouths of volcanoes are safety-valves, and, on account of the obstruction of these outlets, the pent-up lava or vapour shakes the superincumbent layers of the terrestrial envelope, seeking to find some way of issue. This theory has the merit of being very simple, and, in a large number of cases, seems to harmonise very satisfactorily with the facts that have been observed. But we must never forget, however probable this hypothesis as to the volcanic action may be, it has not as yet become a certainty; and the duty for the geographer still is to study events impartially, and to suspend his judgment until a satisfactory conclusion evidently results from the whole body of facts observed.

In the first place, it is important to know if those regions on the surface of the globe in which earthquakes take place with the greatest frequency are distinguished from other parts by any peculiar features in the form of their vertical outline, or in the nature of their rocks. The volcanic districts in Europe, such as the environs of Vesuvius and Etna, the Islands of Santorin and Milo, and the south of Iceland, are not the only places which are subject to severe shocks; the former districts, too, have never been so violently agitated as the mountains of the Abruzzi and Calabria, the Islands of Rhodes and Cyprus, the limestone districts of Carniola and Istria, the Alps of the Valais, the environs of Basle, certain plateaux in Spain, and the hills at the mouth of the Tagus. The mountains of Scotland, and especially those in the county of Perth, have also experienced repeated shocks: out of two hundred and twenty-five earthquakes recorded in the British Isles, eighty-five have taken place in this one county. In Africa, the soil of Algeria, so rich in saline and thermal springs, but devoid of volcanic craters, is sometimes very severely shaken. The districts of the Nile, which are likewise without volcanoes, have also suffered much from subterranean movements. In Asia, the peninsula of Guzerat, a spot in which astonishing modifications of the shape of the coasts were produced during a great earthquake, is situated more than 1,240 miles from the nearest volcanoes, the Demavend and the burning mountains of Thian-Chan; but, on the other hand, the Philippines and Japan, which are volcanic countries, are also frequently agitated by movements of the ground. Again, the sea-coast of Syria, the towns of Aleppo and Antioch, the scenes of some of the most destructive earthquakes that are recorded in history, no longer possess very active volcanoes, and the lavas of the Djebel-Hauran, on the south-east, have long been extinct. In South America most of the great shocks have taken place in the region of the Andes, or not far from their bases. The Argentine town of Mendoza, which was overturned in the violent earthquake of 1861, is comparatively not far from a lava-furnace, since the volcano of Maypu rises at a distance of only 87 miles. The Equatorial Andes are often convulsed by violent oscillations of the soil, and are also the theatre of great volcanic activity, many of their summits being domes of trachyte or craters still vomiting out ashes, mud, or smoke. Nevertheless, according to the testimony of Boussingault, the most energetic shocks experienced in Columbia—those which destroyed the towns of Latacunga, Riobamba, Honda, Merida, and Barquesimeto, and were simultaneously felt over a very extended area—presented no coincidence whatever with any volcanic phenomena, and their centre of agitation was situated at a considerable distance from the smoking peaks.* The plateau of Caraccas, celebrated for the catastrophe of 1812, is situated more than 600 miles eastward of the Grenadini volcanoes of Huila and Tolima, and at rather a less distance from the craters of the Antilles, from which it is separated by wide arms of the sea. Finally, the region in North America where oscillations of the ground are most frequent and most severe is the alluvial plain of the Mississippi, far distant from any volcanic district, and even from any great chain of mountains. Thus, although the history of earthquakes is known only for a few centuries, and over but a small portion of the earth's surface, it is certain that severe oscillations of the ground are felt in countries the most diverse, which bear no resemblance to one another either in their formation or their aspect. The only fact which seems well established is, that shocks are more frequent in mountainous than in flat countries. Nevertheless, if, as Mr. Mallet thinks, all earthquakes not followed by an eruption are "incomplete efforts to open a volcano," if they are produced by the endeavours made by the planet to get rid either of gas or the molten matter within, the ground ought to be most frequently agitated in continental plains, far from volcanoes and mountains; for in those localities there would be no natural "vent-holes" through which to discharge the overflow of the interior fluids, and there, too, according to the common theory, the terrestrial layers must be the thinnest.

Those who look upon every volcano as a safety-valve for the adjacent regions put forward in favour of their theory certain facts

* *Annales de Chimie et de Physique*, 1855.

which have attained to the dignity of legends, although their reality is far from being certain, as M. Otto Volger has more than sufficiently proved.* Thus, it is said that, at the time of the earthquake at Lisbon, Vesuvius, which was vomiting out a considerable quantity of vapour, suddenly sucked in the cloud which it was throwing out, and that the current of lava issuing from its sides was suddenly stopped. But these statements are founded solely on a much less precise expression in the account which Kant the philosopher devoted to the catastrophe at Lisbon. Humboldt tells us "that, after having vomited out for three months a high column of smoke, the volcano of Pasto ceased to throw out vapour at the exact moment when, at a distance of 248 miles, the earthquake of Riobamba and the mud-eruption of Tunguragua caused the death of 30,000 to 40,000 Indians.† Even the great name of Humboldt must not, however, lead us to forget that, on the plateaux of the Andes, communication is both rare and difficult, and that the population scattered over that space of 248 miles does not afford all the guarantees which would be considered requisite in scientific observation. Lastly, the assertion that Stromboli relaxed its incessant activity during the Calabrian earthquake in 1783 is based on nothing but the vaguest information. According to the pamphlets of that date, all the Lipari Isles were to be swallowed up in the sea, leaving but a few shoals to mark the places they once filled! As may be seen, the facts which have been brought forward as the foundation of the theory that all earthquakes are caused by the movements of lava or vapours are deficient in the requisite authenticity, and cannot be looked upon as exempting geologists from the necessity of direct observations.

* *Erdbeben in der Schweiz*, vol. iii. p. 385.
† *Tableaux de la Nature*, vol. ii.

CHAPTER LXXIV.

EARTHQUAKES OF VOLCANIC ORIGIN.—SUBTERRANEAN DOWNFALLS.—EXPLOSIONS OF MINES AND POWDER-MILLS.

THERE are a certain number of cases in which, independently of any theory, we can without difficulty substantiate a relation of cause and effect between a volcanic eruption and an earthquake. Thus, when the sides of a smoking mountain, such as Etna or Kilauea, are suddenly cleft asunder to pour forth a stream of lava, and at the same time the ground is strongly agitated, it is evident that the earthquake is caused by the fracture of the volcano. This local phenomenon is precisely analogous to that produced by the explosion of a mine or a powder-mill. When the fissure is of a considerable length and the fractured sides of the volcano present a great thickness, the shock is a violent one, and reverberates in long oscillations in all the adjacent districts. When, on the contrary, the rocks of the volcano, having been diminished in thickness and partially melted by the rising lava, yield more easily to the pressure which bursts them, the explosion is only felt in the immediate vicinity of the fissure. Thus, at the time of the last great eruption of Etna, the trepidations of the ground, which coincided with the formation of the fissure, were in general very slight, and the sharpest of them, which was, however, perceptible in the town of Aci Reale, was not felt beyond the Etnean region properly so called.* History also affords several examples of volcanic eruptions during which the ground was not perceptibly shaken. In May, 1855, Vesuvius vomited out a considerable quantity of lava without the least trace of any motion of the ground being perceived either in the observatory on the volcano, or at Naples.

When the ground of a volcanic region is convulsed by shocks, and we are unable to observe the least connection, either as regards coincidence or immediate succession, between these phenomena and the eruption of a cone of ashes or the emission of a current of lava, we evidently have no scientific reason for asserting, with any certainty,

* Mariano Grassi, *Eruzione dell' Etna*.

that these shocks originate in the subterranean furnace of burning matter, or that they are caused by vapour endeavouring to break through the terrestrial crust. For still stronger reasons, a similar assertion would be contrary to all the rules of scientific observation when applied to earthquakes which take place far from any volcano. Certainly, according to the "safety-valve" hypothesis, oscillations of the ground ought to take place just in those very localities of the planetary envelope in which there exists no orifice communicating with the lava. But how is it, then, that undulations of the soil are not constantly produced at places far from these gigantic vent-holes situated on the sea-coast? Why is it, too, that frequent eruptions of Vesuvius and Etna did not precede the Calabrian earthquakes, and afford an issue to the pent-up vapour and lava?

The hypothesis that volcanic forces are the cause of earthquakes, being one that cannot be justified in every case, we must have recourse, in respect to many of these phenomena, to some other theory—one, in fact, which in all time has suggested itself to the minds of various people, and was taught by the Greek philosophers. Some two thousand years ago, Lucretius propounded this idea in magnificent language—an idea which has now been scientifically adopted by Boussingault, Virlet, Otto Volger, and other geologists.

"Learn, now, the cause of earthquakes, and first be assured that the interior of the globe is, like the surface, filled with winds, caverns, lakes, precipices, stones, rocks, and a large number of rivers, the impetuous waves of which hurry along in their course numerous submerged blocks. The shakings of the surface of the globe are occasioned by the falling in of enormous caverns which time has succeeded in destroying. Whole mountains thus sink in ruin, and the violent and sudden shock is spread far and wide in terrible vibrations. Thus, a chariot, the weight of which is not, however, very considerable, makes all the houses near tremble as it passes, and the fiery steeds, drawing behind them the iron-armed wheels, shake all the places round. It might well happen that an enormous mass of earth should fall by reason of decay into some great subterranean lake, and that the globe should tremble in a series of undulations. In like manner, on the surface of the earth, a vessel filled with liquid in a state of agitation cannot resume its equilibrium until the water contained in it has found its level."

Various *savants* have recently collected a large number of facts which are in favour of the theory of earthquakes once propounded by

Lucretius, although only generally and without the necessary proofs. In a vast number of cases this theory is certainly the true one, for it is often possible to catch in the act, so to speak, the phenomena which give rise to the oscillations of the ground and subterranean thunders. Thus, great landslips, such as those of the Diablerets, Rossberg, and other mountains of the Alps, have caused real earthquakes, the waves of which were felt at a considerable distance from the scene of the catastrophe. Even the falls of *moraines*, *séracs*, and avalanches of snow shake the ground very severely over considerable areas, so much so that in the mountains of Allemont, in Dauphiny, the inhabitants consider all the vibrations of the ground as the reverberations of distant downfalls of snow or rocks.*

The subsidence of rocks or shifting soil is accompanied by similar phenomena. In September, 1814, near Alais, for twenty-four hours a series of detonations were heard like a cannonade, and then, after a formidable cracking noise, the ground sank 13 feet for a breadth of more than 88 yards. Not far from the town of Wagstadt in Silesia, in 1827, a tract of more than two acres in area sank in a similar way with a great crash. In Carniola, where earthquakes are not unfrequent, heaps of fallen rocks are noticed in the numerous caverns, which heaps correspond with the funnel-shaped cavities on the surface of the ground. These subsidences which man sometimes personally witnesses, either in districts hollowed out by natural caves, or in mining regions pierced with subterranean galleries, may cause local shocks, or, in proportion to the mass of rocks falling, give rise to earthquakes felt simultaneously over vast extents of country. In fact, certain rocky strata sometimes leave intervals between them of very considerable depth, as may easily be noticed on the sides of mountains, and they may, besides, be composed of substances which are more or less easily dissolved and washed away by the infiltrated water. When these voids extend so far that the rocks above, sometimes hundreds or even thousands of yards in thickness, can no longer maintain their position by means of their own cohesion, the whole mass must necessarily fall down on the beds beneath. It is, indeed, impossible to imagine that it can be otherwise; the enormous quantities of salt, gypsum, lime, silex, and other substances which springs bring up to the surface must of necessity leave great voids in the depths below, and, in consequence, the subsidence of the rocks above these vacant spaces becomes inevitable. Only imagine, if it is

* Fournet, *Annales des Sciences Physiques de Lyon*, vol. viii., 1845.

possible, the potency of the shock produced by the sudden fall of several millions of cubic yards.*

Earthquakes produced by artificial causes in no way differ from natural shocks in the effects which they produce; they, indeed, furnish us with excellent terms of comparison. The explosions of mines and the passage of heavily-laden trains cause a motion of the ground over areas increasing in extent as the initial pressure is more considerable, and as the rocks present a greater degree of elasticity. Cannonades, the reaction of which on the earth is, however, trifling in proportion to the effect produced, are heard at distances which seem prodigious, if the ear is applied to the surface of the agitated ground. Vibrations of the layers of the earth—in fact, real earthquakes—are thus prolonged for more than 250, or, indeed, for 375 miles. Thus, in 1832, the bombardment of Antwerp was heard, it is said, in the

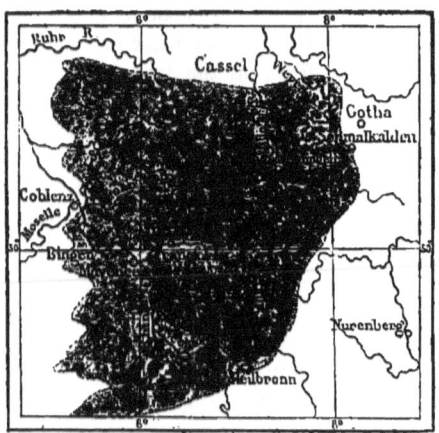

Fig. 210.—Area over which was felt the Explosion of Gunpowder at Mentz.

Erzgebirge, in the centre of Germany. Twenty-five years ago, at the time of the explosion of the powder-mills of Mentz, a very large area of ground was made to tremble. More than 10,000 lbs. weight of the gunpowder exploded (the greater part) blew up in open vaults, and could not, therefore, react on the ground; yet the shock was felt either as a slight trembling of the ground or as distant thunder as far as 100 to 125 miles away, far beyond the districts to which the sound of the detonation was carried by the wind. The shock was felt at several towns in Swabia, at Wurzburg, at Kissingen, in

* Otto Volger, *Erdbeben in der Schweiz*.

the countries of Fulda and Meiningen, in Thuringia, near Cassel, and at Wildungen.* Doubtless the observations made on earthquakes in the various countries of the world, whether these phenomena are produced with or without the intervention of man, will some day enable us to estimate approximately the force that is necessary for shaking some given area of the earth's surface. These would be comparative studies which could be made without prejudice in favour of the hypothesis of the central fire or any other theory.

* Otto Volger, *Erdbeben in der Schweiz*.

CHAPTER LXXV.

GREAT CATASTROPHES.—EARTHQUAKE AT LISBON.—AREA OF DISTURBANCE.—EARTHQUAKES AT SEA.

As a natural consequence of the fellow-feeling which tends to unite all men together, writers of the earth's history have been prone to give to earthquakes a geological importance bearing a proportion to the number of persons overwhelmed and the quantity of the productions of human labour destroyed. A shock which shakes a vast city filled with stone buildings, in which thousands of individuals are assembled, may cause the most terrible disasters, while a much more violent one felt by the inhabitants of a nearly desert country is soon forgotten and passed over by history. According to Otto Volger, the earthquake at Lisbon was scarcely more violent than was that of the valley of Viége, a hundred years after (?); but it remains memorable on account of the thousands of human beings who perished in it, while the shock of Viége, which only killed two poor mountaineers, was soon forgotten.

The sudden vibrations which overturn cities in a few seconds are, in fact, the most frightful catastrophes that can be imagined by man. All other disasters are announced by some precursory signs. The stream issuing from the volcano advances but slowly, and its progress across villages and cultivated land may be foreseen. The floods of rivers threaten the embankments long before they break through them, and preparation may generally be made for the irruption of water. Even the hurricane is preceded by atmospheric signs; but earthquakes generally happen suddenly and unexpectedly, and in an instant, without a single sign to explain the catastrophe, whole cities are demolished, and the inhabitants destroyed by thousands. The earthquake of San Salvador only lasted ten seconds, and this space of time was sufficient for the destruction of the town. The successive vibrations which devastated Calabria, in 1783, were felt during barely two minutes. The terrible movements of the earth which destroyed Lisbon succeeded each other during the space of five minutes; but it was the first shock, lasting from five to six seconds, which caused the greatest damage. The inhabitants sometimes make use of the brief respite given them by the intervals between the great shocks, not in taking refuge in the open air, but in increasing still more their chances of death: struck with terror, they rush into the churches, the roofs of which fall in upon them. After some of these catastrophes the corpses may be counted by tens of thousands: the shocks in Sicily, in 1693, and Calabria, in 1783, must have caused in each of these two countries the death of 100,000 persons. Lastly, records of more or less credibility speak of earthquakes in Syria, Japan, and the Sunda Archipelago, which resulted in a sudden loss of life still more considerable. In 526, more than 200,000 people met with their death at Antioch and the adjacent towns.

These undulations, which are so terrible in their consequences, are simultaneously felt over vast areas. Thus, the commotion which destroyed Lisbon on the 1st of November, 1755, and demolished the greater part of Oporto and several other places in Portugal, threw part of the walls of Cadiz into the moats, and, it is said, on the testimony of the governor of Gibraltar, that the greater part of the towns of Morocco, Tetuan, Tangiers, Fez, Mequinez, and even the capital itself, were overthrown by the earthquake. Kant, the philosopher, and Hoffmann, who were the historians of the earthquake at Lisbon, mention a great number of other countries in Europe, Africa, and even the New World, which must have participated in the violent disturbance. The vibrations extended over an area of 15,400,000 square miles; that is to say, about a twelfth part of the terrestrial surface. The statements upon which the various accounts of the catastrophe are founded are not always of any great value; it is now proved that the extent of the area over which the undulations of the earth were felt on this occasion has been singularly exaggerated. In the whole of Europe popular imagination was so struck by this event, which, in the course of a few minutes, and, indeed, on a day of festival, destroyed so many thousands of persons under the ruins of a great capital, that it was naturally led to look upon the earthquake at Lisbon as a phenomenon without parallel, the scene of which was, if not the whole world, at least a great part of the terrestrial surface. All the oscillations which were felt in Europe, either on the same day or about the same time, were considered in a general way as the result of the great commotion at Lisbon; and gradually a sort of legend was established attributing to the same cause a considerable number of geological facts of an undecided date, such as the downfall of rocks, the formation of lakes, the breaking up of ice, and changes of temperature in thermal springs. Thus, a shock which, "by a strange chance," was felt at Turin alone on the 9th of November, a week after the catastrophe at Lisbon, was attributed to this vast disturbance. The movements of the soil described as having taken place at New York on the 18th of November are also reckoned among the undulations which were then spread far and

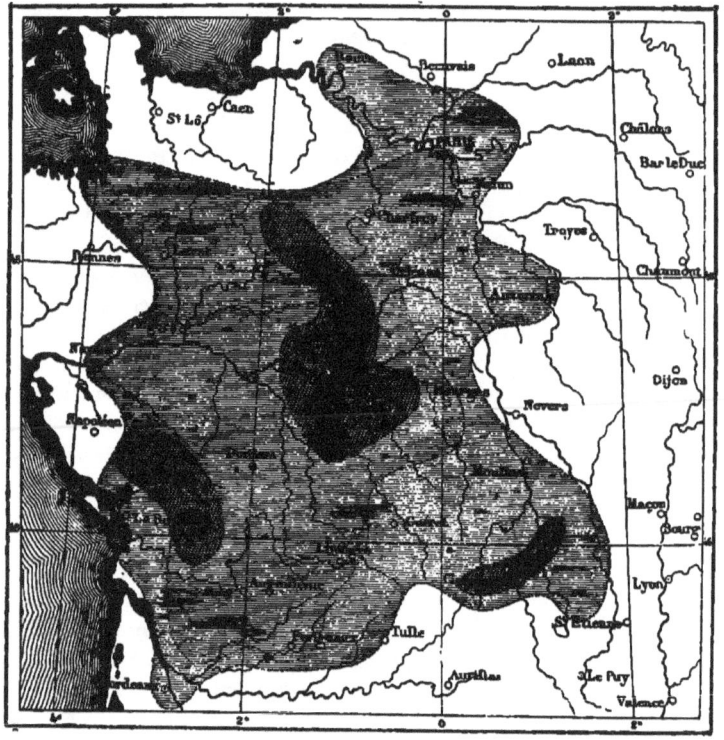

Fig. 211.—Chart of the Earthquake of the 14th September, 1866.

wide. The Lake Ontario was also added to the immense area of disturbance, because strong vibrations agitated its shores during the month of October; that is to say, before the day of the disaster. As a matter of fact, there is no positive proof that the terrestrial wave spread further than Morocco on the south, the Castiles on the east, or in a northerly direction further than Angoulême and Cognac.*

* Otto Volger, *Erdbeben in der Schweiz*, vol. i.

This, however, constitutes an area of 1,118 miles in length, and if, as the diameter of the area of disturbance, the same distance is taken in the direction from east to west, we shall find that the area of the earth shaken by the great terrestrial wave of Lisbon was more than 1,158,000 square miles, or about six times the size of France. As a term of comparison, we will mention the earthquake which was felt in France on the morning of the 14th of September, 1866, the undulations of which were propagated to the north as far as Rouen, and to the south as far as Bordeaux. The area of disturbance of this shock must be estimated at about 77,218 square miles, or the fifteenth part of the surface agitated by the earthquake at Lisbon.

At the time of this latter event there was one fact which contributed much to extend the apparent area of disturbance; this was, that a marine wave, harmonising with the shock of the earth, was spread across the Atlantic in all directions. But the water being more easily moved than the soil, necessarily transmitted the wave to a greater distance than the comparatively rigid beds at the bottom of the sea. At the mouth of the Tagus the wall of water formed by the waves rose, it is said, to a height of nearly 56 feet; then, filling up all the estuary that extends in front of Lisbon, swept over the quays of the city, and rushed among the houses. At Cadiz a wave of nearly equal size rushed above the ramparts, and caused much more havoc than the earthquake itself. On the coasts of Madeira and of Holland, the mouth of the Elbe, the sea-shores of Denmark and Norway, and, lastly, the whole circumference of the British Isles, the sea felt the reaction of the shock communicated to the waves in the waters off Lisbon, and its level underwent rapid fluctuations. The undulations of the wave, variously modified by currents and tides, struck even upon the shores of the New World. At Barbadoes and Martinique, where the flow of the tide never exceeds 28 inches, the wave produced by the Transatlantic earthquake attained a height of 13, 16, and even 19 feet. Thus the marine wave resulting from the shock was carried to a distance of 3,728 miles in a straight line. In 1854, at the time of the earthquake of Simoda, the wave which reached the coast of California had traversed 2,485 miles, the whole width of the Pacific Ocean.

When violent shocks agitate the ground, towns situated on the sea-shore have often suffered much more from the sudden irruption of the water than from the shaking of the earth itself. When the waves receive the shock from the neighbouring coasts, or else when the centre of disturbance is at the bottom of the ocean, the masses of water rise to a formidable height, and dash upon the shores, as if during a hurricane. In 1783, at the time when the shock in Calabria overthrew the towns and villages on the mainland, a terrible bore, after having swept away at once 2,000 persons assembled on the coast of Scylla, rushed into the port of Messina, sank all the ships, and undermined the base of the superb rows of palaces which bordered the shore: more than 12,000 individuals perished, it is said, under the ruins. On the 7th of June, 1692, at the time of the earthquake which shook Jamaica and the neighbouring seas, the waves rushed violently upon the town of Port-Royal, and, in the space of three minutes, covered more than 2,500 houses with a depth of 33 feet of water. The ships were thrown in every direction on to the land, and the frigate *Swan* was stranded upon the roof of a house. In like manner, according to the statement of Acosta, the terrible wave which destroyed Callao in the year 1586, and carried a great ship right up on to the Lima road, at a point 52 feet above the mean level of the sea, must have been altogether 89 feet in height. The Japanese, whose islands have often suffered from earthquakes and sea-bores caused by submarine shocks, say that these frightful phenomena are caused by the blows of the tail of a monster striking against the shore. Thus the Greeks attributed the vibrations of the soil not only to Pluto, the "shaker of the world," but also to Neptune, the "agitator of the waves."

CHAPTER LXXVI.

MOVEMENT OF TERRESTRIAL WAVES.—VARIATIONS CAUSED BY THE INEQUALITY OF VERTICAL OUTLINE AND THE DIVERSITY OF ROCKS.—AREAS OF DISTURBANCE.—NOISE OF EARTHQUAKES.—FRIGHT OF MEN AND ANIMALS.

WHATEVER may be the nature of the first shock, whether it proceed from a sudden swelling up of lava or vapour, or whether it be caused by the falling in of upper strata upon the subjacent beds, the effects produced will always be the same as regards observers who are above the central point where the phenomenon is produced. They will experience a shock tending upwards. Even when falling with the ground, they might well fancy themselves raised, like the aëronaut, whose balloon is falling towards the earth, sees the country mounting up towards him. Around the central point of disturbance, where the shock takes place in all its violence, and is felt vertically, in a manner more or less irregular, according to the number of shocks, the movements become more and more oblique, and are propagated across the strata of the earth in a direction which ultimately becomes perceptibly horizontal.* The phenomenon of undulation which is produced in solid rocks is perfectly analogous to that which may be observed in water when a stone falls into it: a series of concentric waves is formed round the centre of the shock, and gradually disappears in the distance.

Terrestrial waves which are formed thus are very long and very flat, on account of the inflexible nature of the rocks through which the movement is transmitted. There does not, however, exist a single authentic measurement from which the dimensions of each wave may be deduced. The observer feels them pass rapidly under his feet during an earthquake, and is often able to notice the rocking of houses and towers, as well as the to-and-fro motion of church bells; but these movements are much more marked than those of the ground, and the movements of the earth have not been clearly distinguished on any occasion.

As to the direction followed by the waves, the tendency generally

* Robert Mallet, *Observation of Earthquake Phenomena*.

differs much, owing to the inequalities in the relief of the ground, from the regular direction it would otherwise take. This direction is often very difficult to discover, owing to the want of necessary instruments and all the local circumstances which may modify the movement. It appears, however, that in mountainous countries, like Switzerland and the Pyrenees, the great undulations are propagated in the direction of the valleys. In striking against the tilted strata at the base of mountains, the corrugations of the soil act like the waves of a river dashing against the shores; they break up, and, changing their course, run along at the foot of the heights in the same direction as the stream of the valley. After this first rupture in its movement, the undulation is communicated to the rocky masses of the mountain, and traverses their whole thickness. Beyond these lofty groups which disturb the movement, without always opposing to it any insurmountable obstacle, the vibrations corrugate the soil of the

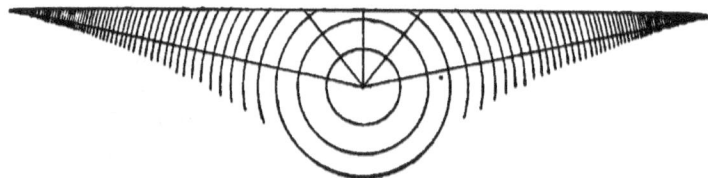

Fig. 212.—Transmission of Earth-waves.

plains in a more regular way; but the intensity weakens proportionately to the square of the distance, and finally ceases to be perceptible to man. It must also be remarked that, at the periphery of the area of disturbance, the various shocks are generally produced at longer intervals than at the centre of the earthquake. The stronger the waves are, the more rapidly are they propagated, and thus it follows that between the different undulations, which generally become weaker and weaker, the interval always increases with the distance. In the centre the shocks all seem to blend together; towards the circumference they succeed each other like waves of slightly-agitated water.

Among the causes which contribute to disturb the regular movement of terrestrial oscillations, the diversity of geological formations must also be reckoned. The swiftness of the transmission of the movement varies considerably, according to the composition of the rocks, the quantity of water they contain, and the hardness and elasticity of their layers. In order to explain the difference that exists between the various strata as regards the propagation of terrestrial waves, Mr. Mallet makes a striking comparison. If a person applies his ear to a railway-rail which is struck with a violent blow at a point about a mile away, the compact iron transmits to him, nearly instantaneously, the wave of sound; immediately after, the observer will feel the undulation which is transmitted through the soil below the rail; then he will hear the noise transmitted by the atmospheric waves. If a canal flows along by the side of the railroad, a man plunged in the water would perceive the sensation of the blow, but not at the same time. In fact, the mean swiftness of the wave is 1,138 feet in the air, 4,692 in the water, and about 11,040 in a bar of iron.

It is a long-established fact that, during earthquakes, the shocks are propagated much more easily through compact rocks than through formations interrupted here and there by faults, fissures, caves, and soft ground. Mr. Mallet has proved these facts by direct and oft-repeated experiments, made not far from the town of Holyhead in Wales. When mines of powder were exploded, the waves of disturbance, which were the more rapid as the charge was stronger, were propagated 951 feet a second in wet sand, 1,283 feet in a rock of friable granite, and 1,640 feet in a compact granite. Subsequently, Mr. Mallet, having made some direct observations as to the speed of transmission of the waves during the earthquake at Calabria in 1857, found that it was about 820 feet a second. According to the same geologist, the starting-point of the shock was nearly three miles below the surface.

Without having made any exact investigations, the Hellenes and the Romans, who inhabited a soil frequently shaken by earthquakes, had found out the fact that caverns, wells, and quarries retarded the disturbance of the earth, and protected the edifices built in their neighbourhood. The town of Capua was, it is said, saved from the effects of earthquakes to a much greater extent than the adjacent cities, on account of the numerous springs in its gardens. Vivenzio also asserts that in building the Capitol the Romans took care to sink several wells in order to weaken the effect of terrestrial oscillations, and this plan succeeded in preserving the building from all damage. In like manner, the great constructions at Naples were built above vast caves, in which the force of the subterranean commotion is lost. Humboldt has described this curious fact—that at St. Domingo the inhabitants of the town spontaneously formed the idea, similar to that of the Greek and Roman naturalists, of digging out deep cavities as the only means of securing the stability of their dwellings.

Besides, as may readily be supposed, the longer, thicker, and lower the walls of the edifices are, the better they resist the shock. In all towns partly destroyed by earthquakes it is said that walls of this form were rarely demolished. When the undulations of the soil are propagated along the whole length of a block of low houses, there is hardly an instance in which the latter have been shattered; in countries, therefore, in which the movements of the soil generally assume the same direction, disasters can nearly always be provided against by setting the principal walls of the edifice in the direction of the terrestrial undulations.

The buildings which always suffer the most are those which have vaulted roofs, elevated in the form of domes or cupolas. The thrusting action of the heavy masses which crown the edifice causes the walls to separate when they are in a state of vibration from the effects of the subterranean shocks; the dome falls down inside, while the walls give way in an outward direction. A considerable area is covered with ruins all round the piece of ground on which the foundations stand, and, in consequence, the danger of being crushed becomes very great to any persons who are near the scene of the catastrophe. It was the earthquakes, and not the barbarians, which, according to the evidence of Mr. Mallet, destroyed so large a number of the palaces and temples of Rome during the period between the fifth and ninth centuries. In like manner, in more modern times, cathedrals and churches have often been overthrown, while other houses were saved. This well-known fragility of vaulted roofs, when shaken by undulations of the soil, will explain the cause of those frightful calamities which took place in various churches at the time of the earthquakes of Lisbon, Calabria, Caraccas, Mendoza, and San Salvador, when kneeling multitudes were crushed in the ruins.

The difference presented by rocks, as regards the speed with which the earthquake wave is propagated through them, and the various obstacles which impede it, has the effect of giving to the areas of perceptible disturbance shapes which are perfectly irregular. The movements, therefore, are not produced round the initial point with a regularity which can be at all compared to that of the wavelets which surround with their regular circles the centre of agitation in a disturbed water. Some earthquakes, as far at least as it is possible to judge from incomplete observations, seem to be propagated in very elongated ellipses. Others appear to have had for their area a space of a polygonal shape; thus the great paroxysm of the valley of Viége, which extended over 108,878 square miles, was felt three times further in a northern than in a southern direction. Sometimes, outside the limits of the ground in a state of vibration, a region has been remarked which likewise shakes, like a kind of trembling islet surrounded by immovable land. At other times vast tracts have not experienced any apparent disturbance, whilst the ground

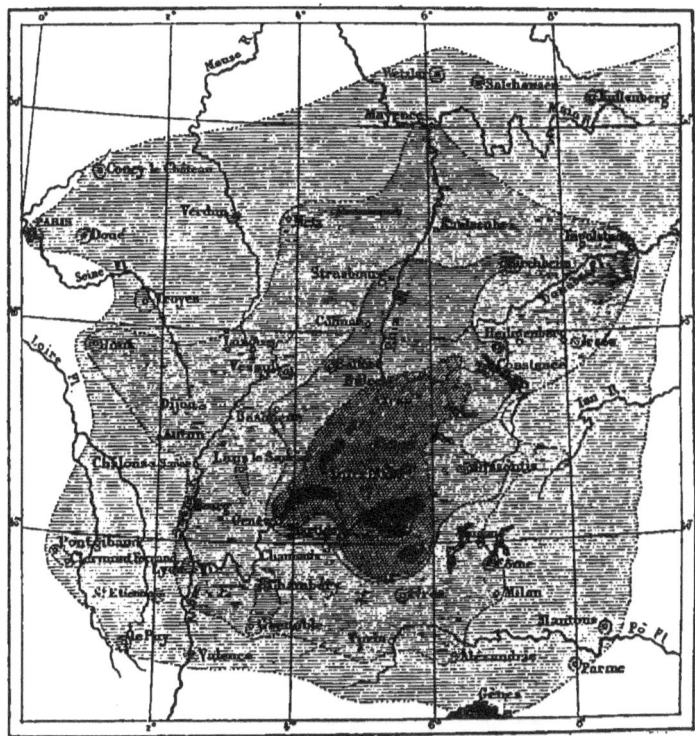

Fig. 213.—Area affected by the Earthquake of Viége in 1855.

all round them was trembling. On the 25th of July, 1846, the shock of which the severest impulse was felt below St. Goar, on the banks of the Rhine, propagated its undulations in France and Germany over a surface estimated at 24,207 square miles; but a belt about 100 yards wide, between Pyrmont (Westphalia) and the right bank of the Rhine, appears to have remained immovable.*

* Daubrée, *Comptes Rendus*, 1847.

According to the testimony of Humboldt, this fact of rocks not participating in the general movement of the surrounding formations has been frequently noted at the time of earthquakes in the Andes. The natives say of these intermediate zones, thus exempted from the vibrations of the ground round them, that they form a bridge (*hacen puente*), meaning by this that the oscillations are transmitted at a great depth below the inactive beds at the surface. It is difficult to imagine how such phenomena could take place, if the oscillations of the soil were caused by the movement of the waves in a subterranean sea of fire: if it were so, the upheaved terrestrial crust would undulate like an object floating on the surface of the water, and the burning waves, spreading out in a circle, would also give a perfectly round periphery to the superficial area of the earthquake.

To the two kinds of movement which have been observed in earthquakes—the upward shock and the long undulations spreading in the manner of marine waves—most of the *savants* since Aristotle also add the rotating or gyratory movement (*vorticosum*). The fact is, that in great cataclysms, when the different shocks cross each other in the ground, it has been thought that proofs of these twisting movements have been felt and even seen. At Quintero, in Chili, three great palm-trees, says Humboldt, twisted round one another like willow-wands, after each had swept a small space round its trunk. Otto Volger, who does not believe in the existence of rotatory movements, mentions, however, the example of the steeple of Grœchen, which twisted during the earthquake in the valley of Viége in 1855: it is true that this twisting may have been caused either by a movement of the soil being communicated to the edifice, or by the want of equilibrium between the different parts of the steeple. Mr. Mallet also explains, by a difference between the centre of form and the centre of gravity, the gyratory movement which the stones of two small obelisks underwent during the earthquake at Calabria in 1783; he absolutely denies that the rotatory movement of the earth could take place, as the Italian naturalists had alleged.

As to the speed of the propagation of terrestrial waves, it was, even recently, very difficult to estimate, owing to the want of precision in the transmission of intelligence and the irregularity of the clocks in the different cities. Since 1853, the period at which the electric telegraph was applied for the first time in describing the shocks of the earthquake of Soleure, almost certain means are at our disposal for noting the passage of terrestrial waves in different localities. But up to our time it has scarcely been employed save in an exceptional way, and too often some of the desirable conditions of exactitude have been neglected.

The incomplete information gathered by Otto Volger about the great earthquake of Viége, in 1855, has warranted him in fixing approximately the speed of the undulation; it was at the rate of 2,861 feet a second from the centre of vibration as far as Strasbourg, and 1,398 feet only in the direction of Turin. Robert Mallet, after having made his celebrated experiments on the speed of transmission of shocks in the rocks at Holyhead, instituted comparative investigations into the speed of the undulations of the great earthquake at Calabria in December, 1857, and found the average rate 774 feet a second. Since that time English observers established at Travancore, in the south of Hindostan, have estimated the movement of the undulations of a local shock at about 656 feet. The result of the calculations varying thus in proportion of one to four, it is impossible to indicate any average figure for the rate of transmission of terrestrial waves; it is certain that the rapidity, as well as the force and direction of the movement, differs according to the nature of the rocks and the position of the chains of mountains and valleys.

The noises which are heard during earthquakes differ in intensity still more than the other phenomena, and are much more difficult to classify, owing to the deep sensation that is felt when the earth is heard to roar, and the part which the imagination never fails to play when memory seeks to recall the past. The sounds, besides, heard at the time of the subterranean downfall sometimes exceed in their violence all known noises, and it is in vain to seek for suitable terms in which to describe them. In a general way, the noises of earthquakes may be compared to explosions of mines, discharges of artillery, peals of thunder, the rolling of carriages, the fall of avalanches, or the roar of cataracts. The diversity of these noises is explained by the difference of the phenomena which may be taking place in the interior of the earth—the falling in and reboundings of rocks, the overflow of subterranean water, the irruption of masses of air through fissures, and distant echoes reverberating far and wide in the abysses. It is a strange thing that sometimes during one and the same earthquake certain persons cannot find terms strong enough to express the frightful noises that they have heard, while others think they have only felt the shock unaccompanied by the least sound. According to M. Otto Volger, this singular difference of sensation proceeds from the fact, well known to natural philosophers, that the scale of sounds perceived by the ear differs in different individuals: just as in a meadow some persons do not hear the cry of the cricket, the note is too shrill for them, so, when the ground is shaken, those who are shaken with it would not all be able to hear the sounds produced by the cataclysm, on account of nothing but the deepness of their tone; the noises would be too deep for their ears. At a distance from the centre of the shock the noise gradually diminishes in intensity, but it always remains difficult to distinguish clearly, because the sound is transmitted with unequal speed both under the ground and in the atmosphere. Through the terrestrial strata the noise of the paroxysm travels with more rapidity than the shock itself; the shock is heard before it is felt; then, when the wave has passed, the sound is heard anew, being transmitted more slowly by the air. This inequality in the passage of the sound through the different mediums results in a great confusion of roaring and rattling, of which it is very difficult to give any just account. Observers have compared the noise of a distant earthquake sometimes to the rumbling of thunder, sometimes to a stormy wind, or the flapping of the wings of a large bird, and sometimes even to a discharge of musketry, the crackling of fire, or the whistle of a locomotive. One might fancy that in this manifestation of its mighty vitality nature makes use of all the sounds known to the human ear.

All this diversified uproar and these frightful noises sufficiently explain the instinctive terror which takes possession of nearly every one during the time of an earthquake, even when the shocks have not caused any fatal consequences. Nevertheless, as Humboldt remarks, a feeling of insecurity is the cause which generally most contributes to upsetting moral force. The earth which we felt so firm under our feet becomes as fluctuating as the waves; one hardly dares to walk a step for fear of being swallowed up in the opening ground. All our ideas as to the nature of things become confounded, and, by a sudden reaction of his physical or his moral nature, the man who feels that he is being deprived of the earth he stands on loses all confidence in his own personal powers.

It is a widely-spread opinion, which, however, is not as yet undeniably confirmed, that animals manifest the greatest uneasiness at the approach of an earthquake. In certain countries, indeed, where these convulsions of the ground frequently take place, care is taken to observe attentively the ways of domestic animals, in order to detect the presentiment of the coming shock, and to prepare for the danger. It may perhaps be the case that slight vibrations, perceptible only to the delicate senses of animals, precede the subterranean downfalls; but very often it is probable that remarks of this kind are made after the catastrophe, and that imagination, excited by the fright, plays no inconsiderable part in the descriptions. Be this as it may, it is said that before an earthquake, rats, mice, moles, lizards, and serpents frequently come out of their holes and hasten hither and thither, as if smitten with terror. Even the crocodiles have fled away from their marshes, and hurried towards *terra firma*, roaring with fear.* At Naples it is said that the ants quitted their underground passages some hours before the earthquake of July 26, 1805, and that the grasshoppers crossed the town in order to reach the coast; also that the fish approached the shore in shoals.† A fact which is better attested is the fright of animals during the catastrophe itself. At the time of the earthquake which shook the valley of Viége, in 1855, the wild birds which most dread the fowler, such as owls, woodpeckers, and peewits, collected on the trees close to the dwelling-houses, and uttered plaintive cries, as if to demand the succour of man. Birds of long flight, swallows and others, at once took wing, and flew away to distant parts. For several days, also, the frogs ceased their croaking.‡

* Humboldt, *Relation Historique*, vol. v.
† Landgrebe, *Naturgeschichte der Vulkane*, vol. ii.
‡ Otto Volger, *Erdbeben in der Schweiz*, vol. iii.

CHAPTER LXXVII.

SECONDARY EFFECTS OF SHOCKS.—SPRINGS.—JETS OF GAS.—FISSURES.—DEPRESSIONS AND ELEVATIONS OF THE GROUND.

EARTHQUAKES very frequently exercise a considerable influence on the discharge of springs rising to the surface of the ground. A great many instances have been brought forward of springs, both thermal and cold, which have suddenly dried up or have increased in volume, accompanied by either an augmentation or diminution of temperature. These phenomena can be easily understood. After each downfall of rocks or fracture of the ground, the conduits through which the subterranean rivulets flow may be either altogether or partially obstructed; the water must then seek some other course, or must flow in a diminished stream through the old channel. Sometimes, also, the breaking down of some barrier which penned back the subterranean water opens up a free passage to it, and the spring is thus augmented in its discharge. Again, the waters of several subterranean currents of various temperatures may become mingled in consequence of some catastrophe, and the springs, therefore, are rendered either warmer or colder. In August, 1854, on the occasion of a violent earthquake in the central Pyrenees, the heat of a spring at Baréges was raised from 64° to 82° (Fahr.), and its discharge, which was 2,729 gallons a day, increased to 6,338 gallons.

The effect more generally produced on springs by the occurrence of an earthquake is to render the water muddy by filling it with the *débris* which has fallen from the rocks disturbed, and been raised by the ascending body of water. During a series of observations made on the artesian well at Passy in 1861 and 1862, M. Hervé-Mangon ascertained that at the time of each of the subterranean shocks of Western Europe which were recorded by M. Perrey during the above-named interval, the water in the well was charged with sediment. On the 14th of November, 1861, the day on which a great earthquake occurred in Switzerland, the sediment in the well at Passy suddenly increased in quantity from 956 grains per cubic mètre to 2,268 grains; the next day it decreased to 1,404 grains. This skilful chemist, by numerous processes, established several other striking coincidences between the impurity of the water and the vibrations of the ground. It is not at all probable that these geological facts are devoid of any mutual relation; it appears, on the contrary, that the artesian spring at Passy, and doubtless also most other springs gushing out to the surface, might be looked upon as actual "scismometers." At the time of earthquake shocks a kind of clearing out takes place of the natural or artificial conduits through which the spring water passes. According to the observations of M. François, who devoted considerable study to the action of earthquakes on the mineral springs of the Pyrenees, the results of a shock are rarely felt for any length of time; after the lapse of two or three days the effects are no longer perceptible.

In all countries frequently affected by earthquakes, the inhabitants never fail to tell numerous stories as to sudden eruptions of water, mud, gas, or flames. Phenomena of this kind may be, in fact, produced in many districts, for shocks violent enough to close up or to enlarge the conduits of springs may equally well open out fresh channels, and thus afford an outlet to water pent up under deep layers of rocks. In like manner, the hydrogen gases which are formed in the ground by the decomposition of organic matter may find an outlet by the breaking down of the rocks above them, and burn spontaneously, like the gases at Baku. Nevertheless, these curious eruptions, however probable they may appear, have not as yet been scientifically observed, and no idea seems to have been formed as to the real importance they might have. Even Mr. Mallet, the great advocate of the constant connection between earthquakes and volcanic phenomena, has not ventured to look upon these sudden jets of water, mud, and gas as facts on which science may depend. With regard to the sudden appearances of flashes and sparks, these may be explained in many cases by the collision of stones, which strike against one another as they fall.

Occasionally, during earthquakes, the ground is rent open for very considerable distances. In 1783, at the time of the terrible shocks which disturbed Calabria, the phenomena of ruptures of the ground ranked among the grandest and most frightful effects of the catastrophe. Whole mountain-sides, undermined by water, slid down *en masse*, and tumbled into the plains below, covering all the cultivated ground. Cliffs fell down in a body, and rocks opened, swallowing up the houses which stood upon them. At the western base of the granitic chain of the peninsula, the ground affected by the shock was cleft open for a length of more than 18 miles, and in some places, especially near Polistena, the fissure was several yards in width. Elsewhere other clefts were opened, one of which, near Cergulli, was no less than 131 feet in depth for a length of more than a mile, and 32 feet wide. In the environs of Rosarno, on the shore of the Gulf of Nicotera, well-like cavities were hollowed out with circular margins, doubtless caused by the gushing out of springs. Finally, districts with a level surface were cleft in every direction by cracks radiating in the shape of a star ; the ground was broken up in a similar manner to mud which has cracked from the loss of its moisture. In February, 1835, on the occasion of the earthquake at Conception, in Chili, similar phenomena were observed : everywhere, says Fitzroy, contact ceased between the shifting ground of the plain and the bases of the rocky hills.

Not only do fissures sometimes form in the ground during earthquakes, but after commotions of this kind the level of the terrestrial surface is occasionally permanently changed. When the catastrophe is caused by some subterranean downfall, it is perfectly natural that the surface layers, being suddenly deprived of their supports, should be included in the subsidence, and, in fact, several instances have been brought forward of these sinkings in the level of the ground. In some countries it is said that phenomena of a directly opposite class have been observed, and whole districts have been suddenly upheaved a few inches, or even several feet. If the facts are certain (and there seems but little doubt about them), they would go to prove that earthquakes, from which these upheavals result, are caused by the pressure of confined vapour.

Amongst shocks of this kind we must mention the great geological catastrophe, unfortunately but too well known, which, in the year 1819, changed the shape of the country of Cutch over a vast area. A part of the Great Runn sank down over an extent of some thousands of square miles, and, in consequence of the inroads of the salt water, it became a tract of land of an undetermined character, during drought a desert without water, and during the monsoons a salt lagoon. A rampart several miles in length, 164 feet broad, and 9 feet in height, was also raised in a straight line across one of the former mouths of the Indus. This rampart is called by the inhabitants of the adjacent districts " God's wall " (*Ullah Bund*), to distinguish it from the Ally and Mora barriers, which the sovereigns of the country had constructed further up stream. According to

Mr. A. Barnes, the earthquake which produced this strange phenomenon of elevation and depression was felt over an area of more than 95,500 square miles.

With regard to the earthquake at Conception in 1835, the affirmative evidence on the point is so abundant, that the raising of the coast at this place must be regarded as a positive fact;* but it must remain unknown whether some enormous depression in the interior of the continent did not compensate for the temporary upheaval of the sea-coast. Still more recently, on the 23rd of January, 1855, at the time of an earthquake which violently shook New Zealand, and especially the two shores of Cook's Straits, the ground on which the

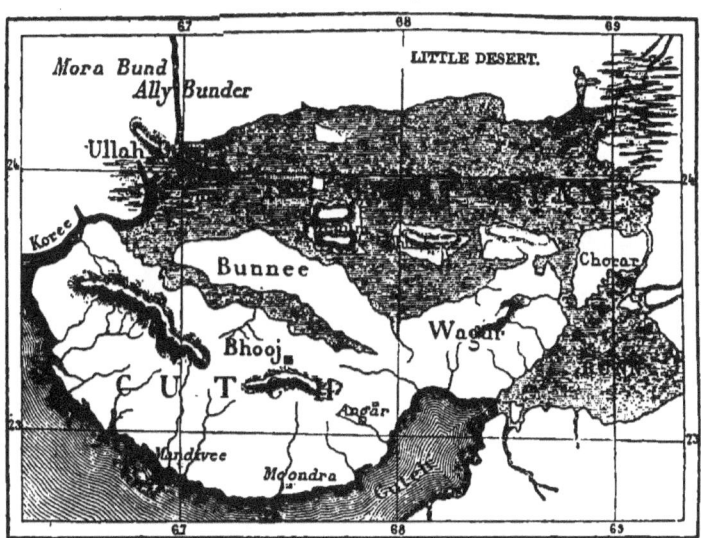

Fig. 214.—The Runn of Cutch and the Ullah Bund.

town of Wellington stands was raised 23 inches, a cape in the vicinity was elevated nearly 10 feet, whilst a fissure of about 40 miles in length opened in the southern island on the other side of the straits, and the alluvial plain of a small stream sank considerably.† Also, in 1861, the coast of Torre del Greco, at the foot of Vesuvius, was suddenly raised 3½ feet for a length of more than a mile. All these are extraordinary phenomena; and a just hesitation must at present be felt in venturing to give any explanation of them.

* *Vide* below, p. 619.
† F. von Hochstetter, *Neu-Seeland*.

There is another geological fact which has been little studied as yet, which, however, must be perhaps attributed to oscillations of the ground: this is the curious arrangement that is assumed in some plains by the boulders, pebbles, and drifts of sand. Thus, in the Desert of Atacama, in many spots, regular figures are to be met with —circles, squares, or lozenges—composed of small fragments of quartz or other stone.* In the entirely uninhabited plains of Safa, at the foot of the former volcanoes of Djebel-Hauran, M. Wetzstein likewise remarked a multitude of small figures with regular angles, formed something like mosaics, over very considerable areas. Must we look upon these designs as some immense symbolical work to be attributed to the ancient inhabitants of the district, or are they, in fact, the sports of nature, and phenomena similar in character to the figures shaped out by grains of sand agitated on vibrating plates? This question remains to be solved by future observers.

* Tschudi, *Ergänzungsheft ; Mittheilungen von Petermann*, 1860.

CHAPTER LXXVIII.

PERIODICITY OF EARTHQUAKES.—THE MAXIMUM IN WINTER.—THE MAXIMUM AT NIGHT.—COINCIDENCE WITH HURRICANES.—INFLUENCE OF THE HEAVENLY BODIES.

FROM time immemorial it has been asserted by the natives of the countries which are most frequently ravaged by earthquakes, that these commotions bear some intimate relation to the movements of the atmosphere, and very generally coincide with certain meteorological conditions, such as rainy seasons, numerous storms, warm and damp winds.* Nevertheless, most geologists, considering these oscillations of the ground to be nothing more than the half-exhausted quiverings of a great ocean of fire, have even recently denied the possibility of any such coincidence.

In 1834, Professor Merian, having classed, according to their appearance in the various seasons of the year, 118 earthquakes which occurred at Basle and in the countries round it, ascertained, to the surprise of the scientific world, that these phenomena are much more frequent in winter than in summer. This fact, which was at first called in question, has since been indubitably established by the investigations of Alexis Perrey and Otto Volger. Only, in proportion as the list of shocks becomes more numerous, the difference between the winter maximum and the summer maximum tends to disappear, for the simple reason that, in the two opposite hemispheres, the seasons follow one another at six months' intervals, and that the various phenomena in connection with the seasons are thus brought into a state of equilibrium on each side of the equator. It is therefore necessary to study, in each region separately, the order in which the occurrence of earthquakes is divided among the different seasons. The comparative frequency of these phenomena during the winter season is then much more easy to be observed. Thus the 656 shocks enumerated in France by Alexis Perrey up to the year 1845 are divided in the proportion of 3 to 2 respectively, for the half-years commencing in November and May. In the region of the

* Luigi Greco; Ferdinand de Luca; Alexis Perrey.

Alps, where there are no volcanoes, the difference between the number of earthquakes in winter and summer is, on the other hand, enormous. By comparing the four months of May, June, July, and August, with December, January, February, and March, we see that

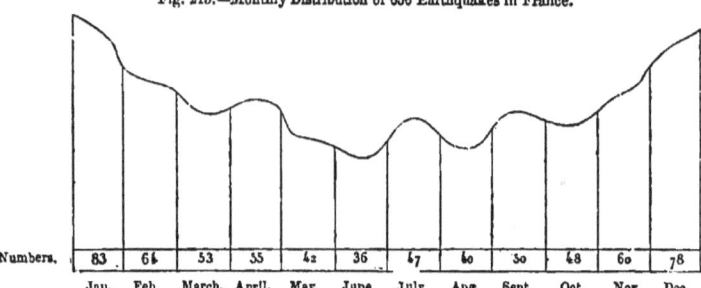

Fig. 215.—Monthly Distribution of 656 Earthquakes in France.

Numbers. 83 | 64 | 53 | 35 | 42 | 36 | 47 | 40 | 30 | 48 | 60 | 78
Jan. Feb. March. April. May. June. July. Aug. Sept. Oct. Nov. Dec.

the shocks are three times as numerous in the second season as in the first. In Italy, according to the same author, the difference was much less perceptible, since out of 984 earthquakes, 453 took place in the summer season, from April to September, and 531 during the

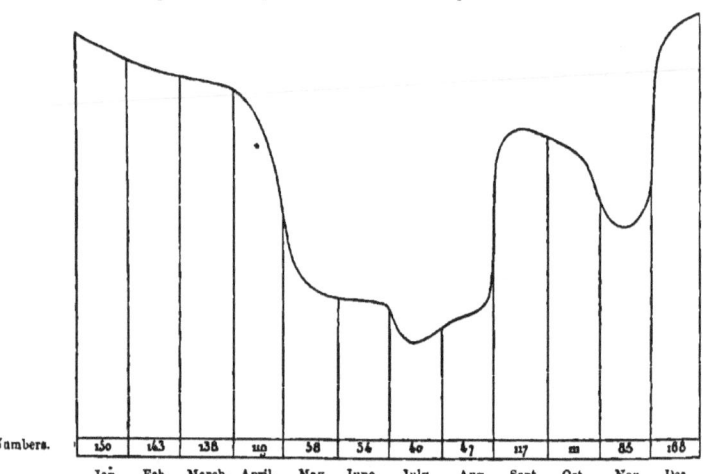

Fig. 216.—Monthly Distribution of 1,230 Earthquakes in Switzerland.

Numbers. 130 | 143 | 138 | 110 | 58 | 54 | 40 | 47 | 117 | 11 | 83 | 188
Jan. Feb. March. April. May. June. July. Aug. Sept. Oct. Nov. Dec.

winter half-year, from October to March. Nevertheless, even in this region, where most of the great subterranean movements are evidently in connection with volcanic action, these phenomena are perceptibly more frequent during the cold portion of the year.

PERIODICITY OF EARTHQUAKES. 613

If we limit ourselves to the study of some particular centre of shocks, the difference observed between the seasons in respect to the frequency of subterranean shocks is still more considerable. As an instance of this we may mention, on the authority of Otto Volger, the remarkable region of the Mid-Valais, where, out of a total number of 98 known earthquakes, one only took place in summer, whilst 72 were felt in winter. The proportion is nearly the same in the region

Fig. 217.—Monthly Distribution of 98 Earthquakes in the Mid-Valais.

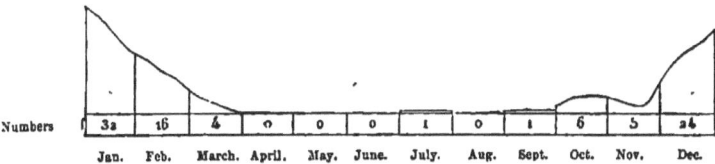

Numbers	32	16	4	0	0	0	1	0	1	6	5	24
	Jan.	Feb.	March.	April.	May.	June.	July.	Aug.	Sept.	Oct.	Nov.	Dec.

of Hohensax, on the southern slopes of Säntis, not far from the former fork of the Rhine.

Not only are subterranean commotions more frequent in winter than in summer, at least in the regions of Central Europe and on the coasts of the Mediterranean as far as Asia Minor and the Caucasus,[*] but this remarkable fact has also been observed—that the shocks are felt more frequently at night than in the day, and this holds good in all seasons of the year. In all earthquake regions this is uniformly the case. In Switzerland, out of 502 earthquakes, the date and time of which are known, only 182 took place between six o'clock in the morning and six o'clock in the evening; 320—that is, nearly double— were felt during the twelve hours of the night. There is, therefore,

Fig. 218.—Monthly Distribution of 98 Earthquakes in Southern Säntis.

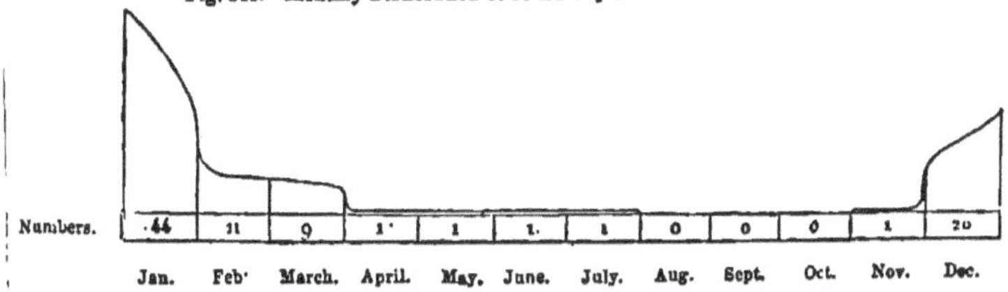

Numbers.	44	11	9	1	1	1	1	0	0	0	1	20
	Jan.	Feb.	March.	April.	May.	June.	July.	Aug.	Sept.	Oct.	Nov.	Dec.

for every diurnal period, a series of alternations resembling those of the annual period; but there is, in truth, no reason for astonishment in this, as, in an entirely general point of view, every day, in its rain, its storms, and all its meteorological phenomena, may be looked upon as an epitome of the whole year. Looking at it in a certain

[*] Moritz, *Bulletin de l'Académie de St. Pétersbourg*, vol. viii.

way, mid-day is summer, and midnight is the winter of each diurnal revolution. Have we not, then, a good right to infer, from the regular fluctuation of subterranean agitations, these fluctuations corresponding with the lapse of seasons and hours, that these great events depend, at least in certain countries, on some external phenomena, and not on the tremblings of a sea of lava? "Are they not," as Volger says, "connected with the whole body of laws which

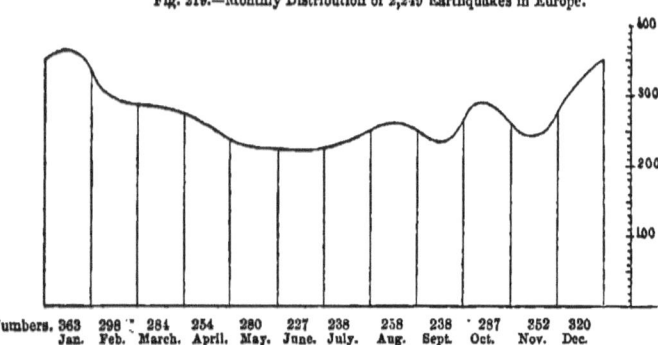

Fig. 219.—Monthly Distribution of 2,249 Earthquakes in Europe.

Numbers. 363 298 284 254 280 227 238 258 238 287 252 320
Jan. Feb. March. April. May. June. July. Aug. Sept. Oct. Nov. Dec.

govern the return of light and darkness, heat and cold, snow and rain, drought and running water?" The greater number of earthquakes would thus be facts essentially connected with the conditions of climate.

It is also related that, during the hurricanes in the West Indies, the ground is severely shaken, as if the wind, which tears down

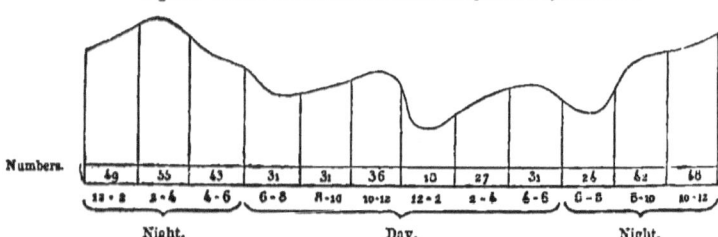

Fig. 220.—Switzerland: Distribution of 435 Earthquakes every other Hour.

trees, overthrows buildings, and drives the waves into immense waterspouts, had also laid hold of the layers of the rocks, and had shaken them on their bases. Is it the fact that the inhabitants, under the influence of terror, fancy that the ground itself participates in the immense convulsion? Such hallucinations would not seem very wonderful when we consider that at each fresh fury of the cyclone, nothing but death is expected. Nevertheless, the evidence relative to this coincidence between hurricanes and earthquakes is so abundant and positive, that it is difficult not to put faith in it. In 1844, on the occasion of a hurricane which destroyed hundreds of vessels in the roads of Havannah, a shock, which was not connected with any volcanic phenomenon, agitated the ground of the island.* Still more recently, during the cyclone of 6th September, 1865, which ravaged Guadaloupe, the Saintes, and Marie-Galante, the earth suddenly shook at the most terrible moment of the tempest, and several houses were thrown down. What is the cause of this coincidence between earthquakes and cyclones? Does the electric storm itself penetrate into the depths of the earth? or do the torrents of rain produce subterranean downfalls? These are questions to which, at present, it is impossible to reply.

However this may be, we must at any rate admit that there are at least two classes of earthquakes; one class proceeding from the pressure and eruption of vapour and lava, the other caused by the downfall of rocks; both, however, producing the same impression on the senses of man, and developing themselves in the same way over vast areas. Perhaps to these two classes of shocks it may be necessary to add a third—those shocks, namely, which originate in the relations existing between our planets and the other bodies in space. Thus, according to Wolf, there is a constant relation between earthquakes and the spots on the sun. Finally, the investigations carried on with so much perseverance by Alexis Perrey prove unquestionably that the successive phases of the moon exercise considerable influence on movements of the ground. Like the ocean, the earth, too, has its tides. The frequency of earthquakes, even of those which are only revealed to us by the delicate instruments of M. d'Abbadie, augments, like the height of the flow of the tide, at the epoch of the syzygies. This frequency increases when the moon approaches the earth, and diminishes when it is further away; in a word, the time when the sea oscillates with the greatest force is also the time when the earth itself most frequently trembles. "Our planet," says M. Boscowitz, "is engaged in a constant exchange of forces and influences with the heavenly bodies which, like it, occupy ethereal space."

* André Poey, *Comptes Rendus de l'Académie des Sciences*, 1865.

CHAPTER LXXIX.

SLOW OSCILLATIONS OF THE GROUND.—DIFFICULTIES PRESENTED IN THE OB-
SERVATION OF THESE·PHENOMENA.—CAUSES OF ERROR : EROSION OF SHORES,
SWELLING AND SINKING OF PEATY SOILS.—INFLUENCE OF TEMPERATURE.
—LOCAL UPHEAVALS.

THE solid ground, which people once considered as the symbol of immobility, is, on the contrary, in a state of constant oscillation. The crust of the earth, carved out incessantly by the various meteorological agents on one side, drawn by the other bodies revolving in space on the other, modified by water and pressed upon by vapours, gases and molten matter, never ceases to undulate as a raft rising and sinking on the waves of the sea. At rare intervals there are great earthquakes; more often this undulation of the ground consists in mere vibrations, which are scarcely perceptible except by the aid of instruments. The terrestrial crust is not only continually shaken by these transient shiverings; it is, besides, actuated by uniform movements of incalculable force, which at some points raise it, and at others depress it, as compared with the level of the sea. Whilst we are busy on its surface, the earth itself is shifting under our feet.

These swellings up and depressions, which recall to mind the phenomena of organised bodies, often take place so slowly, that to verify them with any degree of certainty, successive generations of observers would require the lapse of many years, or even centuries. The " patient earth " seems to revolve inertly in space, and yet it is at work without intermission in modifying the form of its seas and its coasts. During each geological period, the continental surface, motionless in appearance, rises in some spots to a great height above the ocean, and, in others, it sinks beneath the water; everywhere, indeed, the ancient relief and outline of the ground are being slowly modified. In accordance with what law, in what geographical order, and at what comparative rate of progress do these gradual oscillations take place, which result, in the long run, in effecting a change in the general aspect of the globe? Science is not as yet in

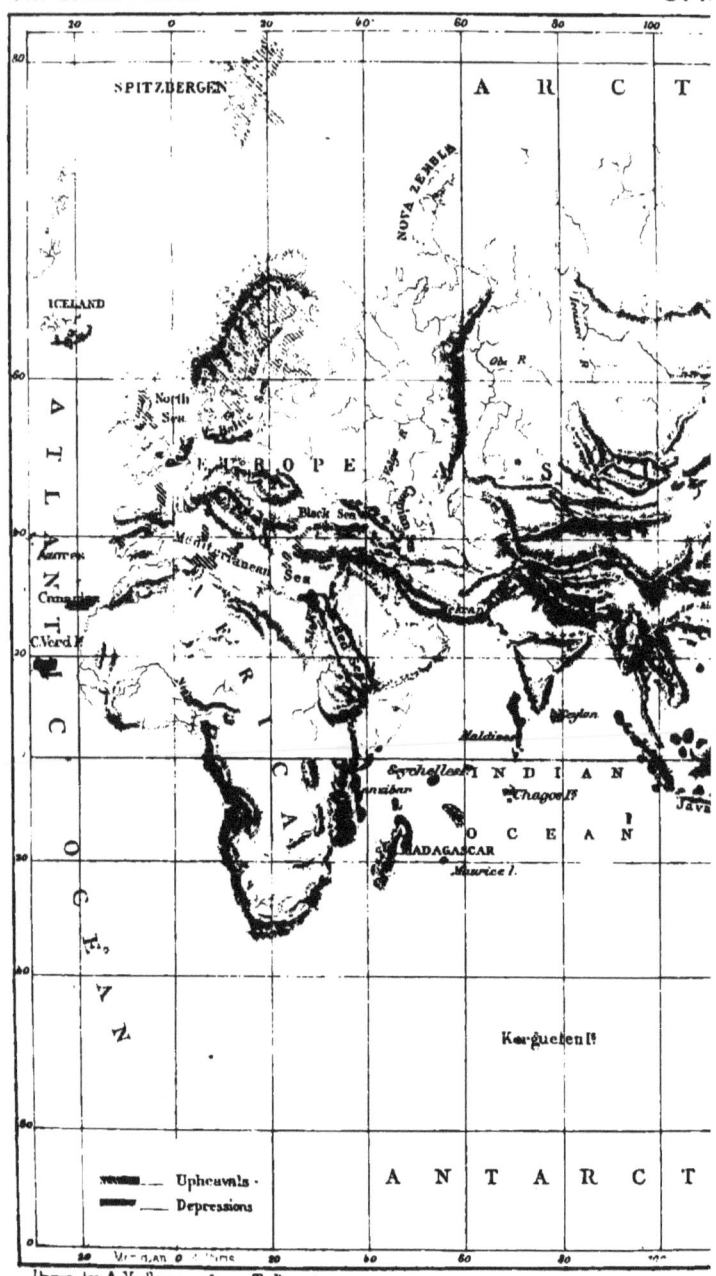

Drawn by A. Vuillemin after Ch. Darwin

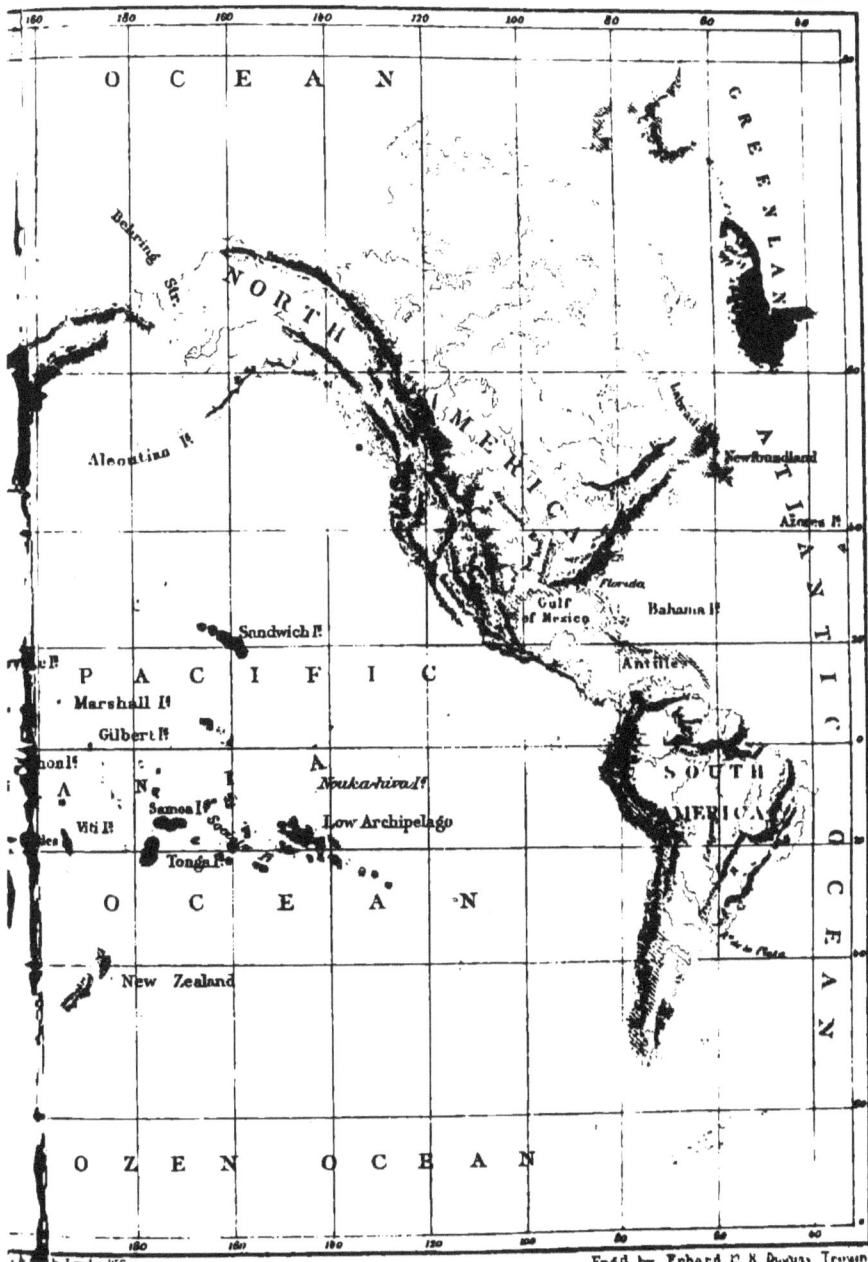

EPRESSIONS   Pl. XXIII

a position to give a positive answer to these questions. But, in the meantime, until geologists are able to estimate exactly the dimensions and progress of each wave of upheaval which is formed by the crust of the earth, it is, at all events, possible to bring together the principal facts which bear upon the subject of the oscillatory movements of continents and the bed of the sea.

Small broken shells, the scattered remains of polypes, almost invisible grooves marked here and there on the sides of rocks—all these signs, which most people would pass by with indifference, are become, thanks to the patience and sagacity of observers, so many undeniable proofs of the regular movements of the ground. It is, however, only on the sea-coast and in the vicinity of seas that *savants* can positively verify these manifestations of the vitality of the globe. Considering the ocean as a fixed level-mark, which is almost motionless in relation to the elevated or depressed masses of the land, they can easily prove the general elevation of a region by noticing on the shore the parallel lines formed at different epochs by the friction of the sea-water. But further inland the marks of the same kind which are traced out on their banks by rivers and lakes are very seldom able to furnish incontestable evidence of vertical oscillations of the soil. The more or less shifting level of lakes and running water depends on several geological circumstances, and it is only when all these circumstances are perfectly well known that the ancient terraces and slopes of erosion which exist in fluvial and lacustrine basins can be made to serve as marks to measure the progress of terrestrial oscillations. As a matter of fact, geologists are, therefore, compelled to limit themselves to bringing under notice those oscillatory movements of the earth's crust which are taking place on sea-coasts.

In studying the oscillations of the earth, it is important to be very carefully on our guard against the numerous causes of error which arise from the eternal combat which is always taking place on the shore between the land and the sea. Neither the gradual encroachments of alluvial shores, nor the progressive erosions which, in so many places, the coast has to undergo from the sea, are to be considered, without due examination, as proofs of the upheaval or subsidence of a region. The sand which is driven up by the waves of the sea, and the alluvium which is drifted down in the currents of rivers, are deposited on most low coasts in sufficiently considerable quantities for the belt of shore to increase constantly in breadth. Even where this zone sinks slowly with the land adjacent, as is taking place at the north of the Adriatic, the alluvium may nevertheless form on the shore a kind of cushion, and may defend the plains of the interior against the waves of the sea. On the other hand, there are a great many steep beaches and cliffs which, being assailed in front by the waves and the tide, and worn away obliquely by lateral currents, recede gradually before the inroads of the sea; but in cases of this kind it is usually impossible to detect the slightest depression in the general level of the country. Gentle geological elevation of the ground may, indeed, actually coincide with a falling back of the shores. The coasts of Aunis and Saintonge present an example of this apparent anomaly.

In movements of the soil it is necessary also to distinguish those which are produced by the slow pressure of subterranean forces, and those which are occasioned by temporary causes, such as the more or less quantity of water contained in the surface layers, the activity of evaporation, and the bringing into cultivation of the country. Thus, when peat-mosses form in the low-lying lands of the valleys, taking the place of lakes and marshes, they hold the water in their masses of moss just as an immense sponge, and, gradually swelling, ultimately rise to a height of several feet above the former level of the soil. On the other hand, those tracts of peat-moss which have been dried by draining operations gradually sink; the mosses wither, die, and are reduced to dust. One might fancy that the soil was slowly sucked towards the interior of the earth by some secret force.

There is, however, no reason to be astonished at these alternate phenomena of swelling and shrinking which are afforded by peaty soil, as a mere variation of temperature is sufficient to produce similar results in the compact strata of mountains. In the day-time the particles of rocks dilate under the influence of the solar rays; in the night-time they become cool, and contract in consequence of radiation, and the total mass sinks to a very slight extent, which is sometimes perceptible by means of instruments. Thus Moesta, the Chilian astronomer, has been enabled to ascertain that the National Observatory of Chili, situated on the hill of Santa Lucia, near Santiago, rises and descends alternately in the space of twenty-four hours. The daily oscillations of the rock, which in turn dilates and shrinks, are, indeed, considerable enough to render it necessary to introduce this element of calculation into the mathematical formulæ devoted to the regular observations. Similar phenomena, but occasioned by different causes, are produced under the Observatory of Armagh, in Ireland.

After heavy rains, the hill on which the edifice stands rises perceptibly; then, after active evaporation has got rid of the extra water contained in its pores, the hill again sinks.

The shocks of greater or less violence communicated to the soil in volcanic districts produce alterations in the level which are much more considerable than the slight oscillations caused by the heat of the sun or the various atmospheric agents. But these alterations of level are none the less merely local phenomena, and although they are doubtless connected with the same class of facts as the slow upheavals and depressions, they must be clearly distinguished from the long-protracted movements which bulge up the crust of the earth under whole continents. As an instance of the local undulations which are mere accidents in the history of the planet, we may mention that which the earthquake of Conception caused to take place temporarily in February, 1835, at the Isle Santa Maria and the adjacent coasts of the Chilian mainland. An enormous mass of earth was suddenly elevated. The shore nearest the town was raised perpendicularly a yard and a half, while the island uprooted, so to speak, by the violence of the subterranean shock was upheaved obliquely 2⅓ yards at its southern point, and 3½ yards at its northern extremity. Two months afterwards the shore at Conception was scarcely 23 inches above its former level, and the island had also sunk in proportion. Finally, towards the middle of the year, every trace of the upheaval had disappeared, and the sea-water came exactly up to the same winding line of *débris* which it washed before the catastrophe.*

The famous columns of the supposed Temple of Serapis, which rise on the shore of the Mediterranean, not far from Puzzuoli, likewise bear on their marble shafts proofs of purely local oscillations. Spallanzani, some time ago, showed that these columns, the bases of which were surrounded with rubbish, must at some former date have been immersed in the water of the sea to a depth of about 21 feet; for up to this height the calcareous cases of the serpulæ may be noticed on the shafts, and also the innumerable holes that the pholades have hollowed out in the thickness of the marble, which is eaten away circularly by the waves. The temple having been repaired in the reign of Marcus Aurelius, must certainly at that time have been above the level of the sea. It is not known at what date it sank, together with the hillock on which it stands. It might, perhaps, have taken place during the year 1198, in which year the *Solfatara* of Puzzuoli produced an eruption. With regard to the

* Fitzroy, *Voyage of the " Adventure" and the "Beagle."*

emergence of the colonnade from the water, it is probable that it happened in the year 1538, at the time when the Monte-Nuovo made its appearance. If these supposed dates are the real ones, the lower half of the Temple of Serapis must have been bathed for 340 years in the waters of the gulf. But this event can only have been caused by some local agitation of the earth; for, during the same period, the adjacent coasts of Naples have not altered their level to any perceptible extent.*

* Lyell, *Principles of Geology*, vol. i.

CHAPTER LXXX.

UPHEAVAL OF THE SCANDINAVIAN PENINSULA; OF SPITZBERGEN; OF THE COASTS OF SIBERIA; OF SCOTLAND; OF WALES.

On all those coasts where heaps of modern shells now left dry, and the ledges cut out at different heights in the faces of the cliffs, furnish unquestionable testimony of a progressive upheaval of the ground, *savants* who desire to study the progress of the phenomenon must, of course, do so both by direct measurements and by comparing the levels taken at longer or shorter intervals. More than a hundred and thirty years ago, Celsius, the Swedish astronomer, formed the idea of resorting to these means, not with the intention of verifying the growth of the Scandinavian peninsula—a fact which to him did not seem at all probable—but in order to prove the supposed alteration in the level of the waters of the Baltic Sea. He was aware, from the unanimous testimony of the inhabitants of the sea-coasts, that the Gulf of Bothnia was constantly diminishing both in depth and extent. Old men pointed out to him various points on the coast over which, during their childhood, the sea used to flow, and, besides, showed him the water-lines which the waves had once traced out further inland. Added to this, the names of places, the position more or less upon the mainland of former ports now abandoned, and of edifices built once upon the shore, the remains of boats found far from the sea; lastly, the written records and popular songs, could leave no doubt whatever as to the retreat of the sea-water. At this epoch, when *savants* still believed in the immovable solidity of the rocky framework of the globe, Celsius was naturally bound to attribute the constant growth of the sea-coast to the gradual depression of the level of the sea. In 1730 he felt himself warranted, by comparing all the evidence he had collected, in propounding the hypothesis that the Baltic sunk about 3 feet 4 inches every hundred years. Then, in the course of the following year, having, in company with Linnæus, made a mark at the base of a rock in the Island of Loeffgrund, situated not far from Jefle, he was able personally to verify, thirteen years afterwards, that the retreat of the Baltic Sea was taking place quite as rapidly as he had supposed. The difference of level observed during these thirteen years was 7 inches, or 4 feet 5 inches for a century. From 1730 to 1839 the upheaval of Loeffgrund amounted to 2 feet 11 inches only.*

Celsius was accused of impiety by the divines of Stockholm and Upsal. The parliament, indeed, wished to cut the matter short by a vote; the two orders of peasants and nobility declared themselves incompetent in the matter; whilst the representatives of the clergy, followed timidly by the burgesses, condemned the new opinion as an abominable heresy.† Nevertheless, the geologists who, since the last century, have visited the coasts of Sweden, have been compelled to verify and complete Celsius's observations. They have, however, been forced to reject the first hypothesis of the gradual subsidence of the water, and to recognise as a fact that the movement, attributed in error to the liquid mass of the Baltic was that of the continent itself. As, indeed, had been already laid down by Antonio Lazzaro Moro, an Italian *savant*, it was the earth, and not the sea, which was the moving and shifting element.‡ In fact, if the sea-level was progressively sinking, as was once supposed, the water, the surface of which, owing to gravity, must always remain horizontal, would retreat equally all round the Scandinavian peninsula, and on all the sea-shores. But this is not the case. At the northern extremity of the Gulf of Bothnia, at the mouth of the Tornea, the continent is emerging at the rate of 5 feet 3 inches a century, but by the side of the Aland Isles it only rises $3\frac{1}{4}$ feet in the same time; south of this archipelago it rises still more slowly; and further down, the line of the shore does not alter as compared with the level of the sea. The terminal point of Scania is gradually being buried under the waters of the Baltic, as is proved by the forests which have been submerged. Several streets of the towns of Trelleborg, Ystad, and Malmoo have already disappeared; the latter has sunk 5 feet 2 inches since the observations made by Linnæus, and the coast has lost, on the average, a belt 32 yards in breadth.

On the west coasts of the Scandinavian peninsula the phenomena proving a recent upheaval of the ground are just as numerous as on the eastern shores; but the rapidity of the ascending movement has not yet been measured, although it is certainly less considerable than in Sweden. The terminal point of Jutland, bounded by an ideal line

* Lyell; Robert Chambers.
† Anton von Etzel, *Die Ostsee*.
‡ Von Hoff, *Veränderungen der Erdoberfläche*, vol. iii. pp. 318, 319.

tending obliquely from Fredericia towards the north-west, rises, according to Forchhammer, 11·70 inches a century. At Christiania the increase is perhaps still less, for, according to Eugène Robert, the pavement of the ancient town appears to have remained stationary for three hundred years. Lastly, further to the north, the present position of several edifices situated in the Island of Munkholm, near Trondhjem, proves, as Keilhau the geologist says, that during a thousand years the elevation of the ground has been less than 20 feet. This is all that is positively known. Nevertheless, a comparison of the various lines of level, and an examination of all the other indications of a slow upheaval, seem to show that, in spite of the numerous inequalities in the rate of progress of the phenomenon, that portion of the coast which is nearest to the pole is rising the most rapidly out of the water. Elevated beaches, which can be traced by the eye like the steps of an amphitheatre, are arranged in stages at various heights on the slopes of the mountains. Heaps of modern shells are found at heights of 500 to 650 feet above the level of the sea, and the great branches of pink coral formed by the *Lophohelia prolifera*, which lives in the sea at a depth varying from 1,000 to 2,000 feet, are now raised up to the base of the cliff.* The pine woods, too, which clothe the summits, are continually being upheaved towards the lower snow-limit, and are gradually withering away in the cooler atmosphere; wide belts of forest are composed of nothing but dead trees, although some of them have stood for centuries.†

The whole body of facts known on the subject of the movements of the ground of Scandinavia will, therefore, warrant *savants* in comparing the whole peninsula to a solid plane turning round a line on which it rests, and raising one of its ends so as to lower the other in the same proportion. The Gulfs of Bothnia and Finland, like vessels tilted up out of the horizontal, slowly pour their waters into the southern basin of the Baltic. Fresh islets and ranges of isles appear in succession, rocks are laid dry, and if the elevation of the bed of the sea continues to take place with the same regularity as during the historic ages, it may be predicted that in three or four thousand years the archipelago of Qvarken, between Umea and Vasa, will be changed into an isthmus, and the Gulf of Tornea will be converted into a lake similar to that of Ladoga. Later still, the Aland Isles will become connected with the continent, and will serve as a bridge between Stockholm and the empire of Russia. It is, besides, very

* Carl Vogt, *Nordfahrt*.
† Keilhau, *Bulletin de la Société Géologique de France*, First Series, vol. vii.

probable, if not certain, that the great lakes and innumerable sheets of water which fill all the granite basins of Finland have taken the place of an arm of the sea which once united the Baltic to the great Polar Ocean. The erratic blocks of granite scattered about all over Russia can only have been carried thither by trains of ice which have made their way over the sea from the mountains of Sweden. The shells belonging to polar waters, which are found as far as the basin of the Volga, also testify in favour of the existence of a former arm of the sea. The name of Scandinavia itself signifies " Isle of Scand," and the name of Bothnia (Botten) proves that these coast provinces were formerly a marine bed.* Here philology steps in to aid geology and tradition.

This is not all. The Baltic Mediterranean also communicated with the North Sea by a wide channel, the deepest depressions of which are now occupied by the Lakes Mäler, Hjelmar, and Wenern. Considerable heaps of oyster-shells are found in several places on the heights which command these great lakes of Southern Sweden. On the rocks now laid dry, which surround the Gulf of Bothnia, banks of these molluscs have also been discovered exactly similar to those of Norway and the western coasts of Denmark. With regard to the celebrated *kjoekkenmœddings* of the Danish islands, they are in great part composed of oyster-shells, which the inhabitants, in the age of stone, evidently used to collect in the bottoms of the neighbouring bays. It has been proved by the investigations of M. de Baer that the oyster cannot live and grow in water holding more than thirty-seven parts in a thousand of salt, or less than sixteen or seventeen parts in a thousand. Now the Baltic Sea, into which its numerous tributaries bring a large quantity of fresh water, does not contain, on the average, more than five parts in a thousand of salt; and, indeed, in some of the gulfs, the water, now devoid of all its former inhabitants, has become entirely fresh. And yet—the heaps of oyster-shells prove it—the Baltic Sea and the inland lakes were once as salt as the North Sea is at the present day. Whence, then, could this saltness proceed, except from some former strait which occupied the depressions in which the Swedish engineers have dug out the Trolhatta Canal? Besides, when the sluices were being constructed, there were found, not far from the cataracts, and at a height of 40 feet above the Cattegat, various marine remains, mingled with relics of human industry—boats, anchors, and piles. According to M. de Baer, it is not, at the most, more than five thousand years before our century that

* Von Maack; Eugène Robert.

ELEVATION OF SCANDINAVIA. 625

we must date the closing up of the straits which used to exist between

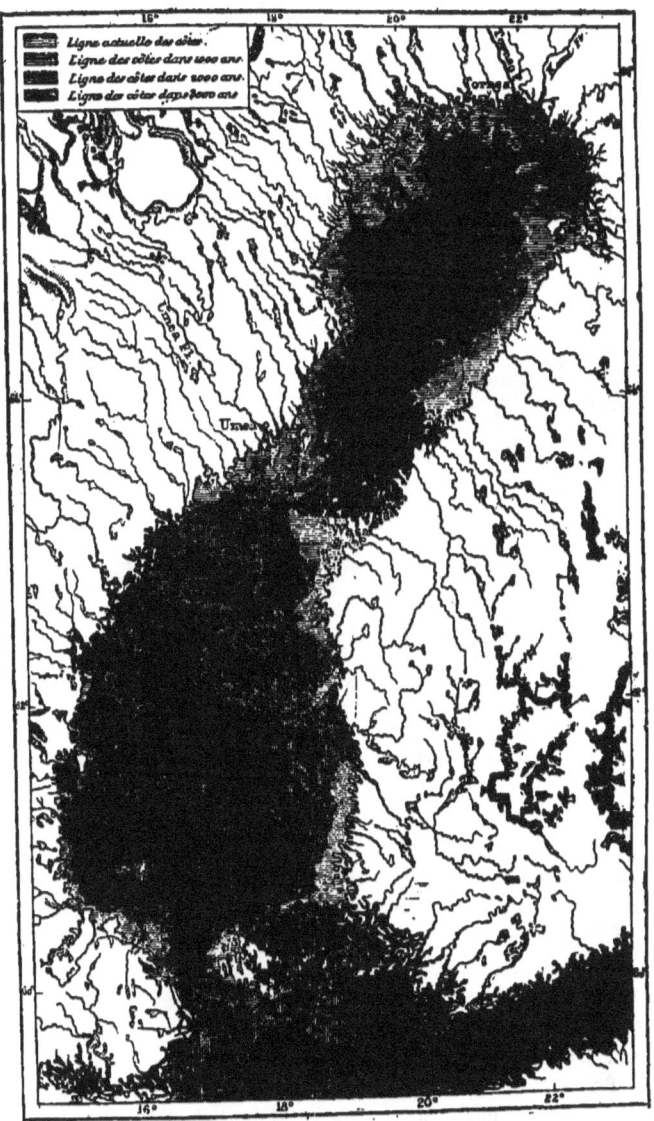

Fig. 221.—Elevation of the Bed of the Gulf of Bothnia.

Southern Sweden and the great mass of the northern plateaux.

Since Leopold von Buch has placed beyond all doubt the important fact of the gradual upheaval of the northern portion of the Scandinavian peninsula, several geologists have ascertained that the elevation does not take place in a mode that is perfectly uniform. During bygone centuries the movement has sometimes been accelerated and sometimes slackened, as is proved by the inequality of the elevated sea-beaches which run along the sides of the mountains on the coasts of Norway. Some of these steps or shelves that the waves have carved out in the rock are wide and gently inclined; others are abrupt, and can scarcely be distinguished from the slopes above them. Lastly, the measurements made by M. Bravais along the

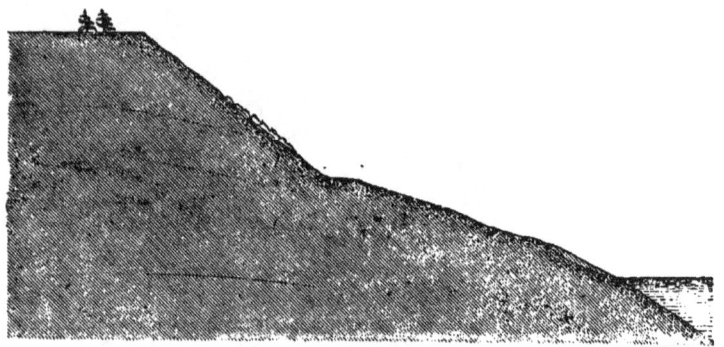

Fig. 222.—Bank of Altenfjord.

lines of erosion of Altenfjord have proved that they are not parallel, and that the rocky masses situated at the ends of the gulfs have been more energetically upheaved than the layers lying nearer the sea. Thus the upper bank of Altenfjord has risen at the eastern end to a height of 219 feet above the level of the sea, but at the entry of the bay it has only risen 91 feet. In like manner, the lower shelf, throughout the whole of its immense circuit round the gulf, presents a slight inclination towards the sea, being no less than 88 feet high on the east, and only 45 feet at the outer promontories. Thus the action of upheaval is evidently stronger in the vicinity of the mass of mountains than it is on the coast; but this does not afford any sufficient reason for saying that at a certain distance to the west, under the bed of the sea, the movement of the ground entirely ceases.*

M. Carl Vogt has propounded an ingenious hypothesis to account for this inequality of elevation. According to his theory, rocks of

* Bravais, *Voyages en Scandinavie*, vol. i.

various natures—schists, sandstones, or limestone—which compose the mountains of the Scandinavian peninsula, incessantly swell in consequence of the infiltration of snow-water, and owing to fresh crystallisations taking place, by means of moisture, are gradually converted into masses of stratified granite. This hypothesis, which has been much discussed by geologists, would certainly explain the raising of the lines of erosion of the Norwegian coast near the groups of mountains, but it would fail to account for the intervals of comparative repose, and especially for the sinking of the ground, which many geological facts prove to have taken place during the Glacial period. It is, therefore, necessary to admit that other forces have been in action in the solid mass of Scandinavia.

Added to this, we must not lose sight of the fact that the upheaval of this peninsula is not an isolated event, and that the other countries of the North of Europe and Asia, notwithstanding the diversity of their rocks, all appear to be actuated by a similar movement of ascension. The islands of Spitzbergen exhibit generally, between the present sea-shore and the mountains, former sea-beaches which are gently inclined, and half a mile to two miles and a half in breadth, on which are found, up to a height of 147 feet, heaps of bones of whales and shells of the present period. These remains, surrounding all the snow-clad slopes of Spitzbergen, prove that this archipelago, like Scandinavia, is gradually emerging from the waves of the Polar Ocean.* The northern coasts of Russia and Siberia are likewise rising, as is attested by popular tradition and the evidence of learned travellers. MM. de Keyserling, Murchison, and de Verneuil have found, at points 250 miles to the south of the White Sea, on the banks of the Dwina and the Vaga, beds of sand and mud containing several kinds of shells similar to those which inhabit the neighbouring seas, and so well preserved that they had not lost their colours. In like manner, M. de Middendorf states that the ground of the Siberian *toundras* is in great part covered with a thin coating of sand and fine clay, exactly similar to that which is now deposited on the shores of the Frozen Ocean: in this clay, too, which contains in such large quantities the buried remains of mammoths, there are also found heaps of shells perfectly identical with those of the adjacent ocean. Far inland, besides, trains of drift wood are seen, the trees forming which once grew in the forests of Southern Siberia; these trees, having been first carried into the sea by the current of the rivers, have been thrown up by the waves on the former coasts,

* Malmgren, *Mittheilungen von Petermann*, vol. ii., 1863.

which are now deserted by the sea. It is this half-rotten wood which is called by the natives "Noah's wood," fancying that they have before them the remains of the ark of the deluge. More than this, there are also direct proofs of the upheaval of Siberia. The Island of Diomida, which Chalaourof noticed in 1760, to the east of Cape Sviatoj, was found to be joined to the continent at the date of Wrangell's voyage, sixty years later. It is, besides, very probable that this upheaval of the ground is prolonged to the east over a great portion of the circumpolar land of North America, as far as Northern Greenland;* for numerous indications of this phenomenon have been recognised in the Arctic isles scattered off the coasts of the continent. At Port Kennedy, Mr. Walker found shells of the present period at a height of 557 feet above the sea; a bone of a whale lay at a height of 164 feet.†

The cliffs of Scotland also present phenomena similar to those of Scandinavia. Parallel water-marks traced out by the waves on the escarpments of rocks, and collections of shells peculiar to the neighbouring seas, attest the gradual elevation of this portion of Great Britain. The elevation, too, must have been of a much more regular character than that of the coasts of Norway; for, according to Robert Chambers, not the slightest variation of level is noticed on the ancient terraces. This ascending movement is still continuing; for it has been ascertained that the marine cliffs which were once situated above the estuaries of the Forth, the Tay, and the Clyde, contain not only organic remains of recent ages, but also heaps of pottery of Roman origin. The former Roman port of Alaterva (Cramond), the quays of which are still visible, is now situated at some distance from the sea, and the ground on which it stands has risen at least $24\frac{1}{2}$ feet. In other places the *débris* scattered on the bank show that the coast has risen about $26\frac{1}{2}$ feet.‡ Now, by a remarkable coincidence, the ancient wall of Antoninus, which, at the time of the Romans, served as a barrier against the Picts, comes to an end at a point 26 feet above the level of high tides. The general upheaval of the region may therefore be estimated at 0·195 inch a year; but since 1810 the movement has become more rapid, as is proved by the tide meters at the port of Leith, and it is, at present, at the rate of 0·546 a year.§ Further to the south, on the sides of the mountains of Wales, there are numerous indications of the

* *Vide* below, p. 651.
† Samuel Haughton, *Natural History Review*, April, 1860.
‡ Arch. Geikie, *Edinburgh New Philosophical Journal*, New Series, xiv.
§ Smith, *Geological Magazine*, September, 1866.

presence of the sea during the present period. Mr. Darbishire lately discovered, not far from Snowdon, at a height of 1,357 feet, a bed of drift containing fifty-four species of shells of similar kinds to those still existing in the northern seas of Europe; the same soil was, however, found at a point 650 feet higher, but devoid of shells.

Thus, from Wales to Spitzbergen and the eastern coasts of Siberia, the ground has continued to rise slowly during a portion of the Glacial period, and also during the present epoch. The area of upheaval includes a portion of the earth's surface which is not less than 160 degrees of longitude. In the face of these facts, are we to consider the phenomena of upheaval as mere local accidents produced by the swelling of rocks and volcanic shocks, or must we look upon them as the results of some general cause acting in various ways over the surface of the whole planet? The latter hypothesis appears to us to be the more probable.

CHAPTER LXXXI.

UPHEAVAL OF THE MEDITERRANEAN REGIONS.—FORMER LIBYAN STRAIT.—
COASTS OF TUNIS, SARDINIA, CORSICA, ITALY, AND WESTERN FRANCE.

THE countries of the South of Europe certainly possess a more gracefully-indented outline than any other regions on the face of the earth. Bays, gulfs, and inland seas penetrate them in every direction. The peninsulas which they throw out present the greatest variety of contour and aspect, and they have thus become, as it were, imbued with vitality, owing to their numerous articulations, similar to those of an organised body. Correspondent with this multiplicity of external shapes are the singular irregularities and exceptional contrasts in the movements of the ground. A certain complication is here and there manifested between the upheaved regions and those which are subsiding. Nevertheless, a sufficient number of observations have been collected to warrant us in admitting, in a general way, the elevation of most of the countries which surround the basin of the Mediterranean. These regions, which in several places have been caused to oscillate by volcanic forces, constitute a considerable area of upheaval, extending from the deserts of the Sahara to Central France, and from the coasts of Spain to the steppes of Tartary. The mountainous peninsula of Scandinavia is situated in the middle of the upheaved regions of Northern Europe, and, by a kind of polarity, the long depression of the Mediterranean occupies the centre of the vast areas in the South of Europe and Northern Africa, which are gradually rising.

This immense space was once bounded, towards the tropical zone, by another sea, or at least by a strait several hundreds of miles in width, which commenced at the Gulf of Syrtes, and, filling up the depressions of the Berber Sahara, joined the Atlantic in front of the archipelago of the Canaries. In 1863 MM. Escher von der Linth and Desor ascertained, according to M. Charles Laurent, that the sands of these regions are entirely identical with those of the nearest Mediterranean shores, and contain the same species of shells. One of these witnesses of the past, the common cockle (*Cardium edule*), is found, not only on the surface of the ground, but also at some depth, and likewise up to a height of 900 feet upon the sides of the hills. The Algerian Sahara has, therefore, risen to this extent during a recent geological period. Various depressions, the surface of which is lower than the level of the Mediterranean, have been gradually separated from the sea, and, in the present day, they exhibit nothing but marshy pools or interminable plains. At a recent, and perhaps historical epoch, Lake Tritonis of the ancients, now the Sebkha-Faraoun, into which flowed the Igharghar, has ceased to be a prolongation of the Gulf of Gabes, and has become a mere marsh. It was the last remnant of the arm of the sea which separated the mountainous regions of Atlas from the African continent, both so distinct in their general character, as well as in their fauna and flora. To the existence of this African Mediterranean, which is now replaced by white sands, beds of salt, and rocks devoid of verdure, MM. Escher von der Linth and Lyell in great part attribute the enormous extent of the former glaciers of Europe. It is, in fact, very natural to think that, before the drying up of this inland sea, the masses of air blown to the north would become saturated with moisture while passing over the water, and, rising gradually to the higher regions, would constantly convey fresh layers of snow to the summits of the Alps, instead of melting the snow, as the *föhn* now does, heated as it is by the reflection from the burning sand of Africa.* It is, however, possible that the Swiss mountains have decreased in height since the Glacial period. The same slow oscillation of the ground which emptied the Libyan Mediterranean has, perhaps, by its reaction lowered the foundations of the Alps, and brought them nearer to the level of the sea.†

On the coasts of the Mediterranean the indications which would lead us to infer the fact of some upheaval of the ground are plentiful enough. Thus the shores of Tunisia are constantly encroaching on the sea. The ancient ports of Carthage, Utica, Mahedia, Porto-Farina, Bizerta, and others, are now filled up.‡ The bays are done away with, the points advance further and further into the sea; and these phenomena take place with a rapidity sufficient to show that we are witnessing the effect of a vertical impulse similar to that which once upheaved the beds of the Saharan seas. In like manner, Sicily appears to be constantly elevated by the forces in action under the beds of its surface. On the heights which command Palermo,

---

* *Vide* vol. ii., the chapter on Winds.
† Lyell, *Inaugural Address to the British Association at Bath*, 1864.
‡ Guérin, *Voyage Archéologique à la Régence de Tunis*.

caves have been observed at an elevation of 180 feet, which have been hollowed out by the sea during a period characterised by existing species of shell-fish.* On the eastern coast of the island, Gemellaro has recorded a recent upheaval of more than 42 feet. In Sardinia, not far from Cagliari, M. De la Marmora points out, as existing at heights of 242 and 231 feet, deposits which contain remnants of pottery mixed with modern shells, which deposits, in his idea, were on a level with the sea at a date when the island was inhabited by man. Certainly, M. Emilien Dumas, an excellent observer, considers that these remnants of pottery and heaps of shells are nothing but the remnants of the cooking of food, similar to the *kjoekkenmœddings* of Denmark. If this be the case, there would be nothing to prove that Sardinia was upheaved at any recent epoch. But are the enormous banks of oyster-shells which cover the ground at the Lake of Diana, 6 feet above the level of the sea, and are continued far under the water, to be considered as nothing but the *débris* of Roman banquets? Such an idea does not, at least, appear at all probable to M. Aucapitaine and a number of other observers.

The facts of upheaval brought forward by geologists as regards the other regions of the coast of the Mediterranean are not as yet sufficiently verified, and it cannot be positively asserted that these shores have risen above the sea during the present period. Nevertheless, the body of evidence is of considerable importance, and merits serious consideration. Thus, round the former Island of Circe, now become a promontory of Tuscany, the rocks, which have much the appearance of a former sea-beach, are pierced by *pholades*.† With regard to the discovery of banks of modern shells made by Risso near Villefranche, at the extremity of the Cape of Saint-Hospice, M. Emilien Dumas disputes its scientific value. Nevertheless, it is plain that this coast, and the whole of the adjacent shore as far as Spezzia, were covered by sea-water at a recent geological epoch. All that is necessary to prove this is to examine the caves of Menton, of Ventimiglia, and of the Cape of Noli, which were hollowed out by the waves at some former date, and open like rows of arched doors and windows along the façade of some palace.

The southern coasts of France do not afford any direct evidence of an upheaval of the soil; but various indications possess a value which cannot be disputed. Astruc, of Languedoc, brings forward a great number of facts, which prove that, at the time of the Romans

---

* Lyell, *Antiquity of Man.*
† Romanelli, Breislak, quoted by Böttger, *Mittelmeer.*

and in the Middle Ages, the marshes extended much further inland. The ancient Roman road from Beaucaire to Béziers describes a wide curve towards the north, doubtless to avoid the plains on the shore, which were then entirely under water. Ancient cities with Gallic names—as *Ugernum* (Beaucaire) and *Nemausus* (Nismes)—are found along the ancient road, whilst all the places situated to the south bear Latin or Roman names—as Aigues-Mortes, Franquevaux, Vauvert, and Frontignan (*Frons stagni*)—and appear, therefore, to be of more modern origin. It is, besides, proved by various documents that ancient ports have filled up, and have been converted into *terra firma*. Astruc also points out the remarkable fact that the Romans, who highly appreciated thermal springs, were not aware of the abundant wells of Balaruc, although the eddies of steam could not have failed to point them out if they had not been covered at that time by the waters of the Lake Thau. This is an important argument in favour of the hypothesis of a gradual elevation of this part of France.

Beyond the Mediterranean basin, this movement of general upheaval appears to continue towards the north and west. Thus, at Seixal, opposite Lisbon, they have been compelled to cease building ships of the line on account of the increasing diminution of the water, which is attributed both to the deposits of mud and also to the upheaval of the rocks. On the Atlantic coasts of France a great number of phenomena of a similar nature have also been adduced. To many geologists, especially to M. Bravais, it seems probable that the whole of France, agitated by a slight and almost imperceptible tremor, is being slowly upheaved on the southern side, and turns on a base-line passing through the peninsula of Brittany. At all events, the coasts of Poitou, Aunis, and Saintonge appear to have risen since the commencement of the historical epoch. Guérande, Croisic, Bourgneuf, and Sables-d'Olonne show upon their shores unquestionable traces of recent elevation. The former Gulf of Poitou, the entrance of which, two thousand years ago, was not less than 18 to 25 miles in width, which, too, penetrated inland as far as Niort, has constantly contracted its dimensions since the above-named date, and is now nothing more than a small bay, known under the name of the Creek of Aiguillon.* The constant deposit of marine alluvium would scarcely be a sufficient cause for this rapid increase of the land; it is, therefore, probable that at this spot the upper layers of the ground are regularly in a course of upheaval. Further

* De Quatrefages, *Les Côtes de la Saintonge*.

to the south, La Rochelle, which owes its name to the position which it once occupied on a rock almost isolated in the midst of the water, now only communicates with the sea by a narrow channel, often choked up with mud. Brouage, another fort which, in the Middle Ages, was a town of important trade, is now nothing more than a ruin, some distance from the sea. The district of Marennes, to which the name of "Colloque des Iles" was once given, is now entirely connected with the mainland; and the arms of the sea which cross it have been converted into draining-channels, salt marshes, and oyster-beds. In like manner, the peninsula of Arvert, situated between the mouth of the Seudre and that of the Gironde, ceased to be an archipelago during the course of the Middle Ages. At Rochefort it has, indeed, been calculated approximately how much the ground has risen, the slips for ship-building, dug out in the time of Louis XIV., having been gradually elevated more than a yard.* "The bank is pushing up," say the inhabitants of the coast, who have long since observed the gradual upheaval of the ground.

* Babinet, *Revue des Deux-Mondes*, September 15, 1855.

CHAPTER LXXXII.

COASTS OF ASIA MINOR.—ANCIENT OCEAN OF HYRCANIA.—COASTS OF PALESTINE AND EGYPT.—THE ADRIATIC GULF.

PHENOMENA of upheaval are also very common in the islands and round the edge of the eastern basin of the Mediterranean. Like Sicily and several parts of the coasts of Italy and Greece, a considerable number of islands—as Malta, Rhodes, and Cyprus—are surrounded with circular terraces more or less elevated above the level of the sea, and composed of calcareous or sandy rocks of recent formation.* The northern portion of the Island of Crete has risen more than 60 feet during the present geological period.† A study of the shores of Asia Minor proves that there also the ground has continued, since man inhabited that region, to rise with a rather rapid movement. During historic times this part of the continent has gained a considerable belt of land at the expense of the Ægean Sea. It is not the alluvium of the rivers, or the matter washed up by the sea, which has caused this increase of the land, for the Anatolian rivers are very inconsiderable, and the water which bathes the coast cannot, on account of its great depth, throw up much sand. It must, therefore, be in consequence of a slow upheaval of the earth's crust that the ruins of Troy, Smyrna, Ephesus, and Miletus have gradually become more distant from the coast, and appear to be receding still further inland. From the same cause so many islands in the Ægean Sea which were once separate are now united, or are connected with the mainland, and form headlands or hills surrounded by low-lying plains. The testimony of ancient authors is unanimous as to these encroachments of the shores. Thus it is said that the two halves of Lesbos, Issa, and Antissa have become united, that the bays have been converted into inland lagoons, and that various islands have joined on to the mainland at Mindus, Miletus, the Parthenion Cape, at Ephesus, also at points near Halicarnassus and Magnesia. At the time of Herodotus, the mountain of Lade, not far

* Albert Gaudry, *Revue des Deux-Mondes*, November 1, 1861; Newbold.
† Raulin; Leycester and Spratt. *Vide* p. 639.

from which the Ionian galleys fought a battle with the Persian fleet, was an island; at the present day it forms part of the mainland, and stands in the midst of the plain of the Meander. Since the date of Strabo and Pliny, several other islands have similarly become headlands. The former Latmican Gulf is converted into a lake, known under the name of Akiz. The encroachments of *terra firma* on this gulf have added to the eastern coast of Asia Minor about 67 square miles in less than two thousand years. This retirement of the sea took place likewise in preceding ages, for the town of Priene (Samsoun), which at the time of Strabo was 4½ miles from the shore, had been built at a previous date on the sea-coast. In like manner, the village of Ayasoulouk, where the ruins of the ancient city of Ephesus may still be seen, is at the present time two leagues from the coast, and the former estuary which was commanded by the town is now converted into a marshy plain. Would the little river Meander, the total length of which is not more than 341 miles, have been able, by nothing but its alluvial deposits, to fill up lakes and estuaries of so large an area, and to modify so considerably the outline of the shore? It is therefore important that the discharge of this river and its annual deposit of alluvium should be exactly measured, in order that we may arrive certainly at the real cause of these encroachments on the part of the Anatolian coast, where, according to the ancient saying of Pausanias, "all is unstable and changing." According to M. de Tchihatcheff, this portion of the coast of Asia Minor has gained, since historic times began, an extent of about 185 square miles, equal to the area of the Isle of Wight.*

Similar phenomena are, however, likewise taking place on the southern coast of Asia Minor. Near Adalia, the Lake of Capria, which was very extensive at the time of Strabo, first ceased to communicate with the sea, and then gradually emptied, being now nothing but a marshy hollow: the surface of the peninsula has thus been increased by an area of 154 square miles. On the north of Asia Minor, a great number of signs prove that there also the water has retreated before the shores and rocks of the continent. During the present geological period the area of the Euxine has been diminished, and, according to the traditions of the Crimean Tartars, it is still diminishing. Banks of modern shells have been left by the sea at considerable heights on the hills of Thrace and Anatolia; † round the Crimea, salt lakes, stagnant marshes, now exist far inland

---

* *Asie Mineure.* Von Hoff, *Veränderungen der Erdoberfläche.*
† De Tchihatcheff, *Le Bosphore et Constantinople.*

in the place of former gulfs. Certainly enough, the Black Sea, before the opening of the Bosphorus, received from its tributary rivers more water than the sun and wind could evaporate, and must necessarily have much exceeded the level to which it now reaches. But if the earth itself had remained stationary and had not slowly risen, the sea-water would not have left the marks of its presence at any point higher than the former Strait of Isnik, the site of which is now dotted over with lakes which once formed a part of the sea. It is, perhaps, in consequence of the upheaval of the ground that this strait was closed, and the water of the Black Sea, gradually accumulating in its basin, was compelled to open by force a new outlet through the volcanic fissures which have now become the Bosphorus.

A geological examination of Southern Russia and the plains of Tartary will preclude us from entertaining any doubt as to the fact that the Caspian Sea, the Sea of Aral, and all those innumerable sheets of water which are scattered over the steppes, were separated from the Euxine and the Gulf of Obi by a gradual upheaval of the continent. The plains are still covered with salt and marine remains. The inland seas and the scattered lakes are still inhabited by seals, and their fauna altogether presents an essentially oceanic character. Herodotus, Strabo, Ptolemæus, and all the authors of antiquity attribute to the ancient Hyrcanian Ocean an extent far larger than that of the Caspian of our day; most of them, indeed, considered this inland sea as a prolongation of the Frozen Ocean. This latter opinion, no doubt erroneous as regards a date two thousand years ago, would certainly have been true of some anterior epoch. After Humboldt's profound investigations on Central Asia, we shall not, at the present day, show too great temerity in assuming that, during some portion of the present period, a vast strait, like that which once ran along the southern base of the Atlas, extended from the Black Sea to the Gulf of Obi and the Frozen Ocean.* The vast depression of the Caspian plains which is below the level of the sea, and, according to Halley, was produced by the shock of some wandering comet, is, on the contrary, really owing to a slow elevation of the ground.

The observations of level made on the coasts of the Mediterranean

* In Captain Maury's magnificent studies, in which, however, imagination sometimes has as large a share as science, he endeavours to prove, by very ingenious arguments, that the upheaval of the Andes, by modifying the system of winds and rain over the whole earth, was the cause of the gradual drying up of the plains of the Caspian and Aral.

have enabled us to ascertain not only that the larger portion of this inland basin of the ancient world, and many of the regions bordering on it, have gradually risen, but that they have, in addition, pointed out the limits of the area of upheaval. They may be distinguished with some degree of precision on the coast of Syria and Palestine; it may be noticed that in this region the surface of the ground is corrugated like that of water, and forms a series of waves and depressions fluctuating in contrary directions. The shores of the Gulf of Iskanderoun are regularly gaining in width by means of the elevation of the ground as well as of the matter thrown up by the sea; but at Beyrout a tower is shown which is sinking further and further into the water. More to the south, the former Isle of Tyre is now connected with the continent, and several parts of the peninsula bear traces of the sojourn of the sea during some recent epoch. Lastly, Kaisarieh and some other towns in Palestine are included in an area of subsidence, as is proved by the remains of fortifications which are now visible below the level of the Mediterranean.

On the east, all the Egyptian coasts were still rising at a comparatively very recent epoch; for the Bitter Lakes and the banks of the Nile exhibit former sea-beaches on which modern shells are found. But at the present day the ground is sinking continuously and imperceptibly. Ruined towns are situated in the midst of the marshy plain of Lake Menzaleh, which is covered by the sea during the greater part of the year. Further on, a former branch of the Nile, with the banks which bordered it, is entirely hidden by the waters of the Mediterranean. The same phenomenon occurs on the other side of the Delta. In 1784 the sea made an irruption into the interior of the land, and formed the Lake of Aboukir in the midst of a plain on which important towns once stood. In like manner it may be inferred, from the ancient descriptions of Alexandria and its environs, that a considerable subsidence of the ground must have taken place there during the centuries of our era. Artificial caves and catacombs, dug out in the days of the Ptolemies and incorrectly known as "Cleopatra's Baths," are now invaded by the waves.*
On the shores of the Red Sea, not far from Suez, some sepulchral caves hollowed out in the calcareous rock have likewise been inundated, owing to the subsidence of the ground. Perhaps this movement in the ground of Egypt is common to all that portion of the Mediterranean which may be called the Egyptian Sea; for in the

* Lyell, *Antiquity of Man;* Pococke; Wilkinson; Schleiden.

Island of Crete the western point has constantly risen during the modern epoch, but the side nearest to Egypt is gradually sinking under the water. As Strabo himself expressly said, nature is seeking to destroy the Isthmus of Suez, which she once formed between the two continents. Man, in his operations of cutting through it, is only anticipating the geological labour of centuries to come.

Along the shores of the Adriatic Sea, to the north of Zara and Pesaro, geographers have noted other phenomena of depression which mark the northern limit of the great Mediterranean area of upheaval. In the middle of the sixteenth century Angiolo Eremitano propounded the opinion that the isles of Venice were sinking at the rate of about a foot a century; and this hypothesis, which was based upon a comparison of the buildings and the pavement of the streets with the water, has since been fully confirmed. In the Isle of St. George, Roman constructions are now found below the level of the lagoons; in other places paved roads are covered by the water, and churches and bridges have sunk in comparison with the surface of the sea. In 1731 Eustache Manfredi noted the same subsidence of the soil under the edifices of Ravenna; but he, in error, attributed it to the gradual elevation of the level of the Adriatic. In addition, an entire town—Conca—once situated not far from Cattolica, at the mouth of the Crustummio, has been entirely submerged for some centuries; and, when the sea is calm, the remains of two of its towers may still be seen under the waves. M. Giacinto Collegno thinks that all these alterations of level are produced by the deposit of the alluvium brought down by the Po, and the other rivers descending from the Apennines and the Alps. This is a cause which must certainly contribute in no small extent to the general depression of the Venetian coasts of the Adriatic Sea, but probably it is not the only cause, for the opposite coasts of Dalmatia and Istria are also sinking, in spite of the compact nature of their rocks. At Trieste, at Zara, and in the Isle of Poragnitza, various works of man—such as pavements, mosaics, and sepulchres—may be seen below the level of the sea.* Moreover, as Lyell remarks, the artesian borings which have been made in the delta of the Po, to a great depth below the sea, have brought up nothing but river-alluvium, which unquestionably establishes the fact of a gradual depression of the ground. The earth which is penetrated by the boring-rod at the bottom of the artesian wells was once situate above the level of the sea.

* Donati, *Histoire Naturelle de la Mer Adriatique*; Schleiden, *La Plante*.

CHAPTER LXXXIII.

SUBSIDENCE OF THE SHORE OF THE CHANNEL, OF HOLLAND, OF SCHLESWIG, OF PRUSSIA.

WHETHER it be that the whole of Central Europe participates in the movement of depression to which the shores of the Adriatic are subject, or whether the latter be merely a local phenomenon, it is nevertheless the case that the southern coasts of the Channel and North Sea are also sinking, although very slowly. On the coast of Brittany and Normandy, numerous forests, which have been submerged, and buildings surrounded by the sea-water, prove that the ground has sunk during the present period. In 709 the Monastery of Mount St. Michael was built in the midst of a forest, ten leagues (?) from the sea; it now stands, like an island, in the midst of sandbanks. The inroads of the sea are still continuing, especially in the Bay of La Hougue and in the harbour of Carteret.* It appears, however, that various undulations, like those on the coast of Syria, also exist on these coasts; for in several places banks of sand and modern shells have been discovered at heights from 40 to 50 feet above the level of the sea. At some remote epoch, but nevertheless contemporary with man, the valley of the Somme was also upheaved; but, for thousands of years, it has been slowly subsiding, as submarine forests are found along the coast; and the peat-bogs of Abbeville, the bottom of which is situated below the Bay of Somme, afford no other *débris* but the remains of animals and vegetables which lived on the earth or in fresh water. When these peat-mosses began to grow, the ground of the valley must have been higher than the surface of the neighbouring seas.†

In Flanders and Holland the phenomena of subsidence have been, if not more considerable, at least more important in their results, on account of the very low level of these countries in comparison with the sea. A mere enumeration of the successive catastrophes brought about by this gradual depression constitutes a terrible history. The

* Bonissent, *Congrès Scientifique de Cherbourg*, 1860.
† Lyell, *Antiquity of Man*.

plains of Dordrecht have become a forest of reeds (Biesbosch). The Zuyderzee itself—once a marsh, next a lake, and then an arm of the sea—is still continuing to sink: at the present time there is, it is said, sufficient depth for ships of heavier burden than those which used to navigate it in former centuries. Like a raft gradually sinking under the waves, Holland would be slowly swallowed up in the abyss if it were not that the inhabitants, accepting the contest with the elements, have walled in their country by means of dikes, and laid it dry by immense operations of drainage, which will never cease to be a subject of astonishment. Several *savants*, at the head of whom

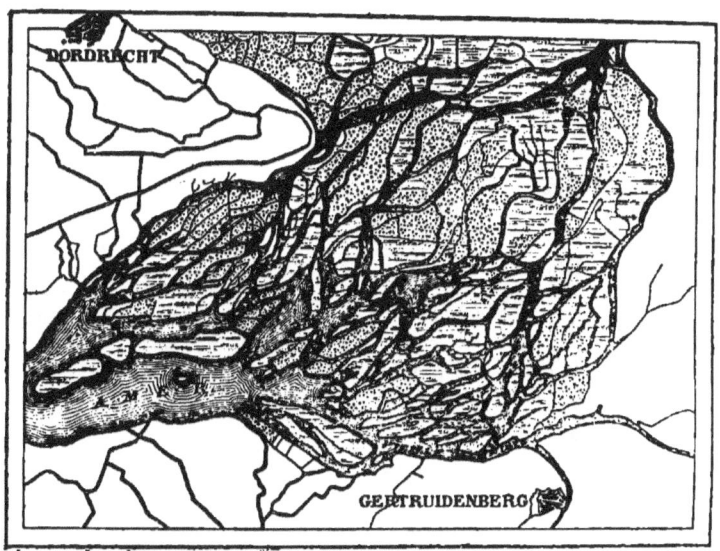

Fig. 223.—Biesbosch.

stands the eminent geologist, M. Staring, are of opinion that the gradual depression of the land which is thus embanked is caused only by the subsidence of the alluvial ground, the weight of the dikes, and the incessant passage of men and cattle. However great may be the importance of these combined causes, the phenomena of subsidence which have been noted for the last fifteen centuries are considerable enough to warrant us in accepting M. Elie de Beaumont's hypothesis as to the depression of the ground of Holland. If, at least, we may judge by the mean level both of the pavement of

T T

the towns and of the fields under cultivation, the movement of depression is most rapid at the mouths of the rivers—the Scheldt, the Meuse, and the Rhine. At Calais the streets are more than a yard above the high tide, whilst the cultivated ground descends to the limits of the tide. At Dunkirk the height of the streets is not more than 23 inches, and the fields are ploughed at a level of a yard below the sea. At Furnes and Ostend the streets are still lower, and the level of the *polders* is always sinking. Near the mouths of the Scheldt it is $11\frac{1}{2}$ feet below the high tide. Further to the north

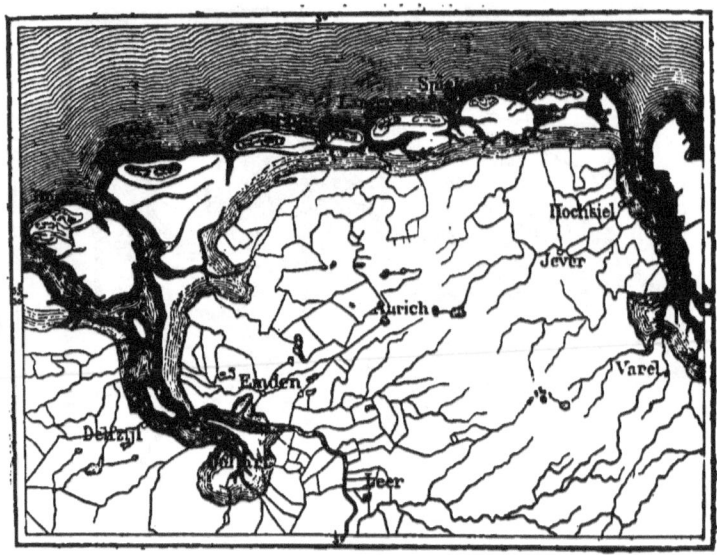

Fig. 224.—Coast of Friesland.

the ground gradually rises; but the streets of Rotterdam and Amsterdam are lower than the level of the equinoctial tides.*

All the adjacent coasts—those of the south of England, and of Cornwall and Yorkshire, as well as those of Hanover and Schleswig —likewise afford certain proofs of a considerable subsidence, by their submarine peat-mosses, their submerged forests, and their former coasts now become islands. On the western coasts of Schleswig the subsidence has been, on the average, 13 feet during the present period. In this locality, at the bottom of the port of Husum, there was dis-

* Bourlot, *Variations de Latitude et de Climat*.

covered, in the midst of a submerged forest of birches, a tomb of the Age of Stone, necessarily dating from a period anterior to the subsidence of the ground on which it stood. On the eastern coasts of Schleswig and Holstein many phenomena of the same nature have been remarked, which can only be explained by a gradual subsidence of the shore. The remains of an old castle, situated at the mouth of the Schlei, are covered by the water. Further on, the stumps of the trees of an ancient deer-forest of the Middle Ages may be seen under the water, about half a mile from the shore. In the Straits of Fehmarn are found the remains of an ancient wall; and, lastly, near Travemunde, two blocks of stone, which stood on the beach at the end of the last century, are now surrounded by water.* These are facts which will not permit us to attribute, solely and wholly, to the action of the waves the transformation of several peninsulas into islands, and lakes by the sea-shore into arms of the sea. According to John Paton, Denmark and Schleswig-Holstein have lost, since the year 1240, an area of about 1,225 square miles, or one-eighteenth of the whole surface of the territory.

Further to the east, all round the southern basin of the Baltic, the inroads of the water have led several geologists to admit that the ground of these countries is slowly subsiding. Rugen is broken up into islets and peninsulas; Bornholm is surrounded by submarine forests, one of which, according to Forchhammer, is 26 feet below the line of the shore. Other submerged forests fringe the coasts of Pomerania and Eastern Prussia. The Islands of Wollin and Usedom, situated in front of the mouths of the Oder, have gradually been eaten away by the waves. The sandy bar which impedes the entrance to the port of Swinemunde used to form a peninsula of Usedom, indeed, during historical periods.† Lastly, according to the evidence of Barth, the point of Samland has been overwhelmed by the water, as may be easily recognised by the fact that the church of St. Adalbert, which was built about the end of the fifteenth century, at a point 4¼ miles from the sea, is now found in a state of ruin only 100 paces from the beach.

These are all facts which cannot be questioned. Nevertheless, we are not yet warranted in considering them as positive proofs of the subsidence of the ground, for Voigt, and several other scientific observers, class them among the simple phenomena of erosion and deposit. However this may be, there are very strong reasons for

* Von Maack, *Das urgeschichtliche Schleswig-Holstein sche Land*, 1860.
† Anton von Etzel, *Die Ostsee*.

considering the channel and the southern waters of the North Sea and the Baltic as a trench of depression, an elongated valley, 1,100 miles in extent, separating the area of upheaval of Northern Europe from that which is bounded at its northern extremity by the coasts of Poitou.

CHAPTER LXXXIV.

UPHEAVAL OF THE COASTS OF CHILI AND PERU.—PROBABLE DEPRESSION OF THE COASTS OF LA PLATA AND BRAZIL.—COASTS OF NORTH AMERICA AND GREENLAND.

THE New World—that double continent the architecture of which is distinguished by general features of such simple grandeur—likewise exhibits a remarkable regularity in the action of its gentle oscillations. The latter are much more easy to study than the movements of the more indented and irregular peninsulas of Europe, and are also better known: since the epoch when the illustrious Darwin noted the fact that a great part of South America was constantly rising, *savants* and observers have only had to confirm the result of his investigations.

It is principally on the coasts of Chili that the traces of the general upheaval of the country are quite self-evident. Round

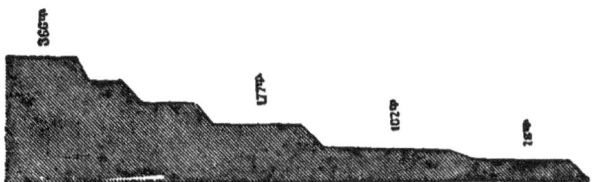

Fig. 225.—Coasts of Puerto San Jorge.

every headland, at the outlets of many of the valleys which cut deep into the mountainous masses on the coast, former sea-beaches may be distinguished, on which shells of the present epoch, like those of the creatures which are now living in the neighbouring bays, are scattered about or even heaped up in thick layers. These beaches, which are separated from one another by cliffs or escarpments of various heights, resemble the steps of a gigantic staircase. From these it may be readily seen that the coast was not raised by any uniform movement, and that intervals of comparative repose have elapsed between each of the stages furnished by the growing mass of rocks. On the hills of the Isle of Chiloe, Darwin found heaps of modern shells at a height of 347 feet. On the north of Conception, several lines of level cut out by the waves during the present period are found at an elevation of 600 to 1,000 feet. Near Valparaiso these levels are no less than 1,295 feet above the level of the sea; but north of this town they become lower. At Coquimbo they scarcely exceed 110 feet, and on the frontier of Bolivia they are only 65 to 80 feet above the sea-level. Thus the rising action of the rocks is especially developed in those regions of the sea-coast which are in the same latitude as the loftiest summits of the Chilian Andes —Aconcagua, Maypu, and Tupungato. We may infer from this that these high peaks indicate the axis of the portion of the upheaved crust, and that the mountains themselves tend to increase more rapidly than the plateaux and shores situated below them. In fact, in Chili, as in Norway, the terraces which overlook former bays or the mouths of valleys are not horizontal, as they appear to be; they rise gradually towards the mountains, and increase in height as they

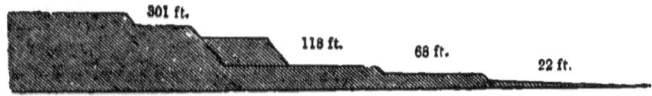

Fig. 226.—Coasts of Coquimbo.

recede from the present coast. The upheaving force acts, therefore, with more energy under the Chilian Andes than under the rocks of the adjacent coast. The white summits are gradually mounting up into the sky.

Trigonometrical measurements carried on for a long series of years will some day enable us to recognise this increase in the giants of Chili, and their upward progress into the regions of eternal snow; but, up to the present time, the only calculations which have been made on the subject of the rapidity of the upheaval of the Andes are based merely on the study of the sea-shores extended at their base. Comparing the present state of things with the evidence derived from history, Darwin proves that, during seventeen years, between 1817 and 1834, the ground at Valparaiso has risen 10 feet 7 inches, or about $7\frac{1}{2}$ inches a year. This rather rapid movement was preceded by a state of comparative inaction, for, from 1614 to 1817, more than two centuries, the elevation of the shore, as proved by an examination of the localities, certainly could not have exceeded 5 feet 11 inches. At Coquimbo, Conception, and the Island of Chiloe, the emergement of the shore has taken place still more slowly. But, however imperceptible the phenomenon may be, it is none the less taking place during the course of ages, and must ultimately completely change the aspect of the American coasts. Several ancient ports which were once frequented are now inaccessible; other harbours have been formed, thanks to the fresh protecting points which have emerged. Numerous islands, always designated by the Indian name *huapi*, have become promontories.

The indications of a gradual upheaval are equally visible on the coasts of Bolivia and Peru. In the eastern zone of the Desert of Atacama the ground is covered at considerable heights with shells and saline efflorescence, and seems as if it had been abandoned by the ocean only the day before. Above Cobija, Iquique, and several other coast towns, stages are marked out, similar to those at Coquimbo, and, like the latter, were once washed by the water of the Pacific. In front of Arica the sea has receded 165 yards in the

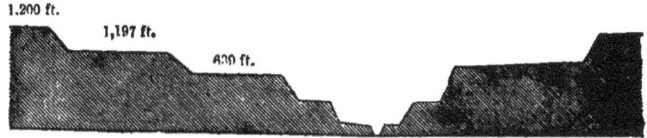

Fig. 227.—Valley of Rio Santa Cruz.

space of forty years, and the merchants of the town have been, in consequence, compelled to lengthen their landing-stage. But in front of Callao, on one of the cliffs of the Island of San Lorenzo, a most interesting proof has been found of the elevation of the shore during the period it was inhabited by man. At a height of 85 feet above the sea Darwin discovered, in a bed of modern shells deposited on a terrace, roots of sea-weed, bones of birds, ears of maize, plaited reeds, and some cotton thread almost entirely decomposed. These relics of human industry exactly resemble those which are found in the *huacas* or burial-places of the ancient Peruvians. There can be no doubt that the Island of San Lorenzo, and probably the whole of the adjacent coast, have risen at least 80 feet since the Red man inhabited the country. It nevertheless appears that in our days the ground on which Callao stands has again sunk, for the place where the ancient town stood is now in great part under water. This depression is doubtless merely local, and only temporarily affects the general ascending movement of the coast; for, further to the north, at Colon and at Santa Marta, and several other points of the coast of New

Granada, the ground has visibly risen since Europeans first landed on the continent. By admitting, however, that Callao forms, in fact, the northern limit of the area of upheaval, it results that the mass raised presents from north to south a length of at least 2,480 miles. It is almost equal to the distance from London to Tobolsk.

The movements of the eastern coast of South America which are at present going on have not been recognised so certainly as those of the western shore, doubtless on account of the extremely slow rate at which they proceed. An investigation of geological facts proves that the ground rose during the Post-pliocene period; that is, during the age of shell-fish still existing, and of the great animals which were the contemporaries of our ancestors, the megatherium, the mastodon, and the glyptodon. The Argentine pampas have preserved the uniform appearance of the ocean which once covered them. The parallel terraces of Patagonia, extending for more than 500 miles, vary but a few yards in height along all the points of their immense development, and the arms of the sea which wind in between the terminal promontory of America and the Tierra del Fuego retain all their ancient outlines. At the present time this portion of the continent appears to be oscillating in a contrary direction, and to be sinking by an imperceptible movement towards the level of the Atlantic. At the foot of the high cliffs of Patagonia the sea is incessantly increasing at the expense of the continent, and although the breakers are not possessed of force sufficient to demolish the rocky beds for more than 13 to 16 feet below the surface, the depth of the sea nevertheless augments with an even slope, in proportion to the distance from the shore, even over the site of the former cliffs. The bed of the sea must, therefore, be sinking, and with it the enormous mass of the plateaux which, during the recent period of the great mammals, rose with such marvellous regularity.

On the coast of Brazil, especially at Bahia, various recent depressions seem to indicate that there also the surface of the continent is regularly sinking. The ascertained facts were not, however, sufficiently numerous to justify any categorical assertion, until Professor Agassiz, in company with some other geologists, undertook his recent exploration of the River of the Amazons. In the first place, he verified the remarkable fact, that, in spite of the enormous quantity of sediment drifted down by its current, this river does not form any deposits at its mouth. Instead of throwing out into the ocean a long peninsula of alluvium, like that of the Mississippi, or

OSCILLATIONS OF SOUTH AMERICA.

at least forming, beyond the regular coast-line, a delta similar to those of the Rhone, the Nile, or the Po, the Amazon, on the contrary, widens out in a large gulf towards the sea, and, in the great estuary, it is difficult to say exactly where the *mouth*, properly so called, commences. The banks of the river, and the islands which lie in its outlet, are not composed of alluvium brought down by the current of fresh water, but are all formed of rock with horizontal strata deposited by the water of the river at some former epoch, and long since solidified. Thus, in the contest which takes place in the estuary of the Amazon, as in every other river-mouth, between the currents of fresh and salt water, between the fluviatile alluvium and the erosions of the sea, the latter always prevail. Instead of encroaching on the ocean, the valley of the Amazons has been invaded by the latter for at least 300 miles; for the geological study of the ground on the two shores of the estuary proves that rocky layers, exactly similar to those further up-stream, exist on the coast as far as the valleys of the Itapicuru and the Parnahyba. These two rivers once fell into the Amazon; but, in consequence of the erosion of their shores and those of the great current which they joined, the sea has advanced, as it were, to meet them, and they have thus by degrees become entirely independent of the Amazon system. In a similar way, the stream of the Tocantins is now only indirectly connected with the great central river, and, sooner or later, it must ultimately become isolated, as the Itapicuru and the Parnahyba already are. The action of erosion, caused doubtless by a constant sinking of the ground, is still continuing. The shores are noticed to recede all round the estuary at Maranhao, at Piauhy, at Macapa, and on the coasts of Marajo. On the shores of the latter island, near Soure, a wide gulf, into which flows the Igarape Grande, has recently been formed across a forest for a space of more than 18 miles from bank to bank. The rocks in the vicinity, which once rose above the sea-level, are gradually becoming covered. At Bragança, the bay, which used to advance scarcely a mile and a half into the land, now penetrates for nearly five miles. The lighthouse of Vigia, which was built at some distance from the sea, was a very few years afterwards washed by the waves. A signal-mast, which was set up in December, out of reach of the water, was surrounded by the sea in the June following. Facts of this kind render very probable the existence of a see-saw movement, which is upheaving all the western coast of America, from the Island of Chiloe to Callao, and depressing the eastern side of the Argentine Andes, of Patagonia, and Brazil. Thus a large portion of the South American continent is constantly gaining on one side that which it loses on the other, and is gradually making its way through the ocean in a westward direction. Agassiz assigns to this phenomenon of displacement a very ancient origin, for, in his view, the Antilles which once formed the isthmus joining the two Americas, have been gradually submerged, and the rivers of Guiana, once tributaries of the Orinoco, have become independent rivers.

In North America the vertical oscillations of the ground have not been recognised over so considerable a length as in the southern continent; but the few observations which have been already made on some points of the coast, in California as well as round the Gulf of Mexico, render the hypothesis very probable that a general upheaval is taking place, to which one of the parallel chains of the Rocky Mountains, or of the Sierra Nevada, serves as axis. The shore-belt of Tamaulipas and Texas increases so rapidly in width, not only because the south wind—which here blows almost the whole year through—throws up a large quantity of sand, but also because the ground itself is rising. During eighteen years, from 1845 to 1863, the shores of the Bay of Matagorda have risen 11 to 22 inches. In consequence of this gradual increase in the land, which is also proved by the heaps of shells left far from the shore, it has been found necessary to transfer the port of Indianola to Powderhorn, a place $4\frac{1}{2}$ miles nearer the entry.* The peninsula of Florida is likewise being upheaved by some subterranean forces, as is proved by the coral-banks which are rising above the level of the sea. Those mysterious elevations, the "mud-lumps," which are scattered around the coast near the mouths of the Mississippi, the origin of which M. Thomassy has tried to explain, by attributing them to the pressure of subterranean water,† also appear to testify in favour of a general upheaval of the region.

The eastern side of North America is not rising uniformly, for, although it is proved that the coasts of Labrador and Newfoundland are being slowly elevated, it is also proved that other regions are sinking. Lyell has ascertained that certain parts of the coasts of Georgia and South Carolina are subject to a movement of subsidence, and it is in consequence of a gradual depression, no less than from a constant action of erosion, that Sullivan and Morris Islands, at the entrance of Charlestown Roads, are incessantly diminishing in

---

* Adolf Douai, *Mittheilungen von Petermann*, April, 1864.
† *Vide* above, p. 288.

area. In like manner, all that portion of the coast which has as its centre the Bay of New York, and is bounded on the north by Cape Cod, and on the south by Cape Hatteras, has gradually sunk under the water of the Atlantic; and the subsidence has not yet ceased, at least as regards the coasts of New York and New Jersey. An isle which, on a map of 1649, is marked down as possessing an area of 290 acres, is at the present day not more than 100 square yards in extent at low tide, and at high tide is entirely submerged. If we are to put faith in tradition, the Straits of Hill-gate, which form the entry to the port of New York, are of recent origin. Two centuries ago the natives related to the Dutch colonists established in the Island of Manhattan, that, at the time of the fathers of their grandfathers, it was possible to cross dry-shod from one bank to the other, and that the sea only penetrated into the straits at the time of the great equinoctial floods. The land-surveyors employed on the survey have calculated that the shores of the Bay of Delaware lose, on the average, nearly eight feet every year. As far as it is possible to judge from the observations made since the colonisation of the country, the slow subsidence of this portion of the American coast may be estimated at $23\frac{1}{2}$ inches every century.*

In the vast island of Greenland, which lies in the axis of North America, the progress of gradual subsidence, succeeding to a period of upheaval, appears to be much more rapid. For a long time the Esquimaux have been acquainted with this phenomenon, and the Danish colonists on the western coast have also been enabled to verify the facts since the last century, by noticing, for a length of more than 620 miles, rocks, advanced promontories, and, indeed, their own dwellings, gradually disappearing under the inroads of the water. According to Wallich, this receding movement is still going on as regards the bed of the sea to the south of Iceland, and the sunken land of Bass, marked out on all the old charts, has really existed. On the north of Greenland, from latitude 76°, and in Grennell's Land, as well as in the other polar regions of the New World, the directly contrary phenomenon is taking place. In his voyage undertaken to discover the open sea, Hayes noticed on all the coasts the existence of ancient sea-beaches, which had gradually risen to a height of 100 feet; added to this, all the cliffs of the headlands had been polished up to this height by the action of the ice.

* Cook, *Geological Survey*; Arnold Guyot, *American Journal*, March, 1861.

CHAPTER LXXXV.

REEFS OF THE SOUTH SEA.—DARWIN'S THEORY AS TO UPHEAVALS AND DEPRESSIONS.

THE study of shores has not only enabled us to note the upheaval and subsidence of great continental masses; it has also made known to *savants* the oscillations of the tracts of ocean, for the numerous islands which lie either alone or in groups in the Indian Ocean have served as evidence to prove the movement of the ground on which they stand. Lines of erosion, parallel terraces, banks of modern shells, and all the other marks of the former presence of the water, point out, in each of the islands of the Pacific, as well as on the coasts of Europe and the New World, the various upheavals which have taken place. But, in addition, most of these islands are surrounded with living girdles of corals, which exactly measure all the changes in level, either elevation or depression, to which the shores are subject. The discovery of this fact, that the terrestrial oscillations are, so to speak, rendered visible by the work of polypes, is doubtless one of the most important achievements of modern geography; and it is to the patient investigations and sagacity of Darwin that science is indebted for it. By comparing his own observations with those of the explorers who preceded him, the English geologist has been enabled to point out, just as if he had witnessed them with his own eyes, the various movements which raise or depress the bed of the ocean, over an area as great as that of the two continents of Europe and Asia.

All the travellers who have crossed the South Seas have been struck with astonishment at the sight of the reefs raised by the polypes in the midst of the water. Some of these reefs circle round at a distance from islands, or even archipelagoes; they then form barriers of coral. Others, situated far from any land, are distributed in the shape of rings, or of more or less elongated crescents, round lagoons or bays remarkable for their pale green water; these are the *atolls*. In those parts of the ring where the constructions of the polypes and madrepores have not yet reached the surface, the waves which flow

ATOLLS, OR CORAL REEFS.

over the submarine bar rise in foamy breakers; in other parts of the reef there are just visible above the waves rocks of a dazzling white or a delicate pink hue. Next comes a semicircular range of islets, like Druidical stones set up by giants in the open sea. Lastly, upon the emerged ground which occupies that portion of the *atoll* which is most exposed to the violence of the waves and wind, cocoa-nut trees and other tropical growths wave in the air, either in mere groups or in regular groves. This is the most common form of the reefs among all the thousands of *atolls* which are dotted over the

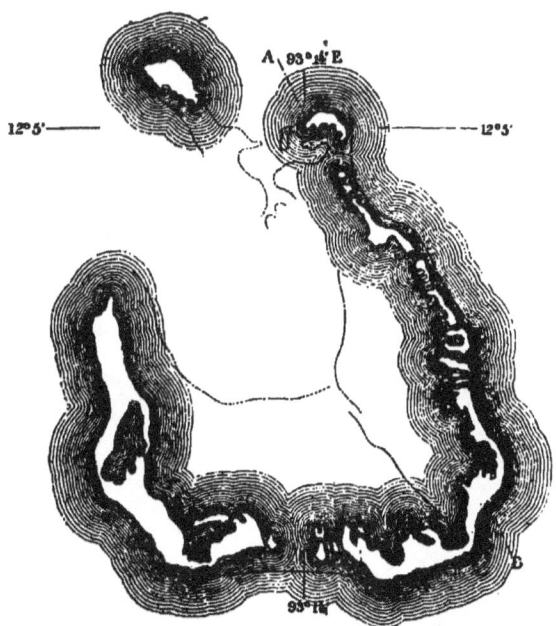

Fig. 228.—Keeling Atoll.

South Seas. When these coral-banks have not as yet reached the surface, their position is only pointed out by a circle of breakers; those that have attained the last stage of their development form a circular grove, which, seen from above, would look like a coronal of leaves floating on the blue water.

How have these wonderful reefs been raised? As was long ago proved by Chamisso, the French traveller, the polypes love to build in the midst of water which is breaking into foam; it may therefore be readily understood that wherever a submarine bank exists, the coral-reefs assume, like the breakers themselves, a more or less annular form. But in spots where the sounding-line reveals no shallows near the *atolls*, how have the polypes been able to raise from the bottom of the abyss their calcareous habitations? In order to explain this phenomenon, scientific men once hit upon a strange hypothesis; they looked upon every *atoll* as the circumference of a crater, which the submarine

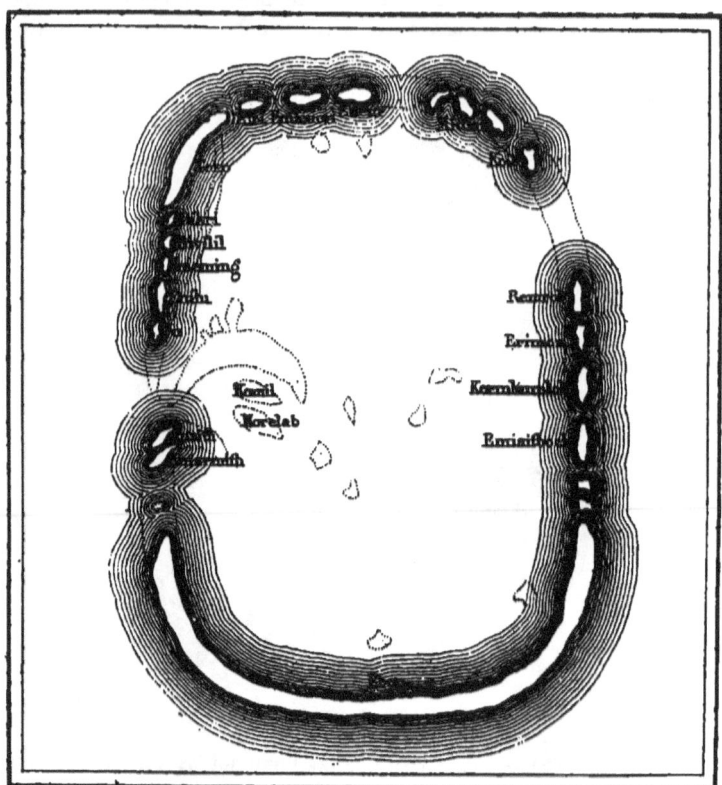

Fig. 229.—Atoll of Ebon.

forces of the globe had raised up to a distance of a few yards from the surface, so as to form a base to the operations of the polypes. Although this explanation might be true enough for a very limited number of *atolls*, it would be incomprehensible how thousands and thousands of volcanoes should have been elevated to the same height below the level of the sea. Nor could it be understood why the craters of these supposed volcanoes should so often assume very elongated forms. Lastly, it would be impossible to conceive why, taking into account the multitudes of annular reefs, of which several archipelagoes are composed, and especially the double range of the Maldives, 465 miles long, by 50 miles broad, no *atoll* has ever distinguished itself by an eruption of lava or ashes.

The form of these reefs, therefore, has no connection with volcanic phenomena, properly so called; and, like so many other facts in the terrestrial history, can only be explained by being attributed to slow movements of the surface. The subsidence of the bed of the sea will account for the formation of *atolls* and barriers of reefs; on the other hand, a gradual elevation of the ground explains the position of the corals which fringe the shore at a certain height above the waves. Thus, whether they rise or sink, the reefs formed by the polypes may serve as a measure of the vertical oscillations to which continental coasts, islands, and even the abysses of the sea are subject.

It is easy enough to verify the movement of land which is rising, as, in this case, we see the banks of coral resting upon the beach, and sprinkling with their *débris* that portion of the shore which is above the level of the sea. Often, too, the channels can be distinguished which once separated them from the coast, and on the high ground of several islands calcareous banks are found, which evidently owe their origin to polypiers. With regard to the coral islands which are not included in the area of upheaval, they are surrounded by annular reefs, constructed in the midst of the water at some distance from the shore. When this distance is but small, and the coral-banks are not very thick, there is nothing to prove that the level of the coast has changed; for the observations of *savants* prove that polypes can live and build their rocky habitations at a depth of from 100 to 150 feet. Generally, however, the walls of coral and calcareous sand, which form the outer sides of the reef, descend much lower. Most of them lie on slopes composed of their own *débris*, and are immersed in the sea at a slope of 45° to depths of several hundreds and even thousands of feet. It is evident that in a case like this the bed of the ocean must have subsided. The polypes commenced their work of construction a few yards below the surface, and then, in proportion as the ground sank on which their coral edifice stood, they continued to build upwards and upwards, in order to approach the light. The mountainous islands, which they surround with their reefs, gradually diminish in height, leaving between themselves and the barrier of coral a channel of increasing width and depth. The time is approaching when, first having been reduced to the state of islets, they will become divided into isolated peaks, which one after the other will be submerged, and disappear in the sea. Then all that remains will be an *atoll*, enclosing within its increasing walls a lagoon, in which calcareous *débris* is slowly gathered; narrow beaches and reefs, like the pieces of wreck still floating above a foundering ship, surround the spot where the island has been swallowed up. The natives of the *atolls* of Ebon relate that

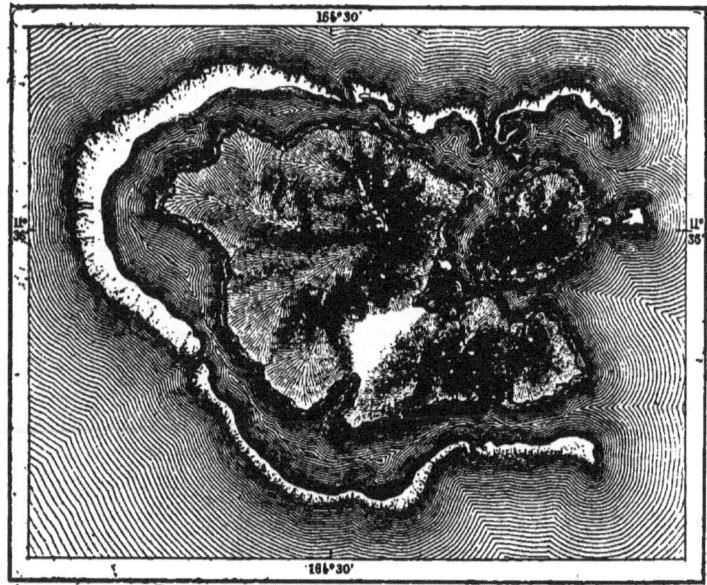

Fig. 230.—Isle of Vanikoro.

they have heard their fathers say that an elevated island, the hills of which were shaded by cocoa-nut trees and bread-fruit trees, once occupied the greater part of the lagoon. The isles disappeared; but the reefs are still maintained just above the level of the water.*

When the subsidence of a whole archipelago of submarine peaks takes place slowly and regularly, it may often happen that the sea, striking forcibly against the outer walls of the reefs, breaks this barrier, and hollows out for itself a free passage across the central

* Doane, *Nautical Magazine*, Sept., 1863.

lagoon. Then the banks of coral rise in the midst of the breakers on both sides of the newly-formed channel, and the original *atoll* is thus divided into two annular isles. In proportion as the submarine mass sinks, other ruptures of the same kind take place in each of the isolated fractions of the *atoll*, and the archipelago of reefs is ultimately composed of a considerable number of islets, which, in their turn, are

Fig. 231.—Section of Isle Vanikoro.

also broken up. In this way are formed these wonderful groups of annular banks arranged in immense ovals.

The Atoll Ari of the Maldives is an example of this astonishing formation of coral islands. If we could, represent in a drawing the shapes that the *ensemble* of the groups has successively assumed during the course of centuries, we should obtain a series of curves similar to those which geographers avail themselves of to delineate

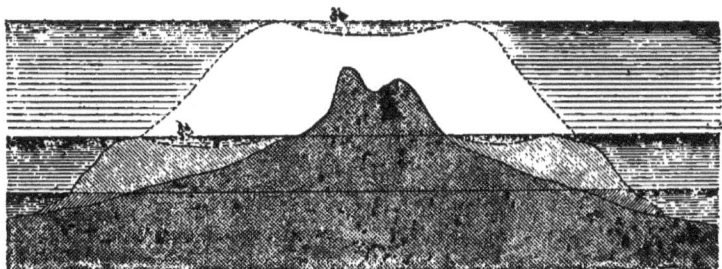

Fig. 232.—Growth of Coral on a Mountain slowly subsiding.

the slopes of a mountainous group. Sometimes, however, the movement of depression is so rapid, that the sea does not confine itself to opening channels here and there across the *atolls*. The coral insects do not build fast enough to maintain their dwellings on a level with the surface; they gradually perish, and the *atolls*, which layer by layer have been raised by innumerable generations of constructers, disappear, and form annular shoals. Of this kind is the great bank of Chagos, south of the Maldives, which the sounding-line shows to have once been one of the largest *atolls* in the Indian Ocean. In the group, indeed, of the Maldives, several islets, which were recently

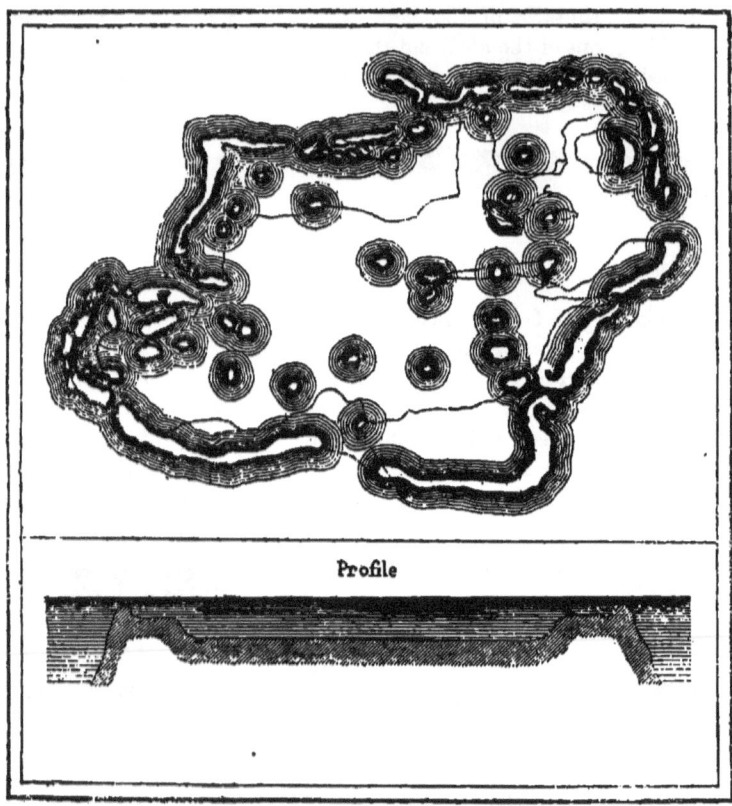

Fig. 233.—Great Bank of Chagos.

inhabited and green with vegetation, are now slowly being swallowed up beneath the surface of the water.*

* Darwin, *Coral Reefs*.

The Earth. Vol. I   ATOLL ARI   PL. XXIV.

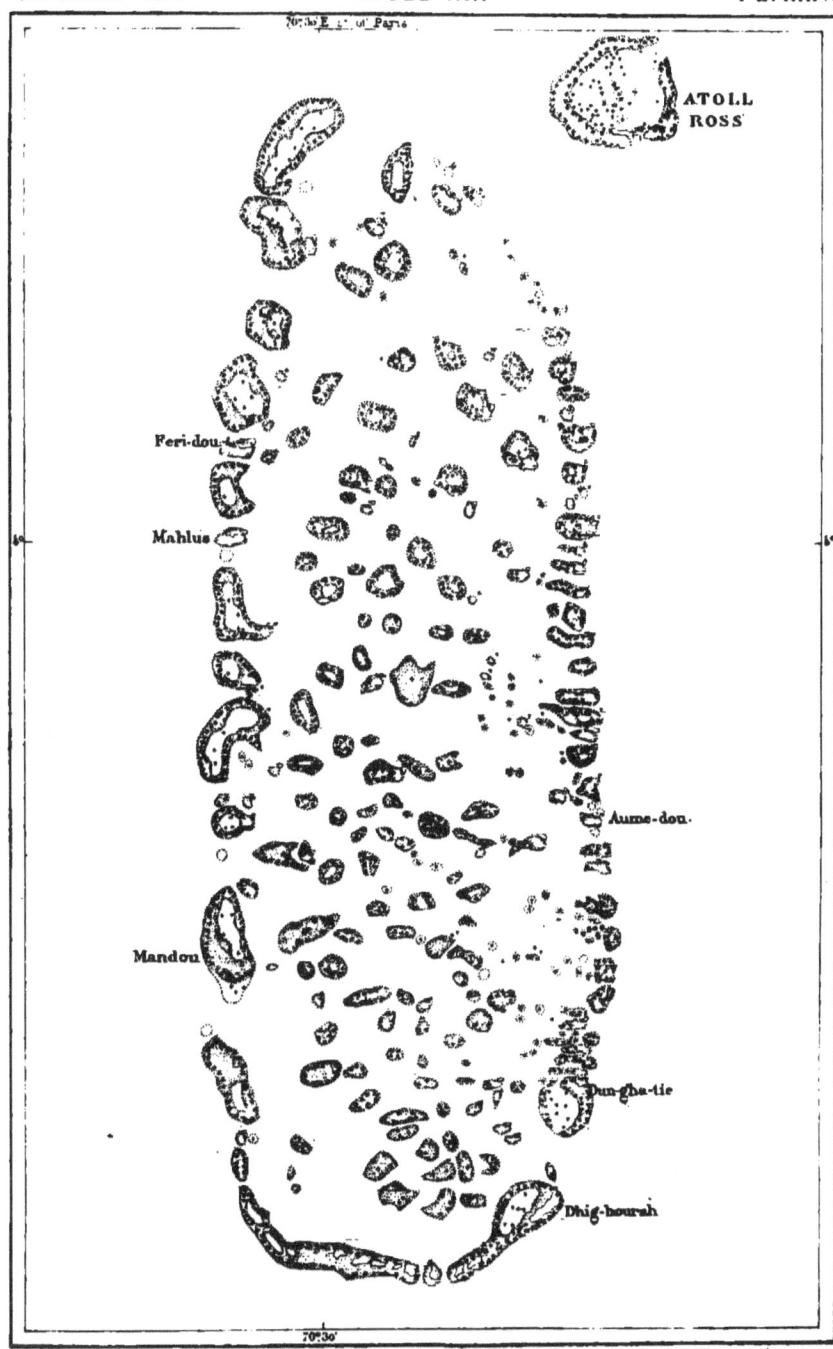

Engd by Erhard   Vincent Brooks, Day & Son, Lith London, W.C.

CHAPMAN & HALL LONDON

CHAPTER LXXXVI.

THE GREAT AREAS OF UPHEAVAL AND DEPRESSION.—MOBILITY OF THE SO-
CALLED RIGID CRUST OF THE EARTH.

OWING to the evidence which is furnished by coral reefs, which evidence, however, is supplemented at a large number of points by other indications, it is now possible to fix almost exactly the limits of the areas both of upheaval and subsidence, which divide between them the hemispheres included within the coasts of South America and Africa. Whilst the group of the Sandwich Islands is rising, as if it were still subject to the forces which are elevating the American continent, a gradual subsidence may be noticed in the archipelagoes of the central basins of the South Seas, the Bass and Society Islands, and also the Gilbert, Marshall, and Caroline Islands; in one word, all this "milky way" of islands, islets, and reefs, which extends diagonally across the Pacific for a length of more than 9,000 miles, and with an average width of 1,200 miles. These islands are the remnants of a former continent, which has sunk down with the people who inhabited it. Since the first European navigators visited these seas, several islands have disappeared, and others, such as Whit-Sunday Island, have considerably diminished in size.

In a parallel line with this great area of depression, which is at least twice as large as Europe, lies a wave of upheaval which coincides with the semicircle of volcanoes running round the western side of the basin of the South Sea. New Zealand, which is situated on the southern end of this rising, and is based on a long furrow of fire, is rising in certain places so considerably, that English colonists, who have only arrived there a few years, have been able to notice that the headlands increase in height, and that banks of rocks are gradually obstructing the entrance of the ports. At the commencement of the present geological epoch, the mountains of New Zealand were at least 1,900 feet lower than they now are, and the ice-floes, from a continent situated to the east, floated with their load of erratic rocks on to the incipient islets, and were there stranded. But since that time the New Zealand Alps have risen ten successive times, as is proved by the ten terraces lying in stages on the sides of the mountains.* Even at the present day the latter are still increasing. In ten years the shores at Lyttleton have risen three feet. The New Hebrides, the Salomon Islands, the northern and western coasts of New Guinea, the numerous islands which compose the great Sunda Archipelago (which latter are proved by their altogether Asiatic fauna, as studied by Wallace, to have once formed a portion of the neighbouring continent), are all now rising, after having quite recently subsided, and banks of coral emerging from the sea are incessantly being added to the shores.

At the angle of the Asiatic continent the wave of elevation divides in a fork, so as to run round the Chinese Sea, which is bordered by the gradually-depressed coasts of Cochin-China and Tonquin. To the north the upheaved region is continued towards America by the Philippine, Formosa,† Liou-Kieou Islands,‡ and Japan; that is, all the islands from Borneo to Kamtschatka, through which passes the fissure of eruption of the volcanoes of the Western Pacific. Quite recently Russian travellers have discovered, on the coast of the great island Saghalien, heaps of modern shells, lying not far from the shore, on beds of marine clay, and also former bays, which are now converted into lakes or salt marshes. In like manner, it has been proved that the regions of the Amoor are gradually being upheaved; for, in order to maintain its level, the river has constantly to hollow out its bed between the cliffs, and on the plateau by the river-side, semicircular sheets of water may still be seen, which are evidently former windings of the Amoor.

On the west of the Sunda Archipelago, Sumatra, fringed on its eastern coasts by peninsulas, which once were islands, and, indeed, still bear the name (*poulo*), seems to be the starting-point of another movement of upheaval, which embraces all the coasts situated round the Bay of Bengal. The Nicobar and Andaman Archipelagoes are gradually rising. Ceylon is likewise rising; at least, part of it, as is proved by the banks of coral lying in gradations one above another on the hills, and also by the traditions of the natives. But it is probable that the extremity of the isle is undergoing a slight see-saw movement, for Adam's Bridge, the chain of shoals which joins Ceylon to the Coromandel coast, which, too, according to the legend,

---

* Julius Haast. F. von Hochstetter also asserts that the eastern coast is being upheaved, but he thinks that the western coast has sunk, and that the axis of a see-saw movement passes through the two islands.
† Ferd. de Richthofen, *Zeitschrift der Geologischen Gesellschaft*, vol. xii.
‡ Swinhoe, *North China Branch of Asiatic Society*, No. 11, May, 1859.

formerly served as a road to the triumphal army of Hanouman, the monkey, appears to have once been a perfect isthmus. Scarcely three centuries have elapsed since the peninsula of Rameseram, whither thousands of Hindoos went every year in pilgrimage, has become detached from the mainland, and formed an islet, like the ruins of a fallen pier.* Further to the north there has been an upheaval. If we may put faith in the Brahmin legend, Veruna, the god of the sea, two thousand three hundred years ago, ordered the waves to abandon the low plain of Malayala, which extends along the Malabar coast between Mangalore and Cape Comorin.† With regard to the basin of the Lower Ganges, it appears to form a part of the area of upheaval of the Bay of Bengal, and, with all its southern part, to be gradually rising; for the tribu-

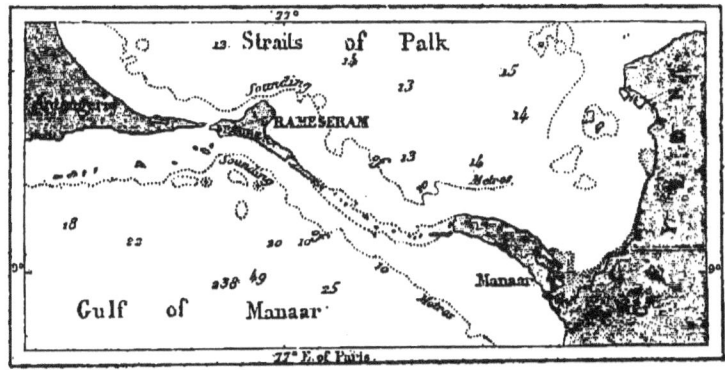

Fig. 234.—Bridge of Adam or Rama.

taries of the river which traverse this region—the Coosy, the Mahanada, and the Soane—are constantly shifting their mouths further up-stream. The last-named water-course has retreated more than four miles during the last eighty years. According to Mr. Ferguson, at the confluence of the Ganges and Gogra we find the western limit of the wave of upheaval, which commences at the islands of New Zealand, 8,000 miles away towards the south-east.

Almost the whole of the space occupied by Australia and the Indian Ocean, properly so called, is situated, like the central basin of the Pacific, in an area of gradual depression. Whilst, from New Guinea to Sumatra and the Philippines, a new continent is emerging

* Ritter, *Erdkunde.*
† Duncan, *Asiatic Researches,* vol. v.; Von Hoff, *Veränderungen.*

from the water, the old Australian continent, so remarkable for its fauna and its flora, which seem to belong to some former geological period, is gradually sinking down, together with the surrounding isles—the Louisiade Archipelago, New Caledonia, and the reefs of the Coral Sea. Up to the present time, one portion alone of Australia is known to be experiencing a continuous movement of elevation—the district of Hobson's Bay, near Melbourne, which, according to M. Becker, is rising at the rate of about four inches a year. Be this as it may, the great mass of the continent is imperceptibly sinking, and the polypes which surround the coast are compelled to heighten their reefs more and more.* The Indian Ocean, to the west of Australia, is almost entirely devoid of islands; but all those that emerge from the depth of the sea over a space of more than 3,700 miles in width are *atolls*, which would be slowly submerged if the polypes were not incessantly building up the edges of the reefs. Among these islands are the celebrated Keeling *Atoll*, which Mr. Darwin has studied so profitably for science, and the Maldive Archipelago, that double chain of submarine mountains, every peak of which is crowned with a coral tiara raised above the water.

Thus the space which extends over two-thirds of the circle of the globe, from the eastern coast of America to the western shores of the Indian Ocean, presents two areas of upheaval and two areas of subsidence succeeding one another from east to west. Next to the American continent, which is slowly rising, we have the innumerable low-lying islands of Oceania, the greater part of which would have long ago disappeared if it were not that the labour of the polypes has maintained them on a level with the waves. Next, there is developed, in a vast semicircle, pointed out from afar by its volcanoes, a large zone of isles and sea-coasts which are gradually rising, as if to replace in the future the old continent of Australia. Lastly, the same causes which are depressing the bed of the Central Pacific are likewise lowering that of the Indian Ocean, with its shallows and its reefs.

Beyond lies the enormous mass of Africa, the coasts of which have not yet been explored by scientific men, except, perhaps, over some small extent. Nevertheless, sufficient observations have been made to warrant us in considering Eastern Africa and the islands adjoining it as a third wave of upheaval, corresponding with those of America and the Sunda Islands. The banks of coral which surround Mauritius Island, Reunion, and Madagascar, and those

* Gregory, *Philosophical Society of Brisbane*.

which border the African coast from Mozambique to Mombaze, bear witness to the elevation of the ground. In like manner, the southern coasts of the Red Sea still exhibit at various elevations evident traces of the recent presence of sea-water. Most of the travellers who have visited these places, Ferret and Gallinier, Rüppel, Salt, Valencia, and Niebuhr, have been struck by the sight of reefs emerged from the sea, former sea-beaches white with salt, and bays which are now left far inland, and are converted into marshes. Quite recently M. Lejean has recognised the fact that, by the upheaval of the ground, the former port of Djeddah is completely separated from the sea, and has become a mere lake; this port, at the time of Niebuhr, was still accessible to ships of small tonnage. The inhabitants of the coast assert that both the bed and the shores of the Red Sea change every twenty years.

Not far from the Isthmus of Suez on the north, the slow elevation of the ground is replaced by a contrary movement; but on the western side it is not yet known where the first signs are to be found of any rising of the ground. The observations of M. Eugène Robert on the coast of Senegal would, perhaps, show that this portion of Africa is in an area of upheaval. It is, however, a fact that beyond the African continent, Madeira, St. Helena, and probably the Canary Islands, the remains of the ancient Atlantis, are gradually sinking into the ocean. All these facts tell in favour of the hypothesis according to which the equatorial position of the circumference of the globe presents three waves of upheaval, separated from one another by three intervening depressions. The centre of each depression lies in the middle of an ocean; the three upheaved regions are the great Sunda Archipelago, a kind of continent in process of formation, and the enormous masses of Africa and America.

As will be understood, these regular oscillations must take place in obedience to some general law still unknown, although none the less certain. We cannot consider them, with Berzelius, to be nothing but mere accidents, produced by the subsidence or the rupture of the terrestrial crust. Neither must these regular movements be confounded with volcanic tremblings, for they are distinguished from the latter by their excessive slowness, as well as by their character of generality. Moreover, all these facts, whatever may be their origin, are determined by causes affecting the whole mass of the planet. If earthquakes have their tides, as is said to be proved by the greater frequency of these phenomena at the time of the full and new moon, we cannot doubt that the slow oscillations of the terrestrial envelope also have their regular cycles. Only the reason of these secular tides still remains unknown. Must we seek for it in some alteration of the physical conditions of the globe, or in the revolutions of some astronomical period? As far as regards these points, we are reduced to mere hypothesis. Some day or other, when scientific men have observed all the lines of level from the north to the south pole, and all the *débris* which have been left by the sea, as it were so many precise measurements, on sea-coasts and mountain-sides, we shall be able to exactly specify the dimensions of each wave of upheaval, and also what the impulsive force is which actuates them. We shall then know whether the regions that are elevated are always equal in extent to those that subside, and whether the surface, like that of every vibrating body, presents certain "nodal lines," round which the agitated portions arrange themselves in rhythmical figures. We shall know, too, whether continents and seas, alternately elevated and depressed, as if by some secular tide, are slowly shifting their positions round the planet, assuming therein various harmonic forms. Perhaps even it will be proved that in the bowels of the earth an exchange of solid particles is taking place similar to the circulation of the aërial and liquid particles of the atmosphere and the ocean. The globe, a simple mass as it is of condensed gas, is not entirely congealed; it has retained, like every other body, some remains of its former fluidity, and, just as in a lump of metal coming out of a furnace, the particles which compose it never cease to turn slowly round and round one another.

Be this as it may, it remains an unquestionable fact that an incessant movement is causing an undulation in the so-called rigid crust of our globe. The continental masses are still elevated through a long course of ages; next, they sink only to rise again with slow and majestic oscillations. Scandinavia, which is at present rising, sank during the Glacial period, and the population, which at that time made it their abode, were forced to abandon their valleys step by step as they became converted into fjords. In like manner, the Chilian Andes and the mountains of New Zealand, which are at the present time increasing in height, previously sank by degrees, the former 8,000 and the latter 5,000 feet, before they rose as they are now doing. It is likewise proved that at a great number of other points—in Peru, in Egypt, in North America, and in Greenland—changes of the same nature have taken place during the present era of geological history, without any violent revolution having thrown the earth into confusion. Continents rise and sink as if through some gentle action of respiration; they move in long undulations, which may be compared to the waves of the sea; the far-reaching glance of science can already trace out these undulations through the long lapse of centuries. "The time will come," says Darwin, "when geologists will consider the quiescence of the terrestrial crust through a long period of its history to be as improbable as an absolute calm in the atmosphere during a whole season of the year."

In the universe everything is changing and everything is in motion, for motion itself is the first condition of vitality. In bygone days, men who, through isolation, hatred, and fear, were left in their native ignorance, and filled with a feeling of their own weakness, could recognise in all that surrounded them nothing but the Immovable and the Eternal. In their ideas the heavens were a solid vault, a firmament on which the stars were fastened; the earth was the firm, unshaken foundation of the heavens, and nothing but a miracle could disturb its surface. But since civilisation has connected all the nations of the earth in one common humanity; since history has linked century to century; since astronomy and geology have enabled science to cast her retrospective glance on epochs thousands and thousands of years back, man has ceased to be an isolated being, and, if we may so speak, is no longer merely mortal: he is become the consciousness of the imperishable universe. No longer connecting the vitality either of the stars or the earth merely with his own brief and fleeting existence, but comparing it with the duration of the existence of his own race and of all the beings who have lived before him, he has seen the celestial vault resolve itself into infinite space, and has recognised the earth as nothing but a little globe rotating in the midst of the "milky way." The firm ground which he treads under his feet, and long thought to be immovable, is replete with vitality, and is actuated by incessant motion; the very mountains rise or sink; not only do the winds and ocean-currents circulate round the planet, but the continents themselves, with their summits and their valleys, are changing their places, and are slowly travelling round the circle of the globe. In order to explain all these geological phenomena, it is no longer necessary to imagine alterations in the earth's axis, ruptures of the solid crust, or gigantic subterranean downfalls. This is not the mode in which Nature generally proceeds; she is more calm and more regular in her operations, and, chary of her might, brings about changes of the grandest character without even the knowledge of the beings that she nourishes. She upheaves mountains and dries up seas without disturbing the flight of the gnat. Some revolution which appears to us to have been produced by a mighty cataclysm, has perhaps taken thousands of years to accomplish. Time is the earth's attribute. Year after year she leisurely renews her charming drapery of foliage and flowers; just as, during the long lapse of ages, she reconstitutes her seas and her continents, and moves them slowly over her surface.

THE END.

www.ingramcontent.com/pod-product-compliance
Lightning Source LLC
Chambersburg PA
CBHW020741020526
44115CB00030B/731